Mediterranean-type Ecosystems

Tasks for vegetation science 19

Series Editors

HELMUT LIETH

University of Osnabrück, F.R.G.

HAROLD A. MOONEY

Stanford University, Stanford, Calif., U.S.A.

1 Box. E.O. *Macroclimate and Plant Forms.* An introduction to predictive modelling in phytogeography. ISBN 90 6193 941 0
2 Navin Sen D. & Singh Rajpurohit, K. *Contributions to the Ecology of Halophytes.* ISBN 90 6193 942 9
3 Ross, J. *The Radiation Regime and Architecture of Plant Stands.* ISBN 90 6193 607 1
4 Margis, N.S. & Mooney, H.A. (eds) *Components of Productivity of Mediterranean – Climate, Regions.* ISBN 90 6193 944 5
5 Müller, M.J. *Selected Climatic Data for a Global Set of Standard Stations for Vegetation Science.* ISBN 90 6193 945 3
6 Roth, I. *Stratification in Tropical Forests as Seen in Leaf Structure.* ISBN 90 6193 946 1
7 Steubing, L. & Jäger, H.J. *Monitoring of air Pollutants by Plants: Methods and Problems.* ISBN 90 6193 947 X
8 Teas, H.J. *Biology and Ecology of Mangroves.* ISBN 90 6193 948 8
9 Teas, H.J. *Physiology and Management of Mangroves.* ISBN 90 6193 949 6
10 Feoli, E., Lagonegro, M. & Orlóci, L. *Information Analysis of Vegetation Data.* ISBN 90 6193 950 X
11 Sesták, Z. (ed) *Photosynthesis during Leaf Development.* ISBN 90 6193 951 8
12 Medina, E., Mooney, H.A. & Vázquez-Yánes, C. (eds) *Physiological Ecology of Plants of the Wet Tropics.* ISBN 90 6193 952 6
13 Margaris, N.S., Arianoustou-Faraggitaki, M. & Oechel, W.C. (eds) *Being Alive on Land* ISBN 90 6193 953 4
14 Hall, D.O., Myers, N. & Margaris, N.S. (eds) *Economics of Ecosystems Management.* ISBN 90 6193 505 9
15 Estrada, A. & Fleming, T.H. (eds) *Frugivores and Seed Dispersal,* ISBN 90 6193 543 1
16 Dell, B., Hopkins, A.J.M. & Lamont, B.B. (eds) *Resilience in Mediterranean-type Ecosystems.* ISBN 90 6193 579 2
17 Roth, I. *Stratification of a Tropical Forest as Seen in Dispersal Types.* ISBN 90 6193 613 6
18 Dässler, H.-G. & Börtitz, S. *Air Pollution and its Influence on Vegetation.* Causes/Effects/Prophylaxis and Therapy. ISBN 90 6193 619 5
19 Specht, R.L. *Mediterranean-type Ecosystems.* A Data Source Book. ISBN 90 6193 652 7

Mediterranean-type Ecosystems

A data source book

edited by

R.L. SPECHT

section co-ordinators

R.L. SPECHT
P.W. RUNDEL
W.E. WESTMAN
P.C. CATLING
J.D. MAJER
P. GREENSLADE

KLUWER ACADEMIC PUBLISHERS
DORDRECHT / BOSTON / LONDON

Library of Congress Cataloging in Publication Data

Mediterranean-type Ecosystems: A data source book / edited by R.L. Specht.
000 p. 000 cm. – (Tasks for vegetation science 19)
Includes index
1. Ecology – Handbooks, manuals, etc. 2. Mediterranean climate –
Handbooks, manuals, etc. 3. Bioclimatology – Handbooks, manuals,
etc. I. Specht, R.L. (Raymond Louis), 1924–. II. Series.
QH541.142.M44 1988
574.5'262 – dc19

ISBN 90-6193-652-7

Published by Kluwer Academic Publishers
P.O. Box 17, 3300 AA Dordrecht, The Netherlands.

Kluwer Academic Publishers incorporates
the publishing programmes of
D. Reidel, Martinus Nijhoff, Dr W. Junk and MTP Press.

Sold and distributed in the U.S.A. and Canada
by Kluwer Academic Publishers,
101 Philip Drive, Norwell, MA 02061, U.S.A.

In all other countries, sold and distributed
by Kluwer Academic Publishers Group,
P.O. Box 322, 3300 AH Dordrecht, The Netherlands.

All Rights Reserved
© 1988 by Kluwer Academic Publishers
No part of the material protected by this copyright notice may be reproduced or
utilised in any form or by any means, electronic or mechanical,
including photocopying, recording or by any information storage and
retrieval system, without written permission from the copyright owner.

Printed in The Netherlands

A tribute to the memory of

- the late Professor P.W. Miller who initiated ecosystem models integrating ecomorphological characters and seasonality (San Diego Workshop, June 1981)
- the late Professor R.H. Whittaker who stimulated the study of species-richness in Mediterranean Terrestrial Ecosystems.

Table of contents

Preface ix

List of contributors xi

1. **CLIMATE, VEGETATION, VERTEBRATES AND SOIL/LITTER INVERTEBRATES OF MEDITERRANEAN-TYPE ECOSYSTEMS – DATA-BANKS**

   prepared by P.C. Catling, Ph. Daget, B.J. Fox, P. Greenslade, J.D. Majer, G. Orshan, P.W. Rundel, R.L. Specht and W.E. Westman 3

2. **VEGETATION, NUTRITION AND CLIMATE – DATA-TABLES**

   (1) Natural vegetation – ecomorphological characters
       co-ordinated by R.L. Specht 13
   (2) Foliar analyses
       co-ordinated by P.W. Rundel 63
   (3) Species richness
       co-ordinated by W.E. Westman 81
   (4) Climate
       co-ordinated by R.L. Specht 93

3. **VEGETATION, NUTRITION AND CLIMATE – EXAMPLES OF INTEGRATION**

   (1) Mediterranean bioclimate and its variation in the palaearctic region
       by Ph. Daget, L. Ahdali and P. David 139
   (2) Climatic control of ecomorphological characters and species richness in mediterranean ecosystems of Australia
       by R.L. Specht 149
   (3) Leaf structure and nutrition in mediterranean-climate sclerophylls
       by P.W. Rundel 157

4. VERTEBRATES

   co-ordinated by P.C. Catling   171

5. SOIL AND LITTER INVERTEBRATES

   co-ordinated by J.D. Majer and P. Greenslade   197

Systematic index   227

General index   237

Preface

The regions of the world which experience a mediterranean type climate, with a cool wet season alternating with a hot dry summer, contain some of the world's most attractive landscapes. In the Old World, the mediterranean landscapes became the cradle of civilization; other mediterranean areas of the world have attracted considerable populations for many centuries. These large human populations have exerted considerable stress on the fragile ecosystems which developed in these sunny, but droughted, fire-prone landscapes. The mediterranean landscape has thus become one of the most threatened in the world.

In recent years much has been learned about the structure and function of mediterranean-type ecosystems (Di Castri and Mooney 1973, Mooney 1977, Thrower and Bradbury 1977, Mooney and Conrad 1977, Specht 1979, 1981, Miller 1981, Di Castri et al. 1981, Conrad and Oechel 1982, Quézel 1982, Margaris and Mooney 1981, Kruger et al. 1983, Long and Pons 1984, Dell et al. 1986, Tenhunen et al. 1987). Much of this research has been fostered under the International Biological Program (IBP), UNESCO Man and the Biosphere Program (MAB) and, recently, the International Society of Mediterranean Ecologists (ISOMED). To facilitate intercontinental comparisons, many of these studies have concentrated on a limited number of intensive sites thought to be representative of a general region.

The objectives of this volume are to ascertain the controls on distribution of community types within the major mediterranean climatic regions and to assess the variation in key environmental and community characteristics. This broad approach is intended to aid in the placement of specific sites in a more general framework.

This volume will emphasize four key community variables:
(1) Climatic controls on growth and community distribution.
(2) Regional variations in plant nutrient status.
(3) Species richness of plants, vertebrates, and soil fauna as a function of stand architecture, climate, soil moisture and substrate stability.
(4) Ecomorphological characteristics and seasonality.

Besides describing the variation which occurs in each of the regional MTEs (mediterranean terrestrial ecosystems), the data on ecomorphological characteristics describe aspects of community function. The changes in these parameters along environmental gradients will help understand adaptive responses of vegetation to environment, and change in community function with environmental variation. Taken as a whole, this data set will act as a resource to analyze the level of similarity of community type to varying environmental conditions and the similarity of the vegetation and climate among the various MTEs. The data on ecomorphological characters will be used to analyze the range in plant growth response to varying environmental conditions.

The tremendous wealth of plant and animal species and the variety of forms (often bizarre) which MTEs exhibit are examined across the mediterranean climatic regions. These attributes and the factors which appear to control this diversity are examined on a continental and global basis in a search for an understanding of the resilience of these ecosystems to the tremendous climatic stresses experienced in the recent geologic past and the effect of disturbance by man in recent times.

Acknowledgements: The tabulation of climatic growth factors, plant nutrients, species-richness and eco-morphological attributes of MTEs involved the cooperation of many ecologists throughout the world. The willing assistance of these contributors (listed) ensured the success of this Volume which will form the basis for a global understanding of ecosystem processes and management. R.L. Specht edited the climatic and botanical contributions; M.M. Specht edited the faunal contributions. Mrs. J. Gamack typed much of the manuscript. The cooperation of all who participated in this international exercise is gratefully acknowledged.

References

Dell, B., Hopkins, A.J.M. and Lamont, B.B. (eds.) 1986. Resilience in Mediterranean-Type Ecosystems. Tasks for Vegetation Science. Vol. 16. Junk, the Hague.

Di Castri, F., and Mooney, H.A. (eds.) 1973. Mediterranean Type Ecosystems: Origin and Structure. Springer-Verlag, Berlin.

Di Castri, F., Goodall, D.W. and Specht, R.L. (eds.) 1981. Ecosystems of the World. Vol. 11. Mediterranean Type Shrublands. Elsevier, Amsterdam.

Conrad, C.E., and Oechel, W.C. (eds.) 1982. Dynamics and Management of Mediterranean Type Ecosystems. USDA Forest Service Gen. Tech. Rep. PSW-58.

Kruger, F.J., Mitchell, D.T. and Jarvis, J.U.M. (eds.) 1983. Ecological Studies. Vol. 43. Mediterranean Type Ecosystems. Springer-Verlag, Berlin.

Long, G. and Pons, A. (eds.) 1984. Bioclimatologie Méditerranéenne. Bull. Soc. Bot. France, Act. Bot. 1984-2/3/4.

Margaris, N.S., and Mooney, H.A. (eds.) 1981. Components of Productivity of Mediterranean Climate Regions, Basic and Applied Aspects. Junk, the Hague.

Miller, P.C. (ed.) 1981. Resource Use by Chaparral and Matorral. Springer-Verlag, Berlin.

Mooney, H.A. (ed.) 1977. Convergent Evolution in Chile and California Mediterranean Climate Ecosystems. Dowden, Hutchison and Ross, Stroudsburg, Pa.

Mooney, H.A., and Conrad, C.E. (eds.) 1977. Proceedings of the Symposium of the Environmental Consequences of Fire and Fuel Management in Mediterranean Ecosystems. USDA Forest Service Gen. Tech. Rep. WO-3.

Quézel, P. (ed.) 1982. Définition et Localisation des Ecosystèmes Méditerranéens Terrestres. Ecologia Mediterranea, Tome VIII, Marseille, France.

Specht, R.L. (ed.) 1979, 1981. Ecosystems of the World. Vol. 9A & B. Heathlands and Related Shrublands. Elsevier, Amsterdam.

Tenhunen, J.D., Catarino, F.M., Lange, O.L. and Oechel, W.C. (eds.) 1987. Plant Response to Stress – Functional Analysis in Mediterranean Ecosystems. Springer-Verlag, Berlin.

Thrower, N.J.W., and Bradbury, D.E. (eds.) 1977. Chile-California Mediterranean Scrub Atlas: A Comparative Analysis. Dowden, Hutchison and Ross, Stroudsburg, Pa.

F. di Castri	UNESCO	M. Arianoutsou	Greece
E.R. Fuentes	Chile	F.M. Catarino	Portugal
F.J. Kruger	South Africa	C. Floret	France
W.C. Oechel	California-Arizona	C.A. Gracia	Spain
R.L. Specht	Australia	G. Orshan	Israel

April 1987

List of contributors

L. Ahdali, Départment de Botanique, Université de Damas, Damascus, Syria

D. Allen, C.S.I.R.O. Division of Wildlife and Rangelands Research, P.O. Box 84, Lyneham, A.C.T. 2602, Australia

L. Amandier, Rue du Soleil Levant, Pietralba, Bat. C, Ajaccio 20000, Corsica

M. Arianoutsou, Department of Biology (Division of Ecology), Aristotelian University, Thessaloniki 540 06 Greece

D.H. Ashton, Botany Department, University of Melbourne, Parkville, Vic. 3052, Australia

L. Atkins, C.S.I.R.O. Division of Wildlife and Rangelands Research, Locked Mail Bag No. 4, Midland, W.A. 6056, Australia

D.T. Bell, Botany Department, University of Western Australia, Nedlands, W.A. 6009, Australia

R. Berliner, Department of Agricultural Botany, The Hebrew University, Rehovot 76100, Israel

L. Best, National Parks and Wildlife Service, G.P.O. Box 1782, Adelaide, S.A. 5001, Australia

W. Bond, Saasveld Forestry Research Centre, Private Bag X6515, George 6530, South Africa

G.J. Breytenbach, Saasveld Forestry Research Centre, Private Bag X6515, George 6530, South Africa

W. Breytenbach, Saasveld Forestry Research Centre, Private Bag X6515, George 6530, South Africa

F.M. Catarino, Departamento de Biologia Vegetal, Faculdade de Ciencias, Universidade de Lisboa, Lisbon 1294, Portugal

P.C. Catling, C.S.I.R.O. Division of Wildlife and Rangelands Research, P.O. Box 84, Lyneham, A.C.T. 2602, Australia

P. Christensen, Department of Conservation and Land Management, P.O. Box 104, Como, W.A. 6152, Australia

H.T. Clifford, Botany Department, University of Queensland, St Lucia, Qld 4067, Australia

A.I.D. Correia, Departamento de Biologia Vegetal, Universidade de Lisboa, Lisbon 1294, Portugal

R.M. Cowling, Botany Department, University of Cape Town, Rondebosch 7700, South Africa

Ph. Daget, Institut de Botanique, 163 Rue Auguste Brousonnet, Montpellier F-34000, France

P. David, C.N.R.S. Centre Louis Emberger, B.P. 5051, Montpellier F-34033, France

S.J.J.F. Davies, Royal Australian Ornithologists Union, 21 Gladstone Street, Moonee Ponds, Vic. 3039, Australia

F. di Castri, C.N.R.S. Centre Louis Emberger, B.P. 5051, Montpellier F-34033, France

K. Dixon, Kings Park and Botanic Garden, West Perth, W.A. 6005, Australia

D. Donnelly, Plant Protection Research Institute, Private Bag X5017, Stellenbosch 7600, South Africa

C. Floret, C.N.R.S. Centre Louis Emberger, B.P. 5051, Montpellier F-34033, France

B.J. Fox, Zoology Department, University of New South Wales, Kensington, N.S.W. 2033, Australia

M.D. Fox, National Herbarium of New South Wales, Royal Botanic Gardens, Sydney, N.S.W. 2000, Australia

E.R. Fuentes, Laboratorio de Ecologia, Pontificia Universidad Cátolica de Chile, Casilla 114-D, Santiago, Chile

R. Gajardo, Department de Botanica, Universidad de Chile, Casilla 1004, Santiago, Chile

J.H. Giliomee, Department of Entomology, University of Stellenbosch, Stellenbosch 7600, South Africa

C.A. Gracia, Departament d'Ecologia, Universitat de Barcelona, Diagonal 647, Barcelona 08028, Spain

P. Greenslade, C.S.I.R.O. Division of Entomology, P.O. Box 1700, Canberra City, A.C.T. 2601, Australia

E.R. Hajek, Laboratorio de Ecologia, Pontificia Universidad Cátolica de Chile, Casilla 114-D, Santiago, Chile

R.J. Hobbs, C.S.I.R.O. Division of Wildlife and Rangelands Research, Locked Mail Bag No. 4., Midland, W.A. 6056, Australia

A.J.M. Hopkins, Western Australian Wildlife Research Centre, Wanneroo, W.A. 6065, Australia

R. How, Western Australian Museum, Francis Street, Perth, W.A. 6000, Australia

L.B. Hutley, Botany Department, University of Queensland, St Lucia, Qld 4067, Australia

M.L. Jarman, C.S.I.R. Botany Department, University of Cape Town, Rondebosch 7700, South Africa

S. Kokkini, Department of Biology (Division of Botany), Aristotelian University, Thessaloniki 540 06 Greece

F.J. Kruger, South African Forestry Research Institute, P.O. Box 727, Pretoria 0001, South Africa

B.B. Lamont, School of Biology, Curtin University of Technology, Bentley, W.A. 6102, Australia

A. Le Roux, Department of Nature and Environmental Conservation, Private Bag 5014, Stellenbosch 7600, South Africa

R. Loisel, Laboratoire de Botanique et d'Ecologie Méditerranéenne, Université d'Aix-Marseille III, Rue Henri Poincaré, Marseille 13397, France

J.D. Majer, School of Biology, Curtin University of Technology, Bentley, W.A. 6102, Australia

C. Martinez, Bolognano 20136, Corsica, France

I.J. Mason, C.S.I.R.O. Division of Wildlife and Rangelands Research, P.O. Box 84, Lyneham, A.C.T. 2602, Australia

J. Merino, Departamento de Ecologia, Facultad de Biologia, Universidad de Sevilla, Seville 41080, Spain

J.D. Molina, Laboratorio de Ecologia, Pontificia Universidad Cátolica de Chile, Casilla 114-D, Santiago, Chile

E.J. Moll, Botany Department, University of Cape Town, Rondebosch 7700, South Africa

G. Montenegro, Laboratorio de Ecologia, Pontificia Universidad Cátolica de Chile, Casilla 114-D, Santiago, Chile

M. Narog, Forest Service, United States Department of Agriculture, 4955 Canyon Crest Drive, Riverside, CA 92507, U.S.A.

Z. Naveh, Faculty of Agricultural Engineering, Technion-Israel Institute of Technology, Haifa 32000, Israel

A.E. Newsome, C.S.I.R.O. Division of Wildlife and Rangelands Research, P.O. Box 84, Lyneham, A.C.T. 2602, Australia

I.R. Noble, Research School of Biological Sciences, Australian National University, P.O. Box 4, Canberra, A.C.T. 2601, Australia

W.C. Oechel, Systems Ecology Research Group, San Diego State University, San Diego, CA 92182-0057, U.S.A.

L. Olsvig-Whittaker, Blaustein Institute for Desert Research, Ben Gurion University of the Negev, Sede Boger Campus 84990, Israel

G. Orshan, Department of Botany, The Hebrew University, Jerusalem 91904, Israel

S. Paraskevopoulos, Department of Biology (Division of Ecology), Aristotelian University, Thessaloniki, 540 06 Greece

S.A. Parker, South Australian Museum, North Terrace, Adelaide, S.A. 5000, Australia

D.C. Paton, Zoology Department, University of Adelaide, Adelaide, S.A. 5001, Australia

A.C. Postle, School of Biology, Curtin University of Technology, Bentley, W.A. 6102, Australia

P. Quézel, Laboratoire de Botanique et d'Ecologie Méditerranéenne, Université d'Aix-Marseille III, Rue Henri Poincaré, Marseille 13397, France

R.D. Quinn, Department of Biological Sciences, California State Polytechnic University, Pomona, CA 91768, U.S.A.

A. Rabinovitz-Vyn, Nature Reserves Authority, 16 Hanatziv Street, Tel Aviv, Israel

D. Rankevich, Department of Biology, Technion-Israel Institute of Technology, Haifa 32000, Israel

T.J. Ridsdill-Smith, C.S.I.R.O. Division of Entomology, Wembley, W.A. 6014, Australia

D. Robertson, Botany Department, University of Melbourne, Parkville, Vic. 3052, Australia

A.C. Robinson, National Parks and Wildlife Service, G.P.O. Box 1782, Adelaide S.A. 5001, Australia

E. Rodá, Departament d'Ecologia, Universitat Autónoma de Barcelona, Bellaterra, Barcelona, Spain

F. Romane, C.N.R.S. Centre Louis Emberger, B.P. 5051, Montpellier F-34033, France

P.W. Rundel, Laboratory of Biomedical and Environmental Science, University of California, Los Angeles, CA 90024-1786, U.S.A.

F. Sáiz, Ecology Laboratory, Universidad Catolica de Valparaiso, Casilla 4059, Valparaiso, Chile

G. Scarascia-Mugnozza, Instituto di Biologia Agraria, Universita degli Studi della Tuscia, Via Riello, Viterbo 01100, Italy

N. Scarlett, Botany Department, La Trobe University, Bundoora, Vic. 3083, Australia

G.A. Scheid, Department of Biology, San Diego State University, San Diego, CA 92181, U.S.A.

G. Smith, C.S.I.R.O. Division of Wildlife and Rangelands Research, P.O. Locked Bag No. 4, Midland, W.A. 6056, Australia

M.M. Specht, C/-Botany Department, University of Queensland, St Lucia, Qld 4067, Australia

R.L. Specht, Botany Department, University of Queensland, St Lucia, Qld 4067, Australia

P.A. Stanton, Department of Biological Sciences, California State Polytechnic University, Pomona, CA 91768, U.S.A.

G. Stewart, Department of Biological Sciences, California State Polytechnic University, Pomona, CA 91768, U.S.A.

R. Struthers, School of Biology, Curtin University of Technology, Bentley, W.A. 6102, Australia

F. Taillole, C.N.R.S. Centre Louis Emberger, B.P. 5051, Montpellier F-34033, France

J. Terradas, Departament d'Ecologia, Universitat Autónoma de Barcelona, Bellaterra, Barcelona, Spain

L. Trabaud, C.N.R.S. Centre Louis Emberger, B.P. 5051, Montpellier F-34033, France

T. Uslu, Biology Department, Gazi University, Ankara, Turkey

M.R. Warburg, Department of Biology, Technion-Israel Institute of Technology, Haifa 32000, Israel

L.E. Watson, Botany Department, University of Western Australia, Nedlands, W.A. 6009, Australia

W.E. Westman, Lawrence Berkeley Laboratory, University of California, Berkeley, CA 94720, U.S.A.

J. Wombey, C.S.I.R.O. Division of Wildlife and Rangelands Research, P.O. Box 84, Lyneham, A.C.T. 2602, Australia

C.A. Zammit, Department of Biology, San Diego State University, San Diego, CA 92182 U.S.A.

P.H. Zedler, Department of Biology, San Diego State University, San Diego, CA 92182, U.S.A.

CHAPTER 1

Climate, vegetation, vertebrates and soil/litter invertebrates of mediterranean-type ecosystems – data-banks

Climate, vegetation, vertebrates and soil/litter invertebrates of mediterranean-type ecosystems – data-banks

P.C. Catling, Ph. Daget, B.J. Fox, P. Greenslade, J.D. Majer, G. Orshan, P.W. Rundel, R.L. Specht and W.E. Westman

Introduction

The objectives of the Data-Source Book are to collate relevant parameters which appear to control the structure and function of major ecosystems (and component plant and animal species) in the mediterranean-climate regions of the world (southern Australia, California-Arizona, Chile, the Mediterranean Basin, Cape Province of South Africa).

Four key community variables are examined:
(1) Climatic controls on growth and community distribution
(2) Regional variations on plant nutrient status and the effects on sclerophylly
(3) Species richness of plants, vertebrates and soil fauna as a function of stand architecture, climate, soil moisture and substrate stability
(4) Ecomorphological characters and seasonality – the effect of changes in these parameters along environmental gradients on community function.

These attributes and the factors which appear to control diversity will form the basis of a continental and global understanding of the resilience of these ecosystems to the perturbations experienced in the recent geologic past and the effect of disturbance by man in recent times.

Many scientists have worked with the coordinators to determine the key attributes (and their method of collection) which should be included in the data-banks. Directions for the preparation of the eight data-banks (covering climate, vegetation, vertebrates and soil/litter invertebrates of mediterranean-type ecosystems) are outlined below.

The data-banks are by no means complete for all attributes or for all mediterranean ecosystems. They indicate the type of data required and where gaps still exist, requiring further international cooperation.

Some sections of the data-banks are complete enough to enable preliminary attempts at integration on a regional basis. Three examples of such integrations are presented in Section 3 of the Volume.

Data-bank 1
Major plant communities in mediterranean-type regions (most recent analysis of region)

Data-bank 2
Climate stations in mediterranean-type regions (long-term meteorological data)
– latitude
– longitude
– altitude
– vegetation type

Data-bank 3
Climatic data: Classification of mediterranean-type regions (Köppen 1923 and Emberger 1955) (prepared by Ph. Daget and R.L. Specht)
– mean annual precipitation (P)
– mean temperature: maximum of hottest month (M)
– mean temperature: minimum of coldest month (m)
– Emberger Pluviothermic Quotient (Q)

R.L. Specht (ed.) Mediterranean-type Ecosystems. ISBN 90-6193-652-7.
© Kluwer Academic Publishers.

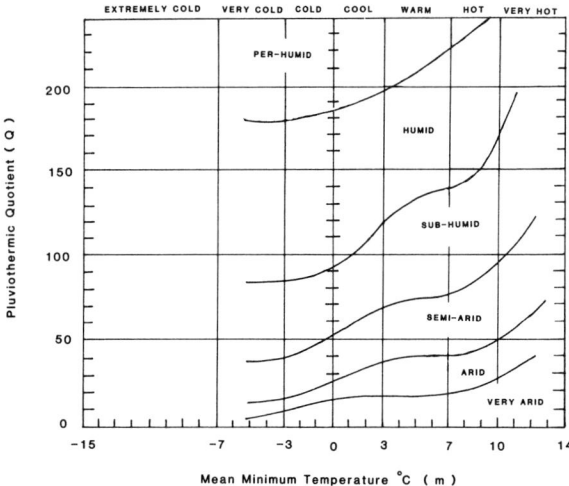

Fig. 1. Climagram showing the Emberger pluviothermic quotient (Q) and the mean minimum temperature (m) of the coldest month for per-humid to very arid mediterranean climates.

$$Q = \frac{2000P}{M^2 - m^2}$$

- Emberger climatic type (from climagram, Fig. 1)
- Köppen climatic type

References

Emberger, L., 1955. Une classification biogéographique des climats. Rev. Trav. Lab. Bot. Fac. Sci., Montpellier 7: 3–43.
Köppen, W., 1923. Der Klimate der Erde. Bornträger, Berlin.
Nahal, I., 1981. The mediterranean climate from a biological viewpoint. In: Di Castri, F., Goodall, D.W. and Specht, R.L. (eds.) Ecosystems of the World. Vol. 11. Mediterranean-Type Shrublands. Elsevier, Amsterdam, pp. 63–86.

Data-bank 4
Climatic data: Growth Indices developed by Specht (1972, 1981) (prepared by R.L. Specht)
- mean annual precipitation
- mean annual pan evaporation (class A pan)
- evaporative coefficient (k)
- mean maximum soil water storage (Smax)
- Moisture Index: mean of 12 months
- Moisture Index: mean of 3 summer months (Northern Hemisphere – June to August; Southern Hemisphere – December to February)
- Drought period: months with Moisture Index <0.10
- Drought period: months with Moisture Index <0.20
- Thermal Index (T.I.) × Moisture Index (M.I.): annual total
  (1) Relative shoot growth of temperate plants (optimum daily temperature 16°C; range 11–21°C)
  (2) Relative shoot growth of subtropical plants (optimum daily temperature 21°C, range 16–26°C)

Rationale

Moisture Index: Monthly values of the Moisture Index of an evergreen plant community may be estimated using the iterative computing technique outlined by Specht (1972).

Moisture Index (M.I.) = E_a/E_o = $k(P - R - D + S_{ext})$

where E_a = actual evapotranspiration (cm per month)
E_o = pan evaporation (cm per month)
P = precipitation (cm per month)
R = runoff (cm per month)
D = drainage (cm per month)
S_{ext} = extractable soil water (cm at beginning of the month)
k = evaporative coefficient

The evaporative coefficient (k), a constant for evergreen plant communities, indicates the resistance to the flow of water through the plant community in the soil-plant-atmosphere continuum. The value of the evaporative coefficient increases as the macroclimate changes from arid to perhumid (Fig. 2).

The Moisture Index provides an integrated estimate of stomatal control of transpiration throughout the month and, *vice versa,* stomatal control on the diffusion of carbon dioxide from the atmosphere to chloroplasts for photosynthesis and growth.

Thermal Index (T.I.) provides an estimate of the monthly shoot growth of species in the upper stratum of the plant community as daily air temper-

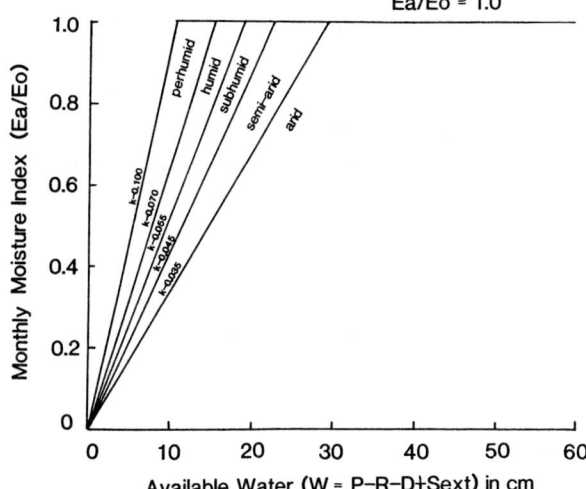

Fig. 2. Relationship of climatic zones (per-humid to arid) to the evaporative coefficient (k), computed from long-term meteorological data (Specht 1972).

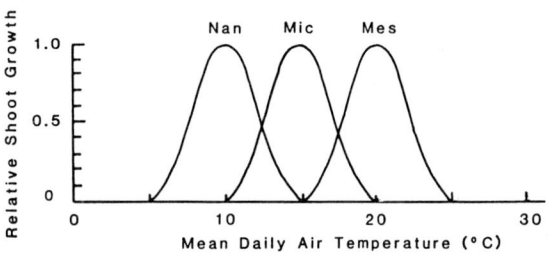

Fig. 3. Relative shoot growth of overstorey species plotted against mean daily air temperature (Specht 1981). (Mes = Mesotherm; Mic = Microtherm; Nan = Nanotherm).

ature oscillates seasonally. Studies on the monthly growth of tagged shoots reveal that overstorey species fall into broad groups with distinctive shoot growth responses to mean daily temperature (Fig. 3 from Specht (1981)).

References

Specht, R.L., 1972. Water use by perennial evergreen plant communities in Australia and Papua New Guinea. Aust. J. Bot. 20: 273–299.

Specht, R.L., 1981. Growth indices – their role in understanding the growth, structure and distribution of Australian vegetation. Oecologia (Berl.) 50: 347–356.

Data-bank 5

Species richness of mediterranean-type plant communities (prepared by W.E. Westman)
- Site location (including latitude, longitude, altitude)
- UNESCO Climatic Zone
- Soil: local name
- Soil: FAO classification
- Minimum age since fire
- Vegetation type
- Number of vascular species (woody + herbaceous + therophyte) in $1 m^2$, $10 m^2$, $100 m^2$, $1000 m^2$
- Reference
- Disturbance

Data-bank 6

Ecomorphological characters of major plant species in mediterranean-type plant communities (prepared by G. Orshan and R.L. Specht)
- Site location (including latitude, longitude, altitude)
- Vegetation type
- Ecomorphological characters and Foliage Projective Cover (%) representative of (1) upper stratum, (2) mid stratum, (3) ground stratum-dicotyledons, (4) ground stratum-monocotyledons
- References

Definition of Ecomorphological Characters: developed from:
Orshan, G. 1981. Monocharacter growth form types as a tool in an analytic-synthetic study of growth forms in mediterranean type ecosystems. Ecologia Mediterranea 8 (1/2): 159–171.

1. Renewal buds
   (1) Phanerophyte (Megaphanerophyte >30 m, Mesophanerophyte – MM 8–30 m, Microphanerophyte – M 2–8 m, Nanophanerophyte – N 25 cm–2 m) (2) Chamaephyte – Ch <25 cm (possibly up to 80 cm) (3) Hemicryptophyte – H (may have evergreen or seasonal foliage) (4) Cryptophyte (Geophyte – G) (5) Therophyte – Th (6) Amphiphytes (more than one type of renewal bud e.g. G-Ch)

2. Trophic type
   (1) Autotrophic only (2) N-fixation (3) Saprophytes (4) Stem semi-parasites (5) Root semi-parasites (6) Stem and leaf parasites (7) Root parasite (8) Carnivorous plants

3. Life duration of plant
   (1) <1 yr (2) 1–2 yr (3) 2–5 yr (4) 5–25 yr (5) 25–50 yr (6) 50–100 yr (7) >100 yr

4. Plant height
   (1) <10 cm (2) 10–25 cm (3) 25–50 cm (4) 50–100 cm (5) 1–2 m (6) 2–5 m (7) 5–10 m (8) 10–20 m (9) 20–30 m (10) >30 m

5. Crown diameter
   (1) <10 cm (2) 10–25 cm (3) 25–50 cm (4) 50–100 cm (5) 1–2 m (6) 2–5 m (7) 5–10 m (8) >10 m

6. Crown density
   (1) <10% (2) 10–25% (3) 25–50% (4) 50–75% (5) 75–90% (6) >90%

7. Photosynthetic organs
   (1) Leaves (2) Phyllodes (3) Cladodes (4) Leaves and Stems (5) Absent

8. 'Leaf size'
   (1) Subleptophyll <0.10 cm^2 (2) Leptophyll 0.10–0.25 cm^2 (3) Nanophyll 0.25–2.25 cm^2 (4) Nano-microphyll 2.25–12.25 cm^2 (5) Microphyll 12.25–20.25 cm^2 (6) Micro-mesophyll 20.25–56.25 cm^2 (7) Mesophyll 56.25–180.25 cm^2 (8) Macrophyll 180.25–1640.25 cm^2 (9) Megaphyll >1640.25 cm^2

9. 'Leaf' colour (upper and lower surface)
   (1) All green (2) All glaucous (3) All white (4) Green and white (5) Glaucous and white

10. 'Leaf' length
    (1) <1 cm (2) 1–2 cm (3) 2–5 cm (4) 5–10 cm (5) 10–20 cm (6) 20–50 cm (7) >50 cm

11. 'Leaf' width
    (1) <1 mm (2) 1–2 mm (3) 2–3 mm (4) 3–5 mm (5) 5–10 mm (6) 10–20 mm (7) 20–50 mm (8) >50 mm

12. 'Leaf' angle
    (1) Mainly horizontal (2) Mainly vertical (3) Between horizontal and vertical (include estimate of angle from horizontal)

13. 'Leaf' margin
    (1) Entire (2) Serrate/toothed (3) Lobed/deeply dissected (4) Rolled (5) Recurved/Revolute (6) Grooved/Incurved

14. 'Leaf' surface resins, oils, mucilage
    (1) Present (2) Absent

15. 'Leaf' glands and ducts
    (1) Present (2) Absent

16. 'Leaf' consistency
    (1) Malacophyll (2) Semi-sclerophyll (3) Sclerophyll (4) Semi-succulent (5) Water-succulent (6) Resin-succulent

17. 'Leaf' tomentosity
    (1) Non-hairy (2) Lower side hairy (3) Upper side hairy (4) Both sides hairy

18. 'Leaf' stomata – location
    (1) Without stomata (2) Lower side only (3) Upper side only (4) Both sides

19. 'Leaf' duration
    (1) <6 mth (2) 6–14 mth (3) 14–26 mth (4) 26–38 mth (5) 38–50 mth (6) many years (estimate number)

20. 'Leaf' seasonality
    (1) Evergreen (2) Winter shedders (3) Summer shedders

21. 'Leaf' area: Assimilating-stem area Ratio
    (1) 1:0 (no assim. stems) (2) 0:1 (no leaves) (3) 1:1 (4) 0:0 (no photosynthetic leaves or stems)

22. Organ periodically shed

(1) Whole plant (2) Shoot (3) Branch (4) Leaf

23. Organ shed – seasonality (see 24)

24. Shoot growth – seasonality
    (1) Spring (Sept.–Nov.*) (2) Summer (Dec.–Feb.) (3) Autumn (March-May) (4) Winter (June–Aug.)
    It is better to indicate actual months of active shoot growth, rather than seasons. (* Southern Hemisphere)

25. Stems – number of stems per plant, arising from ground-level
    (1) Single (2) Several (include range) (3) Absent

26. Stems – height to first branch (in metres)

27. Stems – diameter at breast height – 1.3 metres (trees and tall shrubs)
    Stems – diameter near ground level (small shrubs)

28. Stem lignification
    (1) Holoxyles (all branches lignified) (2) Semixyles (parts of branches not lignified) (3) Axyles (no lignified branches)

29. Bark consistency
    (1) Smooth (may be slightly roughened with lenticels etc.) (2) Fibrous (3) Flaky (4) Papery (5) Horny/hard (6) Corky

30. Bark thickness (from cambium outwards)
    – at breast height (trees and tall shrubs)
    – near ground level (small shrubs)
    (1) <2 mm (2) 2–5 mm (3) 5–10 mm (4) 10–20 mm (5) 20–50 mm (6) >50 mm

31. Bark-shedding rhythm
    (1) After 1 year (2) After 2–5 years (3) After more than 5 years (4) Absent

32. Bark shedding – seasonality (see 24)

33. Spinescence
    (1) Stem (2) Leaves (3) Stem and Leaves (4) Absent

34. Underground stems – morphology
    (1) Rhizomes (with short internodes) (2) Rhizodes (with elongated internodes, of regular architecture) (3) Sobols (with elongated internodes, of irregular architecture) (4) Stem tubers (5) Bulbs (6) Corms (7) Lignotubers (8) Underground caudex (Stemless Cycads, *Xanthorrhoea*, Palms)

35. Root morphology
    (1) Tap root (2) Horizontal roots (3) Vertical-horizontal roots (4) Hemispheric roots (5) Netted roots (6) Fleshy storage roots

36. Root modifications
    (1) Fleshy roots (2) Tubers (3) Water storing woody roots (4) Sucker bearing roots (5) Stilt roots (6) Contractile roots (7) None

37. Rootlet modifications
    (1) Sand-binding rootlets (2) Dauciform (cyperoid) rootlets (3) Proteoid rootlets (4) Restioid rootlets (5) Mycorrhizal rootlets (6) Root nodules (legumes, cycads) (7) Beaded short roots (e.g. *Podocarpus*) (8) None

38. Root depth
    (1) <10 cm (2) 10–25 cm (3) 25–50 cm (4) 50–100 cm (5) 1–2 m (6) 2–5 m (7) >5 m

39. Root spread
    (1) <10 cm (2) 10–25 cm (3) 25–50 cm (4) 50–100 cm (5) 1–2 m (6) 2–5 m (7) >5 m

40. Vegetative multiplication
    (1) By above ground organs (2) By below ground organs (3) By root/shoot splitting (4) None

41. Vegetative regeneration after fire
    (1) Plant killed (2) From epicormic buds below ground (3) From epicormic buds above ground (4) From non-epicormic buds below ground (5) From non-epicormic buds above ground

42. Flowering – seasonality
    (1) Spring (Sept.–Nov.*) (2) Summer (Dec.–Feb.) (3) Autumn (March-May) (4) Winter (June-Aug.)
    It is better to indicate actual months of active flowering, rather than seasons. (* Southern Hemisphere)

43. Flowering – pyrogenic (stimulated by fire)
    (1) Present (2) Absent

44. Fruit dehiscence
    (1) Dry indehiscent (2) Dry dehiscent (3) Fleshy indehiscent (4) Fleshy dehiscent

45. Seed dissemination
    (1) Shed annually (2) Shed after several years (3) Shed after fire

46. Seedling – germination type
    (1) Epigeal cryptocotyly (2) Epigeal phanerocotyly (3) Hypogeal cryptocotyly (4) Hypogeal phanerocotyly

47. Seedling – plumule position
    (1) Above surface of ground (2) Surface of ground (3) Below surface of ground

48. Number of 'leaves' per 10 cm stem

Data-bank 7
Foliar analyses of mediterranean-type plant communities (prepared by P.W. Rundel)
- Site location (including latitude, longitude, altitude)
- Vegetation type
- References
- Foliar nutrients: N, P, K, Ca, Mg
- Leaf structure: lignin, cellulose, other extractives (fats, waxes, oils)
- Sclerophyll Index (SI) of Loveless (1961)

$$SI = \frac{(Lignin + Cellulose)}{Crude\ Protein} \times 100$$

where Crude Protein = (nitrogen × 6.25)

Reference

Loveless, A.R., 1961. A nutritional interpretation of sclerophylly based on differences in the chemical composition of sclerophyllous and mesophytic leaves. Ann. Bot. 25: 168–184.

Data-bank 8
Vertebrates in mediterranean-type ecosystems – composition, species richness, dietary types, source of food (prepared by P.C. Catling and B.J. Fox)
- Site location (including latitude, longitude, altitude, area)
- Aspect
- UNESCO Climatic Zone
- Soils: FAO classification
- Vegetation type
- Abundant plant species: 3 or 4 spp.
- Vegetation strata: height, percent cover
- Disturbance: type, years
- Collecting dates
- References
- Species richness of vertebrates
- Dietary type and source of food (including foraging zone)
    (1) Mammals
        – Type of mammal: A – arboreal, B – bats, F – fossorial, L – large ground, M – medium ground, S – small ground, W – aquatic
        – Dietary type: b – browser, c – carnivore, f – fruit, x – fish, g – grazer, h – herbivore, i – insectivore, l – leaves, n – nectar, o – omnivore, s – seeds
    (2) Birds (resident and non-resident species)
        – Foraging zone: A – above canopy air space, B – below canopy air space, C – outer canopy and branches, G – ground, S – shrubs <2 metres, T – trunks and main branches, U – understorey, W – water
        Dietary type: b – seeds on shrubs and trees, c – carnivore, f – fruit, g – seed on ground, h – herbivore, i – insects, l – leaves, n – nectar, o – omnivore, s – seeds
    (3) Reptiles and Amphibia
        – Families: A – snakes, B – amphibians, C – tortoise, D – Agamidae, E – Iguanidae, F – Teiidae, G – Scincidae, H – Geck-

onidae, I – Pygopodidae, J – Varanidae, K – Typhlopidae, L – Elapidae, M – Boidae, N – Colubridae, O – Anugidae

Data-bank 9
Soil and litter invertebrates in mediterranean-type ecosystems – Composition, species richness and seasonality (prepared by J.D. Majer and P. Greenslade)
- Site location (including latitude, longitude, altitude, area)
- UNESCO Climatic Zone
- Soils: FAO Classification
- Vegetation type
- Disturbance: type, years
- Collecting dates
- References
- Epigaeic invertebrates (pitfall traps)
- Litter/soil invertebrates (Tullgren-type funnels)

Techniques

(1) Pitfall traps (epigaeic invertebrates)
The traps used are glass specimen tubes (such as large McCartney bottles), mouth diameter 1.8–1.9 cm (a number of workers in different parts of Australia have standardised on this size so that their results are comparable).

Traps should be about two-thirds filled with alcohol (90–95% ethanol) as a preservative to which should be added (and well mixed) up to 5% glycerol to reduce evaporation.

The traps are sunk in the ground with the lip absolutely flush with the soil surface so that there is no obstacle to animals falling into them. A circular area about 20 cm in diameter around each trap needs to be cleared of grasses, litter etc., to reduce variability between traps. We usually work with a unit of five traps run for a working week i.e. put in on Monday, checked on Wednesday, taken out on Friday. This gives 20 trapping-days. They often need topping up with alcohol after a couple of days – hence the check on Wednesday. There is trouble if it rains so heavily that the alcohol in the traps is diluted. Each trap then needs dividing between two specimen tubes which are then filled with alcohol to avoid decomposition.

Twenty trap-days can also be achieved by having more traps for a shorter period. But this does cause a problem with digging-in effects where the initial catch is higher and the trapping period is too short to achieve stable conditions around the traps. Preferably we use at least two sets of five traps in each habitat sampled to provide some replication. As a rule we set traps in a rough line (which makes it easier to find them again), about three or four paces apart in a reasonably typical area of whatever habitat is being sampled.

References

Greenslade, P. and Greenslade, P.J.M., 1971. The use of baits and preservatives in pitfall traps. J. Aust. Ent. Soc. 10: 253–260.
Majer, J.D., 1978. An improved pitfall trap for sampling ants and other epigaeic invertebrates. J. Aust. Ent. Soc. 17: 261–262.

(2) Tullgren-type funnels (litter and soil invertebrates)
(a) *Litter* is removed from a number (say 20) of small quadrats (say 20 × 20 cm), located at randomly selected points. The samples are bulked and sealed in a polyethylene bag. In the laboratory, invertebrates are extracted from the samples in a Tullgren-type funnel. The temperature in the funnel is raised from ambient to 40° C over a period of one week.

(b) *Soil cores* (50 mm diameter, 100 mm deep) may be collected from below all litter samples and extruded into plastic sleeves. Invertebrates may be extracted from individual soil cores using a multiple canister heat extractor in which the temperature regime is the same as that used for the litter extraction.

CHAPTER 2

Vegetation, nutrition and climate – data tables

Vegetation, nutrition and climate – data-tables

(1) Natural vegetation – ecomorphological characters

Co-ordinator: R.L. Specht

Contributors

D.H. Ashton, H.T. Clifford, L. Olsvig-Whittaker,
D. Robertson, N. Scarlett, R.L. Specht (Southeastern and South Australia)
L. Atkins, K. Dixon, R.J. Hobbs, A.J.M. Hopkins,
B.B. Lamont, R.L. Specht, R. Struthers (Southwestern Australia)
G.A. Scheid, C.A. Zammit, P.H. Zedler (California and Arizona)
R.A. Gajardo, G. Montenegro (Chile)
L. Amandier, C. Martinez, R.L. Specht (Corsica)
C. Floret, F. Romane, L. Trabaud (France)
M. Arianoutsou, S. Kokkini (Greece)
G. Orshan, A. Rabinovitz-Vyn (Israel)
P. Quezel (Mediterranean Basin)
F.M. Catarino, A.I.V.D. Correia (Portugal)
E.J. Moll (South Africa)
C.A. Gracia, J. Merino (Spain)

Contents

Australia (Tables 1–2)	14
– South Australia (Tables 3–9)	16
– Victoria (Tables 10–14)	23
– Western Australia (Tables 15–18)	28
California and Arizona (Tables 19–27)	32
Chile – Central (Tables 28–32)	41
Mediterranean Basin (Table 33)	46
– Corsica (Tables 34–35)	47
– France (Tables 36–37)	49
– Greece (Tables 38–39)	51
– Israel (Tables 40–43)	54
– Portugal (Tables 44)	57
– Spain (Tables 45–47)	58
South Africa – Cape Province (Table 48)	61

R.L. Specht (ed.) Mediterranean-type Ecosystems. ISBN 90-6193-652-7.
© Kluwer Academic Publishers.

Table 1. Natural Vegetation (TWINSPAN floristic groups) of the mediterranean-region of southeastern Australia – R.L. Specht.

Supra-Mediterranean Zone	
Tall-open forest	*Eucalyptus regnans*
	E. cypellocarpa-E. obliqua-E. radiata
Mediterranean Zone	
Open-forest (sclerophyll*)	*E. dives-E. macrorhyncha-E. radiata*
	E. macrorhyncha-E. polyanthemos
	E. obliqua (approaches savanna in some areas)
	E. obliqua-E. viminalis
Woodland (sclerophyll)	*E. baxteri*
	E. fasciculosa
Woodland (savanna*)	*E. leucoxylon*
	E. microcarpa
	E. odorata (E. porosa)-E. microcarpa
	Allocasuarina luehmannii
	Allocasuarina verticillata-Melaleuca lanceolata
	Callitris preissii
Woodland (wetland)	*E. camaldulensis*
	E. largiflorens
Open scrub (sclerophyll)	*E. diversifolia-E. cosmophylla*
– Mallee	*E. diversifolia-E. socialis*
	E. incrassata-E. foecunda-Melaleuca uncinata
	E. incrassata-E. viridis
Open scrub (savanna)	*E. behriana*
– Mallee	*E. socialis-E. dumosa*
	E. socialis-E. dumosa-E. gracilis
Heathland (dry)	*Acacia bivenosa ssp. wayi*
	Allocasuarina pusilla-Leptospermum myrsinoides-Xanthorrhoea australis
	Banksia marginata-Epacris impressa
	Melaleuca uncinata-Xanthorrhoea tateana
	Pteridium esculentum
Grassland	*Lomandra dura-L. effusa*
	Themeda triandra

* Sclerophyll refers to heath understorey; savanna refers to grassy understorey.

References

Boomsma, C.D. and Lewis, N.B., 1980. The Native Forest and Woodland Vegetation of South Australia. S. Aust. Woods and Forests Dept., Bull. No. 25, 313 pp. + map.

Carnahan, J.A., 1976. Natural vegetation. In: Atlas of Australian Resources. Second Series. Div. National Mapping, Dept. National Resources, Canberra, 26 pp. + map 1:6,000,000.

Specht, R.L., 1972. The Vegetation of South Australia. Second Edition. Govt. Printer, Adelaide, 328 pp.

Specht, R.L. (ed.) in press. Major Plant Communities in Australia: An Objective Assessment. C.S.I.R.O. Aust., Melbourne.

Table 2. Natural Vegetation (TWINSPAN floristic groups) of south-west Western Australia – R.L. Specht.

Supra-Mediterranean Zone	
Tall open forest	*Eucalyptus diversicolor – E. calophylla*
Mediterranean Zone	
Open forest (sclerophyll*)	*E. marginata – E. calophylla*
Woodland (sclerophyll)	*E. gomphocephala – limestone heath*
	E. loxophleba – Acacia acuminata
	E. loxophleba – E. salmonophloia – Melaleuca uncinata
	E. wandoo – Leptospermum erubescens
Woodland (savanna*/sclerophyll)	*E. loxophleba – E. salmonophloia – Acacia acuminata*
Low woodland	*E. marginata – Banksia attenuata – B. menziesii*
Woodland (wetland)	*E. rudis – Melaleuca rhaphiophylla*
Low woodland (wetland)	*Agonis hypericifolia – Banksia littoralis*
Open scrub (sclerophyll)	*E. obtusiflora – Allocasuarina campestris – Melaleuca cordata* (mallee-heath)
	E. gracilis – E. foecunda (mallee)
	E. leptopoda – Baeckea maidenii – Melaleuca cordata (sclerophyll mallee)
	E. diversifolia – E. oleosa – Melaleuca lanceolata (limestone mallee)
Heathy shrubland	*Allocasuarina acutivalvis – Callitris preissii – Hakea falcata*
	Allocasuarina campestris – Melaleuca uncinata
Shrubby heathland	*Calothamnus quadrifidus – Dryandra nivea* (kwongan on sand)
	Dryandra nivea – Xanthorrhoea drummondii (kwongan on laterite)

* Sclerophyll refers to heath understorey; savanna refers to grassy understorey.

References

Beard, J.S. 1975, 1981. Vegetation Survey of Western Australia. Sheet 4. Nullarbor. Sheet 7. Swan. (1:1,000,000 Vegetation Series). University of Western Australia Press, Perth.

Specht, R.L. (ed.) in press. Major Plant Communities in Australia: An Objective Assessment. C.S.I.R.O. Aust., Melbourne.

Table 3. Ecomorphological characters: South Australia.
Location: Mount Lofty Summit, South Australia (34° 59′ S, 138° 43′ E).
Vegetation structure: Open-forest with heath understorey (*Eucalyptus obliqua – E. baxteri* alliance).
Age since last fire: >10 years.
Soil: Shallow grey-brown sandy soil (Uc 6.11) over quartzite.
Reference: Specht, R.L. and Perry, R.A., 1948. Trans. R. Soc. S. Aust. 72: 91–132.
Compiler: R.L. Specht, February 1983.

Character	*Eucalyptus obliqua* (Myrtaceae)	*Eucalyptus baxteri* (Myrtaceae)	*Banksia marginata* (Proteaceae)	*Xanthorrhoea semiplana* (Xanthorrhoeaceae)	*Leptospermum myrsinoides* (Myrtaceae)	*Acacia myrtifolia* (Mimosaceae)	*Pultenaea daphnoides* (Fabaceae)	*Platylobium obtusangulum* (Fabaceae)	*Acrotriche fasciculiflora* (Epacridaceae)	*Hibbertia sericea* (Dilleniaceae)	*Lepidosperma semiteres* (Cyperaceae)
Foliage Projective Cover (%)											
Upper stratum (20–26 m)	←	59	→	–	–	–	–	–	–	–	–
Mid stratum (1–2 m)	–	–	2	1	1	1	1	–	–	–	–
Ground stratum (25–100 cm)	–	–	–	–	–	–	–	15	11	2	6
1. Renewal buds	MM	MM	N	N	N	N	N	Ch	Ch	Ch	Hev
2. Trophic type	1	1	1	1	1	2	2	2	1	1	1
3. Life duration (yrs)	7	7	6	6	5	4	4	4	5	5	6
4. Height (m)	9	9	5	4	5	5	5	4	4	2	2
5. Crown diameter (m)	7	7	5/6	5	4	5	3	2	2	2	2
6. Canopy density (%)	4/5	5	6	6	6	4	5	5	6	5	6
7. Photosynthetic organs	1	1	1	1	1	2	1	1	1	1	4
8. 'Leaf' size	6	6	4	7	2	4	3	4	2	2	4
9. 'Leaf' colour	1	1	4	1	1	1	1	1	1	3	1
10. 'Leaf' length (cm)	4/5	3/4	3/4	7	1	3/4	2/3	2/3	1/2	1/2	6
11. 'Leaf' width (mm)	6/7	7	5	5	2	6/7	5	2/3	3/4	3	2
12. 'Leaf' angle (°)	3 (30°)	3 (30°)	3 (50°)	3 (50°)	3 (50°)	3 (55°)	3 (60°)	3 (20°)	1	3 (45°)	2
13. 'Leaf' margin	1	1	5	1	1	1	1	1	1	5	1
14. 'Leaf' surface resins	2	2	2	2	2	2	2	2	2	2	2
15. 'Leaf' glands	1	1	2	2	1	2	2	2	2	2	2
16. 'Leaf' consistency	3	3	3	3	3	3	2	2	3	2	3
17. 'Leaf' tomentosity	1	1	2	1	1	1	1	1	3	4	1
18. 'Leaf' stomata	4	4									
19. 'Leaf' duration (yrs)	2/3	2/3	2/3	?	2	3/4	2	2	2/3	2	?
20. 'Leaf' seasonality	1	1	1	1	1	1	1	1	1	1	1
21. 'Leaf' area: assim. stem area	1	1	1	1	1	1	1	1	1	1	3
22. Organs periodically shed	4	4	4	–*	4	4	4	4	4	4	–*
23. Organs shed – seasonality	2	2	2	–	2	2	2	2	2	2	–
24. Shoot growth – seasonality	2	2	2	1	1	1	1	1	1	1	1
25. Stem – number from base	1	1	2 (1–3)	3	2	1	1	1	2	2	3
26. Stem – height (m)	6–12	6–10	0.3–0.4	–	–	–	–	–	–	–	–
27. Stem – diameter (cm)	50	50	2–3	–	<1	2	1	1	<1	<1	–
28. Stem lignification	1	1	1	–	1	1	1	1	1	1	–
29. Bark consistency	2	2	1	–	1/4	1	1	1	1	1	–
30. Bark thickness (mm)	10–15	10–15	1	–	1	1	1	<1	<1	<1	–
31. Bark shedding	4	4	4	–	4	4	4	4	4	4	–
32. Bark shedding – seasonality	–	–	–	–	–	–	–	–	–	–	–
33. Spinescence	4	4	4	4	4	4	4	2	2	4	4
34. Underground stems	7	7	–	8	7	–	–	–	–	–	1
35. Root morphology	3	3	3	4	3	3	3	3	3	3	4
36. Root modification	7	7	7	6	7	7	7	7	7	7	6
37. Rootlet modification	5	5	3	8	5	6	6	6	8	8	2
38. Root depth (cm)	7	7	5/6	5/6	5/6	4/5	4/5	2/3	3/4	2/3	2/3
39. Root spread (cm)	7	7	5	5	5	5	4	3	3	4	2
40. Veg. multiplication	4	4	4	2	4	4	4	4	4	4	2
41. Veg. regen. after fire	3	3	2/3	4	2	1	1	1	2	2	4
42. Flowering – seasonality	2/3	2	2/3	1	1	1	1	1	1	1	1
43. Flowering – pyrogenic	2	2	2	1	2	2	2	2	2	2	1
44. Fruit dehiscence	2	2	2	2	2	2	2	2	3	2	1
45. Seed dissemination	2/3	2/3	3	1	2/3	1	1	1	1	1	1
46. Seedling – germ. type	2	2	2	–	2	2	2	2	2	2	–
47. Plumule position	1	1	1	2	1	1	1	1	1	1	2
48. Number leaves per 10 cm stem	5–6	4–5	20–30	–	200	10–12	15–18	3–4	50	25–45	–

* Leaf dead but not shed.

Code of Ecomorphological Characters: See Chapter 1.

Table 4. Ecomorphological characters: South Australia.
Location: Para Wirra Recreation Park, South Australia (34° 43′ S, 138° 50′ E).
Vegetation structure: Woodland with heath understorey (*Eucalyptus goniocalyx-E. fasciculosa* alliance).
Age since last fire: >20 years.
Soil: Hard-setting loamy soil with yellow clayey subsoil – grey-brown podzolic soils (Dy 2.21 and Dy 2.22) over shales.
Reference: Specht, R.L. Brownell, P.F. and Hewitt, P.A., 1961. Trans. R. Soc. S. Aust. 85: 155–176.
Compiler: R.L. Specht, February 1983.

Character	*Eucalyptus goniocalyx* (Myrtaceae)	*Eucalyptus fasciculosa* (Myrtaceae)	*Acacia pycnantha* (Mimosaceae)	*Xanthorrhoea semiplana* (Xanthorrhoeaceae)	*Calytrix tetragona* (Myrtaceae)	*Astroloma conostephioides* (Epacridaceae)	*Spyridium spathulatum* (Rhamnaceae)	*Hibbertia sericea* (Dilleniaceae)	*Acrotriche serrulata* (Epacridaceae)	*Lepidosperma semiteres* (Cyperaceae)
Foliage Projective Cover (%)										
Upper stratum (10 m)	14	20	–	–	–	–	–	–	–	–
Mid stratum (2–4 m)	–	–	5	–	–	–	–	–	–	–
Ground stratum (<2 m)	–	–	–	32	3	2	trace	1	1	trace
Density (mature plants ha^{-1})	400	300	1,100							
1. Renewal buds	MM	MM	M	N	N	N	N	Ch	Ch	Hev
2. Trophic type	1	1	2	1	1	1	1	1	1	1
3. Life duration (yrs)	7	7	5	6	5	5	4	5	4	6
4. Height (m)	7/8	7/8	6	5	5	4	4/5	3	2	2
5. Crown diameter (m)	7/8	7	4/5	6	4/5	4	4	3	3	2
6. Canopy density (%)	3/4	3/4	3/4	6	3	3/4	3/4	3/4	4	6
7. Photosynthetic organs	1	1	2	1	1	1	1	1	1	4
8. 'Leaf' size	6	5	6	6	1	1/2	2/3	2	1	3
9. 'Leaf' colour	1	1	1	1	1	1	4	4	1	1
10. 'Leaf' length (cm)	5/6	4/5	4/5	7	1	1/2	1/2	1/2	1	6/7
11. 'Leaf' width (mm)	6/7	6/7	6/7	5	1	2	4	2	1/2	2
12. 'Leaf' angle (°)	3 (45°)/2	3 (45°)/2	3 (50°)/2	3 (50°)	3 (45°)	3 (30°)	3 (30°)	3 (40°)	3 (40°)	2
13. 'Leaf' margin	1	1	1	1	1	5	1	5	1	1
14. 'Leaf' surface resins	2	2	2	2	2	2	2	2	2	2
15. 'Leaf' glands	1	1	2	2	1	2	2	2	2	2
16. 'Leaf' consistency	3	3	3	3	2	3	2	2	3	3
17. 'Leaf' tomentosity	1	1	1	1	4	1	2	4	4/1	1
18. 'Leaf' stomata	4	4	4							
19. 'Leaf' duration (yrs)	3	3	3	?	2	3	3	3	3	?
20. 'Leaf' seasonality	1	1	1	1	1	1	1	1	1	1
21. 'Leaf' area: assim. stem area	1	1	1	1	1	1	1	1	1	1
22. Organs periodically shed	4	4	4	–*	4	4	4	4	4	–*
23. Organs shed – seasonality	2	2	2	–	2	2	2	2	2	–
24. Shoot growth – seasonality	2	2	1	4/1	1	1	1	1	1	1
25. Stem – number from base	1	1	1	3	1/2	2	1	2	2	3
26. Stem – height (m)	3–4 m	3–4 m	2 m	–	–	–	–	–	–	–
27. Stem – diameter (cm)	25 cm	23 cm	2–5 cm	–	–	–	–	–	–	–
28. Stem lignification	1	1	1	–	1	1	1	1	1	–
29. Bark consistency	2	1	1	–	1	1	1	1	1	–
30. Bark thickness (mm)	6 mm	5 mm	2 mm	–	–	–	–	–	–	–
31. Bark shedding	4	2	4	–	–	–	–	–	–	–
32. Bark shedding – seasonality	–	1/2	–	–	–	–	–	–	–	–
33. Spinescence	4	4	4	4	4	2	4	4	2	4
34. Underground stems	7	7	–	8	–	–	–	–	–	1
35. Root morphology	3	3	3	4	3	3	3	3	3	4
36. Root modification	7	7	7	6	7	7	7	7	7	6
37. Rootlet modification	5	5	6	8	5	8	8	8	8	2
38. Root depth (cm)	7	7	6	5/6	3/4	4/5	4/5	2/3	2/3	2/3
39. Root spread (cm)	7	7	6	5	3/4	4	4	4	3	2
40. Veg. multiplication	4	4	4	2	4	4	4	4	4	2
41. Veg. regen. after fire	3	3	1	4	1	2	2	2	2	4
42. Flowering – seasonality	3/4/1	3/4/1	1	1	1	3/4/1	1/2	1	1/2	1/2
43. Flowering – pyrogenic	2	2	2	1	2	2	2	2	2	2
44. Fruit dehiscence	2	2	2	2	1	3	2	2	3	1
45. Seed dissemination	2/3	2/3	2	1	1	1	1	1	1	1
46. Seedling – germ. type	2	2	2	–	2	2	2	2	2	–
47. Plumule position	1	1	1	2	1	1	1	1	1	2
48. Number of leaves per 10 cm stem	4–5	4–6	3	–	170–210	70–150	20	25–45	?	–

* Leaf dead but not shed.

Code of Ecomorphological Characters: See Chapter 1.

Table 5. Ecomorphological characters: South Australia.
Location: Dark Island Soak, near Keith, South Australia (36° 02′ S, 140° 31′ E).
Vegetation structure: Open-heathland (dry-heathland) (*Banksia ornata – Xanthorrhoea australis – Allocasuarina pusilla* alliance).
Age since last fire: 25 years.
Soil: Deep sandy soil with neutral mottled yellow clayey subsoil (solodized solonetzic) – Dy 5.43.
References: Specht, R.L. and Rayson, P., 1957. Aust. J. Bot. 5: 52–85. Specht, R.L., and Rayson, P., 1957. Aust. J. Bot. 5: 103–114. Specht, R.L., Rayson, P. and Jackman, M.E. 1958. Aust. J. Bot. 6: 59–88.
Compiler: R.L. Specht, December 1981.

Character	*Banksia ornata* (Proteaceae)	*Banksia marginata* (Proteaceae)	*Xanthorrhoea australis* (Xanthorrhoeaceae)	*Allocasuarina pusilla* (Casuarinaceae)	*Leptospermum myrsinoides* (Myrtaceae)	*Phyllota remota* (Fabaceae)	*Calytrix alpestris* (Myrtaceae)	*Hibbertia riparia* (Dilleniaceae)	*Hibbertia sericea* (Dilleniaceae)	*Hypolaena fastigiata* (Restionaceae)	*Lepidosperma carphoides* (Cyperaceae)	*Lepidosperma laterale* (Cyperaceae)
Foliage Projective Cover (%)												
Upper stratum (>2 m)	11.3	10.8	10.0	–	–	–	–	–	–	–	–	–
Mid stratum (30 cm–2 m)	–	–	–	14.0	17.2	trace	–	–	–	–	–	–
Ground stratum (<30 cm)	–	–	–	–	–	–	2.0	4.5	1.2	3.1	2.9	1.6
Density (mature plants m^{-2})	0.02	0.75	0.46	0.16	0.55	trace	0.28	1.04	0.13	?	?	4.48
Biomass tops (kg ha^{-1})	12,899	4,090	3,067	2,886	2,619	trace	128	220	99	413	259	91
1. Renewal buds	N	N	G	N	N	N	Ch	Ch	Ch	Hev	Hev	Hev
2. Trophic type	1	1	1	2	1	2	1	1	1	1	1	1
3. Life duration (yrs)	6	6	6	5	5	4	5	5	5	6	6	6
4. Height (m)	5/6	5	4	5	5	2/3	2/3	2/3	2/3	2/3	2/3	2/3
5. Crown diameter (m)	5/6	4/5	5	4/5	5	2	2	2	2	1	1/2	1/2
6. Canopy density (%)	6	6	5	4/5	5	4	4	3	4	3/4	6	6
7. Photosynthetic organs	1	1	1	3	1	1	1	1	1	1	4	4
8. 'Leaf' size	4	3	6	3	2	2	1	1	2	3	3	4
9. 'Leaf' colour	1	4	1	1	1	1	1	1	1	1	1	1
10. 'Leaf' length (cm)	4	3	7	3/4	1	1	1	1	1	5/6	5/6	5/6
11. 'Leaf' width (mm)	6	4/5	4	1	1	1	1	1	1	1	1	3
12. 'Leaf' angle (°)	3 (55°)	3 (50°)	3 (50°)	3 (65°)	3 (50°)	3 (45°)	3 (30°)	3 (60°)	3 (30°)	2	2	2
13. 'Leaf' margin	2	5	1	1	1	1	1	5	5	1	6	1
14. 'Leaf' surface resins	2	2	2	2	2	2	2	2	2	2	2	2
15. 'Leaf' glands	2	2	2	2	1	2	1	2	2	2	2	2
16. 'Leaf' consistency	3	3	3	3	3	2	2	3	2	3	3	3
17. 'Leaf' tomentosity	1	2	1	1	1	1	1	1	2	1	1	1
18. 'Leaf' stomata												
19. 'Leaf' duration (yrs)	4	3	3/4	3/4	2/3	2/3	2/3	2/3	3/4	?	?	?
20. 'Leaf' seasonality	1	1	1	1	1	1	1	1	1	1	1	1
21. 'Leaf' area: assim. stem area	1	1	1	2	1	1	1	1	1	2	1	1
22. Organs periodically shed	4	4	–*	2	4	4	4	4	4	–*	–*	–*
23. Organs shed – seasonality	2	2	–	2	2	2	2	2	2	–	–	–
24. Shoot growth – seasonality	2	2	4/1	1	1	1	1	1	1	1/2	1/2	1/2
25. Stem – number from base	1	2	3	2	2	2	1	2	2	3	3	3
26. Stem – height (m)	0.05	0.05	–	0.10	0.10	0.05	0.05	0.05	0.05	–	–	–
27. Stem – diameter (cm)	2–3	1	–	<1	<1	<1	<1	<1	<1	–	–	–
28. Stem lignification	1	1	–	1	1	1	1	1	1	2	–	–
29. Bark consistency	1	1	–	1	4	1	1	1	1	–	–	–
30. Bark thickness (mm)	2	2	–	1	1	1	1	1	1	–	–	–
31. Bark shedding	4	4	–	4	4	4	4	4	4	–	–	–
32. Bark shedding – seasonality	–	–	–	–	–	–	–	–	–	–	–	–
33. Spinescence	4	4	4	4	4	4	4	4	4	4	4	4
34. Underground stems	–	–	8	–	–	–	–	–	–	1	1	1
35. Root morphology	3	3	4	3	3	3	3	3	3	4	4	4
36. Root modification	7	4	6	7	7	7	7	7	7	6	6	6
37. Rootlet modification	3	3	8	6	5	6	5	8	8	4	2	2
38. Root depth (cm)	6	6	6	6	6	4	4	4	4	2/3	2/3	2/3
39. Root spread (cm)	7	6	5	5	5	4	4	4	4	1	2	2
40. Veg. multiplication	4	2	2	4	4	2/4	4	4	4	2	2	2
41. Veg. regen. after fire	1	2	4	2	2	2	2	2	2	4	4	4
42. Flowering – seasonality	3/4	3	1	3	1	1	1	4/1	1	1/2	3	3
43. Flowering – pyrogenic	2	2	1	2	2	2	2	2	2	2	2	2
44. Fruit dehiscence	2	2	2	2	2	2	1	2	2	1	1	1
45. Seed dissemination	3	3	1	3	2/3	1	1	1	1	1	1	1
46. Seedling – germ. type	2	2	–	2	2	2	2	2	2	–	–	–
47. Plumule position	1	1	2	1	1	1	1	1	1	2	2	2
48. Number leaves per 10 cm stem	12–16	20–30	–	40–50	200	320	150	60–80	25–45	–	–	–

* Leaf dead but not shed.

Code of Ecomorphological Characters: See Chapter 1.

Table 6. Ecomorphological characters: South Australia.
Location: Keith, South Australia (36° 06′ S, 140° 21′ E).
Vegetation structure: Savanna woodland (*Eucalyptus leucoxylon* alliance).
Age since last fire: Unknown.
Soil: Shallow red sandy-loam soil – sandy terra rossa soil (Um 6.24) over limestone.
Reference: Specht, R.L., 1951. Trans. R. Soc. S. Aust. 74: 79–107.
Compiler: R.L. Specht, August 1983.

Character	*Eucalyptus leucoxylon* (Myrtaceae)	*Allocasuarina muelleriana* (Casuarinaceae)	*Stipa eremophila* (Poaceae)	*Clematis microphylla* (Ranunculaceae)
Foliage Projective Cover (%)				
Upper stratum (10–15 m)	30	–	–	–
Mid stratum (2–4 m)	–	2	–	2
Ground stratum (< 30 cm)	–	–	22	–
1. Renewal buds	MM	N/M	Hev	L
2. Trophic type	1	2	1	1
3. Life duration (yrs)	7	6	6	6
4. Height (m)	8	6	2	6
5. Crown diameter (m)	7	5	2	5
6. Canopy density (%)	4	6	6	3
7. Photosynthetic organs	1	3	1	1
8. 'Leaf' size	6	3	3	2/3
9. 'Leaf' colour	1	1	1	1
10. 'Leaf' length (cm)	5/6	3/4	6	2/3
11. 'Leaf' width (mm)	6/7	1	4	4
12. 'Leaf' angle (°)	3 (60°)/2	3 (65°)	3 (80°)	3
13. 'Leaf' margin	1	1	1	1
14. 'Leaf' surface resins	2	2	2	2
15. 'Leaf' glands	1	2	2	2
16. 'Leaf' consistency	3	3	1	1
17. 'Leaf' tomentosity	1	1	1	1
18. 'Leaf' stomata	4	4	3	4
19. 'Leaf' duration (yrs)	2/3	3/4	2	2/3
20. 'Leaf' seasonality	1	1	1	1
21. 'Leaf' area: assim. stem area	1	2	1	1
22. Organs periodically shed	4	2	–	4
23. Organs shed – seasonality	2	2	–	2
24. Shoot growth – seasonality	2	1	1	1
25. Stem – number from base	1	1	3	1
26. Stem – height (m)	8	0.10	–	?
27. Stem – diameter (cm)	40	5	–	?
28. Stem lignification	1	1	–	1
29. Bark consistency	1	1	–	1
30. Bark thickness (mm)	5	2	–	1
31. Bark shedding	2	4	–	4
32. Bark shedding – seasonality	1/2	–	–	–
33. Spinescence	4	4	4	4
34. Underground stems	7	–	1	–
35. Root morphology	3	3	4	1
36. Root modification	7	7	7	7
37. Rootlet modification	5	6	8	8
38. Root depth (cm)	7	6	4/5	?
39. Root spread (cm)	7	5	4	?
40. Veg. multiplication	4	4	2	4
41. Veg. regen. after fire	3	1/2	4	1
42. Flowering – seasonality	4/1	3	1	4/1
43. Flowering – pyrogenic	2	2	2	2
44. Fruit dehiscence	2	2	1	1
45. Seed dissemination	2/3	3	1	1
46. Seedling – germ. type	2	2	–	2
47. Plumule position	1	1	2	1
48. Number of leaves per 10 cm stem	5–6	40–50	–	2

Code of Ecomorphological Characters: See Chapter 1.

Table 7. Ecomorphological characters: South Australia.
Location: Warrenben Conservation Reserve, Yorke Peninsula, South Australia (35° 08' S, 137° 02' E).
Vegetation structure: Mallee open-scrub (*Eucalyptus diversifolia* alliance).
Age since last fire: Unknown.
Soil: Calcareous sand (Uc 1.11) over limestone (aeolianite).
Reference: Foale, M.R. (ed.) 1977. Nature Conserv. Soc. S. Aust. Publ., Adelaide.
Compiler: R.L. Specht, August 1983.

Character	*Eucalyptus diversifolia* (Myrtaceae)	*Eucalyptus oleosa* (Myrtaceae)	*Melaleuca lanceolata* (Myrtaceae)	*Acacia rupicola* (Mimosaceae)	*Dodonaea humilis* (Sapindaceae)	*Pomaderris obcordata* (Rhamnaceae)	*Beyeria lechenaultii* (Euphorbiaceae)	*Templetonia retusa* (Fabaceae)	*Acrotriche patula* (Epacridaceae)
Foliage Projective Cover (%)									
Upper stratum (4–8 m)	36	10	–	–	–	–	–	–	–
Mid stratum (3–6 m)	–	–	19	–	–	–	–	–	–
Ground stratum (<1 m)	–	–	–	5	7	2	2	3	7
1. Renewal buds	M	M	M	N	N	N	N	N	Ch
2. Trophic type	1	1	1	2	1	1	1	2	1
3. Life duration (yrs)	7	7	7	6	4/6	4/6	4/6	4/6	6
4. Height (m)	6/7	7	6	4	4	4	4	5	3
5. Crown diameter (m)	6	6	5	4	3	3	3	5	3
6. Canopy density (%)	3	3	4	3	3	4	4	3	3
7. Photosynthetic organs	1	1	1	2	1	1	1	1	1
8. 'Leaf' size	4	4	1	2	1	3	2/3	3	1
9. 'Leaf' colour	1	1	1	1	1	1	4	2	1
10. 'Leaf' length (cm)	4	4	1	2	1	2	2	2	1
11. 'Leaf' width (mm)	6	6	2	2	4	5/6	2	5/6	2
12. 'Leaf' angle (°)	3 (50°)/2	3 (60°)/2	3 (45°)	1 (5°)	3 (55°)	3 (40°)	3 (55°)	3 (55°)	3 (25°)
13. 'Leaf' margin	1	1	1	1	2	1/5	5	1	1
14. 'Leaf' surface resins	2	2	2	1	1/2	2	1	2	2
15. 'Leaf' glands	1	1	1	2	2	2	2	2	2
16. 'Leaf' consistency	3	3	3	3	2	2	2	3	3
17. 'Leaf' tomentosity	1	1	1	1	1	1	1	1	1
18. 'Leaf' stomata	4	4							
19. 'Leaf' duration (yrs)	3	3/4	3	3	3	3	2	2	3
20. 'Leaf' seasonality	1	1	1	1	1	1	1	1	1
21. 'Leaf' area: assim. stem area	1	1	1	1	1	1	1	1	1
22. Organs periodically shed	4	4	4	4	4	4	4	4	4
23. Organs shed – seasonality	2	2	2	2	2	2	2	2	2
24. Shoot growth – seasonality	2	2	1	1	1	1	1	2	1
25. Stem – number from base	2 (3–8)	2 (3–8)	1	1	1	2(1–2)	1	1	2
26. Stem – height (m)	3	2–4	1–2	small	small	small	small	small	small
27. Stem – diameter (cm)	2–6	5–8	2–4	1–2	1–2	1	<1	2–3	<1
28. Stem lignification	1	1	1	1	1	1	1	1	1
29. Bark consistency	1	1	2/6	1	1	1	1	1	1
30. Bark thickness (mm)	1–2	1–2	2–3	–	–	–	–	–	–
31. Bark shedding	2	2	4	–	–	–	–	–	–
32. Bark shedding – seasonality	1/2	1/2	–	–	–	–	–	–	–
33. Spinescence	4	4	4	2	4	4	4	4	2
34. Underground stems	7	7	–	–	–	–	–	–	–
35. Root morphology	2/3	2/3	3	3	3	3	3	3	3
36. Root modification	7	7	7	7	7	7	7	7	7
37. Rootlet modification	5	5	5	6	8	8	8	6	5
38. Root depth (cm)	7	7	6	5	5	5	5	5	3/4
39. Root spread (cm)	7	7	6	5	5	5	5	5	3
40. Veg. multiplication	4	4	4	4	4	4	4	4	4
41. Veg. regen. after fire	2	2	2	1	1	1	1	1	2
42. Flowering – seasonality	4/1	3/4	2	1	1/2	3/4	1	4/1	1
43. Flowering – pyrogenic	2	2	2	2	2	2	2	2	2
44. Fruit dehiscence	2	2	2	2	2	2	2	2	3
45. Seed dissemination	2/3	2/3	2/3	1	1	1	1	1	1
46. Seedling – germ. type	2	2	2	2	2	2	2	2	2
47. Plumule position	1	1	1	1	1	1	1	1	1
48. Number of 'leaves' per 10 cm	10–12	7–10	85–90	70–75	16	24–38	14–16	20–24	80

Code of Ecomorphological Characters: See Chapter 1.

Table 8. Ecomorphological characters: South Australia.
Location: Dark Island Soak, near Keith, South Australia (36° 02′ S, 140° 31′ E).
Vegetation structure: Mallee open-scrub (*Eucalyptus incrassata – Melaleuca uncinata* alliance).
Age since last fire: 29 years.
Soil: Shallow sandy soil with neutral mottled yellow clayey subsoil (solodized solonetzic) – Dy 5.43.
Reference: Specht, R.L., 1966. Aust. J. Bot. 14: 361–371.
Compiler: R.L. Specht, February 1983.

Character	*Eucalyptus incrassata* (Myrtaceae)	*Eucalyptus foecunda* (Myrtaceae)	*Melaleuca uncinata* (Myrtaceae)	*Hakea muellerana* (Proteaceae)	*Pultenaea tenuifolia* (Fabaceae)	*Acacia spinescens* (Mimosaceae)	*Baeckea crassifolia* (Myrtaceae)	*Calytrix tetragona* (Myrtaceae)	*Brachyloma ericoides* (Epacridaceae)	*Lepidosperma laterale* (Cyperaceae)
Foliage Projective Cover (%)										
Upper stratum (3–4 m)	25	9.5	–	–	–	–	–	–	–	–
Mid stratum (30 cm–2 m)	–	–	8	6	–	–	–	7.5	2.5	–
Ground stratum (<30 cm)	–	–	–	–	2	trace	6	–	–	2
Density (mature plants m^{-2})	0.21	0.03	0.33	0.02	trace	trace	394	433	355	39
Biomass tops (kg ha^{-1})	13,095	1,254	1,473	1,000						
1. Renewal buds	M	M	N	N	Ch	Ch	Ch	Ch	Ch	Hev
2. Trophic type	1	1	1	1	2	2	1	1	1	1
3. Life duration (yrs)	7	7	6	5/6	4	4	5	5	5	6
4. Height (m)	6	6	5	5/6	2/3	2/3	2	2	2	2
5. Crown diameter (m)	6	6	4	6	2/3	2	2	2/3	3/4	1
6. Canopy density (%)	2/3	5/6	3/4	5/6	3/4	2	5	4	5/6	6
7. Photosynthetic organs	1	1	1	1	1	3	1	1	1	1
8. 'Leaf' size	5	4	2	3	1	3	1	1	1	4
9. 'Leaf' colour	1	1	1	1	1	1	1	1	1	1
10. 'Leaf' length (cm)	4	3/4	2/3	3/4	1	2/3	1	1	1	5/6
11. 'Leaf' width (mm)	7	4/5	1	1	1	2	1	1	1	4
12. 'Leaf' angle (°)	3 (50°)	3 (55°)	3 (70°)	3 (25°)	3 (60°)	3 (45°)	3 (55°)	3 (60°)	3 (75°)	2
13. 'Leaf' margin	1	1	1	1	6	1	1	1	1	1
14. 'Leaf' surface resins	2	2	2	2	2	2	2	2	2	2
15. 'Leaf' glands	1	1	1	2	2	2	1	1	2	2
16. 'Leaf' consistency	3	3	3	3	2	3	2	2	3	3
17. 'Leaf' tomentosity	1	1	1	1	4	1	1	1	1	1
18. 'Leaf' stomata	4	4								
19. 'Leaf' duration (yrs)	3	3	3	3	3	?	2	2	3	?
20. 'Leaf' seasonality	1	1	1	1	1	1	1	1	1	1
21. 'Leaf' area: assim. stem area	1	1	1	1	1	2	1	1	1	1
22. Organs periodically shed	4	4	4	4	4	–	4	4	4	–*
23. Organs shed – seasonality	2	2	2	2	2	–	2	2	2	–
24. Shoot growth – seasonality	2	2	1	1	1	1	1	1	1	1/2
25. Stem – number from base	2 (2–3)	2 (2–3)	2	2 (1–2)	2	1/2	1	1/2	2	3
26. Stem – height (m)	2.4 m	2.0 m	0.10 cm	0.10 cm	–	–	–	–	–	–
27. Stem – diameter (cm)	1.6–6.4 cm	1.6–3.8 cm	0.6–1.2 cm	7.12 cm	–	–	–	–	–	–
28. Stem lignification	1	1	1	1	1	1	1	1	1	2
29. Bark consistency	1	1	1	1	1	1	1	1	1	–
30. Bark thickness (mm)	1.5 mm	1.5 mm	1 mm	1 mm	1 mm	1 mm	1 mm	1 mm	1 mm	–
31. Bark shedding	2	2	4	4	4	4	4	4	4	–
32. Bark shedding – seasonality	1/2	1/2	–	–	–	–	–	–	–	–
33. Spinescence	4	4	4	2	4	1	4	4	2	4
34. Underground stems	7	7	7	–	–	–	–	–	–	1
35. Root morphology	3	3	3	3	3	3	3	3	3	4
36. Root modification	7	7	7	7	7	7	7	7	7	6
37. Rootlet modification	5	5	5	3	6	6	5	5	8	2
38. Root depth (cm)	7	7	6	6	3/4	5	2/3	3/4	2/3	2/3
39. Root spread (cm)	7	7	6	6	3/4	5	2	3/4	2	2
40. Veg. multiplication	4	4	4	4	4	4	4	4	4	2
41. Veg. regen. after fire	2	2	2	1	1	1	1	1	1	5
42. Flowering – seasonality	4/1	1/2	1	1/2	1	1	1	1	1	3
43. Flowering – pyrogenic	2	2	2	2	2	2	2	2	2	2
44. Fruit dehiscence	2	2	2	2	2	2	2	1	3	1
45. Seed dissemination	2/3	2/3	2/3	2/3	1	1	1	1	1	1
46. Seedling- germ. type	2	2	2	2	2	2	2	2	2	–
47. Plumule position	1	1	1	1	1	1	1	1	1	2
48. Number leaves per 10 cm stem	4–5	6–7	25	20–25	18–24	–	150	170–210	150	–

* Leaf dead but not shed.

Code of Ecomorphological Characters: See Chapter 1.

Table 9. Ecomorphological characters: South Australia.
Location: Brookfield Conservation Park, South Australia (34° 20′ S, 139° 23′ E).
Vegetation structure: (1) Mallee open-scrub (*Eucalyptus gracilis* – chenopod association) (2) Mallee-open-scrub (*E. oleosa* – *Triodia irritans* association).
Age since last fire: Unknown.
Soil: (1) Brown calcareous earth-Solonized brown soil (Gc 1.12 and Gc 1.22) (2) Dunes of brown sands (Uc 5.1) over calcarenite limestone.
Reference: Jessup, R.W. 1948. Trans. R. Soc. S. Aust. 72: 33–68.
Compilers: L. Olsvig-Whittaker and R.L. Specht, October 1983.

Character	(1.)					(2.)			
	Eucalyptus gracilis (Myrtaceae)	*Atriplex suberecta* (Chenopodiaceae)	*Maireana pentatropis* (Chenopodiaceae)	*Zygophyllum ovatum* (Zygophyllaceae)	*Stipa nitida* (Poaceae)		*Eucalyptus oleosa* (Myrtaceae)	*Melaleuca acuminata* (Myrtaceae)	*Triodia irritans* (Poaceae)
Foliage Projective Cover (%)									
Upper stratum (5–6 m)	46	–	–	–	–	(3–5 m)	40	–	–
Mid stratum (40–50 cm)	–	3	–	–	–	(1.5–2 m)	–	10	–
Ground stratum (< 15 cm)	–	–	6	7	2	(40 cm)	–	–	7
Density (mature plants m^{-2})	0.1	0.1	2.0	3.5	5.1		0.07	0.10	0.17
1. Renewal buds	M	N	Ch	Th	Th		M	N	Hev
2. Trophic type	1	1	1	1	1		1	1	1
3. Life duration (yrs)	7	4	2/3	2/3	1		7	6	7
4. Height (m)	7	3	2	2	1		6	5	3
5. Crown diameter (m)	7	2	1/2	1	1		5/6	5/6	5/6
6. Canopy density (%)	2/3	2/3	2/3	2	2		2/3	5/6	4/5
7. Photosynthetic organs	1	1	1	1	1		1	1	1
8. 'Leaf' size	3/4	3/4	1	1	3		4/5	2	
9. 'Leaf' colour	1	4	1	1	1		1	1	1
10. 'Leaf' length (cm)	4	2/3	1	1	5/6		4/5	1	5
11. 'Leaf' width (mm)	5	6/7	1	1	2		6	3/4	4
12. 'Leaf' angle (°)	3 (50°)	3	3	3	3		3 (50°)	3 (60°)	3
13. 'Leaf' margin	1	2	1	1	1		1	1	1
14. 'Leaf' surface resins	2	2	2	2	2		2	2	1
15. 'Leaf' glands	1	2	2	2	2		1	1	2
16. 'Leaf' consistency	3	4	4	5	1		3	3	3
17. 'Leaf' tomentosity	1	1	1	1	1		1	2	1
18. 'Leaf' stomata	4						4		
19. 'Leaf' duration (yrs)	3	2	2	1	1		3	3	?
20. 'Leaf' seasonality	1	1	1	2	2		1	1	1
21. 'Leaf' area: assim. stem area	1	1	1	1	1		1	1	1
22. Organs periodically shed	4	4	4	–	–		4	4	–*
23. Organs shed – seasonality	2	2	2	–	–		2	2	–
24. Shoot growth – seasonality	2	1	1	4/1	4/1		2	1	1
25. Stem – number from base	2 (5–10)	1	1	3	3		2 (2–5)	2 (10–15)	3
26. Stem – height (m)	4–5	short	short	–	–		2–4	short	–
27. Stem – diameter (cm)	5–7	<1	<1	–	–		4–5	1	–
28. Stem lignification	1	2	2	–	–		1	1	–
29. Bark consistency	1	1	1	–	–		1	1	–
30. Bark thickness (mm)	1–2	<1	<1	–	–		1–2	1	–
31. Bark shedding	2/3	4	4	–	–		2/3	4	–
32. Bark shedding – seasonality	1/2	–	–	–	–		1/2	–	–
33. Spinescence	4	4	4	4	4		4	2	2
34. Underground stems	7	–	–	–	–		7	–	1
35. Root morphology	3	3	3	1	4		3	3	4
36. Root modification	7	7	7	7	7		7	7	7
37. Rootlet modification	5	8	8	8	8		5	5	8
38. Root depth (cm)	7	4/5	4/5	1	1/2		7	6	5/6
39. Root spread (cm)	7	4/5	4/5	1	1/2		7	6	5/6
40. Veg. multiplication	4	4	4	–	–		4	4	2
41. Veg. regen. after fire	2	1	1	1	1		2	2	4
42. Flowering – seasonality	3/4	1/2	1/2	4/1	4/1		3/4	1	1
43. Flowering – pyrogenic	2	2	2	2	2		2	2	2
44. Fruit dehiscence	2	1	1	2	1		2	2	1
45. Seed dissemination	2/3	1	1	1	1		2/3	2/3	1
46. Seedling – germ. type	2	2	2	2	1		2	2	–
47. Plumule position	1	1	1	1	2		1	1	2
48. Number leaves per 10 cm stem	6–7	10–20	?	20	–		7–10	20–24	–

* Leaf dead but not shed.

Code of Ecomorphological Characters: See Chapter 1.

Table 10. Ecomorphological characters: Victoria.
Location: Mount Dandenong, Victoria (37° 50′ S, 145° 21′ E).
Vegetation structure: Grassy open-forest (*Eucalyptus macrorhyncha* – *E. radiata* association).
Age since last fire: Unknown.
Soil: Hard-setting acidic loamy soil, with yellow clayey subsoil – yellow podzolic soil (Dy 3.41), often skeletal on slopes.
Reference: Clifford, H.T., 1953. Proc. R. Soc. Vict. 65: 30–55.
Compilers: D.H. Ashton and H.T. Clifford, September 1982.

Character	*Eucalyptus macrorhyncha* (Myrtaceae)	*Eucalyptus radiata* (Myrtaceae)	*Lomatia ilicifolia* (Proteaceae)	*Acacia stricta* (Mimosaceae)	*Danthonia pallida* (Poaceae)	*Themeda triandra* (Poaceae)	*Lomandra filiformis* (Xanthorrhoeaceae)	*Dianella tasmanica* (Liliaceae)
Foliage Projective Cover (%)								
Upper stratum (18–23 m)	28	27	–	–	–	–	–	–
Mid stratum (2–8 m)	–	–	5	5	–	–	–	–
Ground stratum (<30 cm)	–	–	–	–	55	10	17	12
Density (mature plants m^{-2})	0.026	0.021	0.7	0.4	1.1	0.6	0.7	0.5
1. Renewal buds	MM	MM	N	N	Hev	Hev	Hev	Hev
2. Trophic type	1	1	1	1	1	1	1	1
3. Life duration (yrs)	7	7	4	4	6	6	6	6
4. Height (m)	9	8	5	5	2	2	2	3
5. Crown diameter (m)	8	8	3	3	2	2	2	2
6. Canopy density (%)	5	5	4	4	6	6	6	6
7. Photosynthetic organs	1	1	1	2	1	1	1	1
8. 'Leaf' size	6	5	5	4	3	4	3	4
9. 'Leaf' colour	1	1	1	1	1	1	1	1
10. 'Leaf' length (cm)	4/5	4/5	4/5	4/5	4	4	5	5
11. 'Leaf' width (mm)	7	7	7	5	3	4	3	5
12. 'Leaf' angle (°)	3 (60°)	3 (60°)	3	3	3 (80°)	3 (80°)	3 (80°)	3 (80°)
13. 'Leaf' margin	1	1	2	1	1	1	1	1
14. 'Leaf' surface resins	2	2	2	1	2	2	2	2
15. 'Leaf' glands	1	1	2	2	2	2	2	2
16. 'Leaf' consistency	3	3	3	3	1	1	3	3
17. 'Leaf' tomentosity	1	1	1	1	1	1	1	1
18. 'Leaf' stomata	4	4	?	4	3	3	?	?
19. 'Leaf' duration (yrs)	2/3	2/3	2/3	2/3	2	2	4	4
20. 'Leaf' seasonality	1	1	1	1	1	1	1	1
21. 'Leaf' area: assim. stem area	1	1	1	1	1	1	1	1
22. Organs periodically shed	4	4	4	4	–	–	–	–
23. Organs shed – seasonality	2	2	2	2	–	–	–	–
24. Shoot growth – seasonality	2	2	1	1	1	1	1	1
25. Stem – number from base	1	1	1	1	3	3	3	3
26. Stem – height (m)	5	5	0.5	0.5	–	–	–	–
27. Stem – diameter (cm)	50	50	3	3	–	–	–	–
28. Stem lignification	1	1	1	1	–	–	–	–
29. Bark consistency	2	2	1	1	–	–	–	–
30. Bark thickness (mm)	6	5	2	2	–	–	–	–
31. Bark shedding	4	4	4	4	–	–	–	–
32. Bark shedding – seasonality	–	–	–	–	–	–	–	–
33. Spinescence	4	4	2	4	4	4	4	4
34. Underground stems	7	7	7	abs.	1	1	1	1
35. Root morphology	3	3	3	3	4	4	4	4
36. Root modification	7	7	7	7	7	7	7	7
37. Rootlet modification	5	5	3	6	8	8	8	8
38. Root depth (cm)	7	7	5	5	4/5	4/5	4/5	4/5
39. Root spread (cm)	7	7	5/6	5/6	4	4	4	4
40. Veg. multiplication	4	4	4	4	2	2	2	2
41. Veg. regen. after fire	3	3	2	1	4	4	4	4
42. Flowering – seasonality	2/3	2/3	2	1	2	1/2	1/2	1/2
43. Flowering – pyrogenic	2	2	2	2	2	2	2	2
44. Fruit dehiscence	2	2	2	2	1	1	2	3
45. Seed dissemination	2/3	2/3	1	1	1	1	1	1
46. Seedling – germ. type	2	2	2	2	–	–	–	–
47. Plumule position	1	1	1	1	2	2	2	2
48. Number leaves per 10 cm stem	6–7	5–6	6–8	6–8	–	–	–	–

Code of Ecomorphological Characters: See Chapter 1.

Table 11. Ecomorphological characters: Victoria.
Location: Brisbane Ranges National Park, Victoria (37° 55′ S, 144° 20′ E).
Vegetation structure: Heathy open-forest (*Eucalyptus obliqua* – *E. macrorhyncha* association).
Age since last fire: Unknown.
Soil: Sandy acidic mottled-yellow soil, containing ironstone gravel – lateritic podzolic soil (Dy 5.61).
Reference: Bridgewater, P.D., 1976. Proc. R. Soc. Vic. 88: 43–48.
Compilers: R.L. Specht and N. Scarlett, August 1983.

Character	*Eucalyptus obliqua* (Myrtaceae)	*Eucalyptus macrorhyncha* (Myrtaceae)	*Xanthorrhoea australis* (Xanthorrhoeaceae)	*Hibbertia stricta* (Dilleniaceae)	*Lepidosperma semiteres* (Cyperaceae)	*Isopogon ceratophyllus* (Proteaceae)
Foliage Projective Cover (%)						
Upper stratum (20–30 m)	← 45 →		–	–	–	–
Mid stratum (1–1.5 m)	–	–	23	–	–	–
Ground stratum (< 50 cm)	–	–	–	4	4	trace
1. Renewal buds	MM	MM	N-G	Ch	Hev	Ch
2. Trophic type	1	1	1	1	1	1
3. Life duration (yrs)	7	7	7	5	7	5
4. Height (m)	9	9	5	4	3	3
5. Crown diameter (m)	7	7	5	3	3/4	2
6. Canopy density (%)	3/4	3/4	4	3	5	5
7. Photosynthetic organs	1	1	·1	1	4	1
8. 'Leaf' size	6	6	6	2	3	3/4
9. 'Leaf' colour	1	1	1	1	1	1
10. 'Leaf' length (cm)	4/5	4/5	7	1	6	4
11. 'Leaf' width (mm)	6/7	7	4	2	2	4
12. 'Leaf' angle (°)	3 (30°)	3 (60°)	3	3	2	3
13. 'Leaf' margin	1	1	1	1	1	3
14. 'Leaf' surface resins	2	2	2	2	2	2
15. 'Leaf' glands	1	1	2	2	2	2
16. 'Leaf' consistency	3	3	3	2	3	3
17. 'Leaf' tomentosity	1	1	1	2	1	1
18. 'Leaf' stomata	4	4	?	?	?	?
19. 'Leaf' duration (yrs)	2/3	2/3	2/3	2/3	?	?
20. 'Leaf' seasonality	1	1	1	1	1	1
21. 'Leaf' area: assim. stem area	1	1	1	1	3	1
22. Organs periodically shed	4	4	–*	4	–*	–*
23. Organs shed – seasonality	2	2	–	2	–	–
24. Shoot growth – seasonality	2	2	4/1	1	4/1	1
25. Stem – number from base	1	1	3/1	2 (2–6)	3	1/2 (1–5)
26. Stem – height (m)	10–15 mm	10–15 mm	0/50 cm	25 cm	–	10 cm
27. Stem – diameter (cm)	40 cm	40 cm	0/15 cm	1 cm	–	2 cm
28. Stem lignification	1	1	–	1	–	1
29. Bark consistency	2	2	–	1	–	1
30. Bark thickness (mm)	4–7 mm	4–7 mm	–	1 mm	–	1 mm
31. Bark shedding	4	4	–	4	–	4
32. Bark shedding – seasonality	–	–	–	–	–	–
33. Spinescence	4	4	4	4	4	2
34. Underground stems	7	7	8	–	1	–
35. Root morphology	3	3	4	3	4	3
36. Root modification	7	7	6	7	6	7
37. Rootlet modification	5	5	8	8	2	3
38. Root depth (cm)	7	7	5/6	4/5	2/3	4/5
39. Root spread (cm)	7	7	5	4	2	3
40. Veg. multiplication	4	4	4/2	4	2	4
41. Veg. regen. after fire	3	3	4	2	4	2
42. Flowering – seasonality	2/3	2/3	1	1	1	1
43. Flowering – pyrogenic	2	2	1	2	2	2
44. Fruit dehiscence	2	2	2	2	1	2
45. Seed dissemination	2/3	2/3	1	1	1	3
46. Seedling – germ. type	2	2	–	2	–	2
47. Plumule position	1	1	2	1	2	1
48. Number leaves per 10 cm stem	5–6	6–7	–	6–8	–	12

* Leaf dead but not shed.

Code of Ecomorphological Characters: See Chapter 1.

Table 12. Ecomorphological characters: Victoria.
Location: Gellibrand Hill State Park, Victoria (37° 40′ S, 144° 48′ E).
Vegetation structure: Savanna woodland (*Eucalyptus microcarpa* alliance).
Age since last fire: Unknown.
Soil: Hard-setting loamy soil, with alkaline, mottled-yellow clayey subsoil – solodic soil (Dy 3.43).
Reference: Robertson, D., 1985. Ph.D. Thesis, University of Melbourne.
Compilers: R.L. Specht and D. Robertson, August 1983.

Character	*Eucalyptus microcarpa* (Myrtaceae)	*Lolium rigidum* (Poaceae introd.)	*Stipa curticoma* (Poaceae)	*Danthonia racemosa* (Poaceae)	*Agropyron scabrum* (Poaceae)	*Arctotheca calendula* (Asteraceae introd.)
Foliage Projective Cover (%)						
Upper stratum (15–25 m)	65	–	–	–	–	–
Mid stratum (2–8 m)	–	–	–	–	–	–
Ground stratum (<30 cm)	–	12	9	4	2	2
1. Renewal buds	MM	Th	Hev	Hev	Hev	Th
2. Trophic type	1	1	1	1	1	1
3. Life duration (yrs)	7	1	6	6	6	1
4. Height (m)	8	2	2	2	2	2
5. Crown diameter (m)	7	1	2	2	2	2/3
6. Canopy density (%)	4	6	6	6	5	6
7. Photosynthetic organs	1	1	1	1	4	1
8. 'Leaf' size	5	3	3	3	3	6
9. 'Leaf' colour	1	1	1	1	1	4
10. 'Leaf' length (cm)	4/5	5	6	4/5	4/5	4/5
11. 'Leaf' width (mm)	6/7	5	4	2	2	7
12. 'Leaf' angle (°)	3 (60°)	3 (80°)	3 (80°)	3 (80°)	3 (80°)	1/3
13. 'Leaf' margin	1	1	1	1	1	3
14. 'Leaf' surface resins	2	2	2	2	2	2
15. 'Leaf' glands	1	2	2	2	2	1
16. 'Leaf' consistency	3	1	1	1	1	1
17. 'Leaf' tomentosity	1	1	1	1	1	4
18. 'Leaf' stomata	4	3	3	3	3	3
19. 'Leaf' duration (yrs)	2/3	1	2	2	2	1
20. 'Leaf' seasonality	1	2	1	1	1	2
21. 'Leaf' area: assim. stem area	1	1	1	1	3	1
22. Organs periodically shed	4	–	–	–	–	–
23. Organs shed – seasonality	2	–	–	–	–	–
24. Shoot growth – seasonality	2	1	1	1	1	1
25. Stem – number from base	1	3	3	3	3	3
26. Stem – height (m)	8	–	–	–	–	–
27. Stem – diameter (cm)	40	–	–	–	–	–
28. Stem lignification	1	–	–	–	–	–
29. Bark consistency	2	–	–	–	–	–
30. Bark thickness (mm)	5	–	–	–	–	–
31. Bark shedding	4	–	–	–	–	–
32. Bark shedding – seasonality	–	–	–	–	–	–
33. Spinescence	4	4	4	4	4	4
34. Underground stems	7	abs.	1	1	1	abs.
35. Root morphology	3	4	4	4	4	1
36. Root modification	7	7	7	7	7	7
37. Rootlet modification	5	8	8	8	8	8
38. Root depth (cm)	7	3/4	4/5	4/5	4/5	3/4
39. Root spread (cm)	7	1	4	4	4	1
40. Veg. multiplication	4	4	2	2	2	4
41. Veg. regen. after fire	3	1	4	4	4	1
42. Flowering – seasonality	3/4	1	1	1	1	1
43. Flowering – pyrogenic	2	2	2	2	2	2
44. Fruit dehiscence	2	1	1	1	1	1
45. Seed dissemination	2/3	1	1	1	1	1
46. Seedling – germ. type	2	–	–	–	–	2
47. Plumule position	1	2	2	2	2	1
48. Number of leaves per 10 cm stem	5–6	–	–	–	–	–

Code of Ecomorphological Characters: See Chapter 1.

Table 13. Ecomorphological characters: Victoria.
Location: Melton, Victoria (37°41′, 144°35′).
Vegetation structure: Mallee open-scrub (*Eucalyptus behriana* association).
Age since last fire: Unknown.
Soil: Sandy surface soil over mottled-yellow clay – solodized solonetz (Dy 5.43).
Reference: Myers, B.A., Ashton, D.H. and Osborne, J.A., 1986. Aust. J. Bot. 34: 15–39.
Compiler: R.L. Specht, August 1983.

Character	*Eucalyptus behriana* (Myrtaceae)	*Rhagodia parabolica* (Chenopodiaceae)	*Calandrinia calyptrata* (Portulacaceae)	*Crassula sieberana* (Crassulaceae)
Foliage Projective Cover (%)				
Upper stratum (8–12 m)	51	–	–	–
Mid stratum (1.0–1.3 m)	–	36	–	–
Ground stratum (<5 cm)	–	–	1	1
1. Renewal buds	M/MM	N	Th	Th
2. Trophic type	1	1	1	1
3. Life duration (yrs)	7	5/6	1	1
4. Height (m)	7/8	5	1	1
5. Crown diameter (m)	6	4	1	1
6. Canopy density (%)	4	5	2	2
7. Photosynthetic organs	1	1	1	1
8. 'Leaf' size	5/6	4	3	1
9. 'Leaf' colour	1	3	1	1
10. 'Leaf' length (cm)	4	2/3	2/3	1
11. 'Leaf' width (mm)	7	6/7	4/5	2/3
12. 'Leaf' angle (°)	3 (60°)	3 (60°)	3	3
13. 'Leaf' margin	1	1	1	1
14. 'Leaf' surface resins	2	2	2	2
15. 'Leaf' glands	1	1	2	2
16. 'Leaf' consistency	3	4	5	4
17. 'Leaf' tomentosity	1	1	1	1
18. 'Leaf' stomata	4	4	4	4
19. 'Leaf' duration (yrs)	2/3	2/3	1	1
20. 'Leaf' seasonality	1	1	2	2
21. 'Leaf' area: assim. stem area	1	1	3	3
22. Organs periodically shed	4	4	–	–
23. Organs shed – seasonality	2	2	–	–
24. Shoot growth – seasonality	2	1/2	1	1
25. Stem – number from base	1–6	1	3	3
26. Stem – height (m)	5	10	–	–
27. Stem – diameter (cm)	8–20	4	–	–
28. Stem lignification	1	1	–	–
29. Bark consistency	1	1/2	–	–
30. Bark thickness (mm)	5	2	–	–
31. Bark shedding	1	4	–	–
32. Bark shedding – seasonality	?	–	–	–
33. Spinescence	4	4	4	4
34. Underground stems	7	abs.	abs.	abs.
35. Root morphology	3	3	1	4
36. Root modification	7	7	1	7
37. Rootlet modification	5	8	8	8
38. Root depth (cm)	7	6	1	1
39. Root spread (cm)	7	6	1	1
40. Veg. multiplication	4	4	4	4
41. Veg. regen. after fire	2/3	?	1	1
42. Flowering – seasonality	4–1	4–1	1	1
43. Flowering – pyrogenic	2	2	2	2
44. Fruit dehiscence	2	3	2	2
45. Seed dissemination	2/3	1	1	1
46. Seedling – germ. type	2	2	2	2
47. Plumule position	1	1	1	1
48. Number leaves per 10 cm stem	5–6	10	10	40

Code of Ecomorphological Characters: See Chapter 1.

Table 14. Ecomorphological characters: Victoria.
Location: St Albans, Victoria (37° 45′, 144° 48′ E).
Vegetation structure: Tussock grassland (*Themeda triandra* association).
Age since last fire: 2–3 years.
Soil: Cracking grey clay soil (Ug 5.2).
Reference: Groves, R.H., 1965. Aust. J. Bot. 13: 291–302.
Compiler: R.L. Specht, August 1983.

Character	*Themeda triandra* (Poaceae)	*Danthonia auriculata* (Poaceae)	*Eryngium rostratum* (Apiaceae)	*Plantago varia* (Plantaginaceae)	*Wahlenbergia stricta* (Campanulaceae)	*Schoenus breviculmis* (Cyperaceae)
Foliage Projective Cover (%)						
Upper stratum (10–15 cm)	60	7	6	–	–	–
Mid stratum (5–10 cm)	–	–	–	7	8	–
Ground stratum (< 5 cm)	–	–	–	–	–	16
Biomass tops (kg ha^{-1})	1,509	88	72	58	38	20
1. Renewal buds	Hev	Hev	Ch	G	Ch	Hev
2. Trophic type	1	1	1	1	1	1
3. Life duration (yrs)	6	6	4	4	4	6
4. Height (m)	2	2	2	1	1	1
5. Crown diameter (m)	2	2	2	1	1	1
6. Canopy density (%)	6	6	2	3	1	5
7. Photosynthetic organs	1	1	4	2	4	1
8. 'Leaf' size	4	3	4	5	2	1
9. 'Leaf' colour	1	1	2	1	1	1
10. 'Leaf' length (cm)	4	4	4–5	4–5	1–2	1
11. 'Leaf' width (mm)	4	3	4	5–6	2–4	1
12. 'Leaf' angle (°)	3 (80°)	3 (80°)	3 (60°)	1	3 (60°)	2
13. 'Leaf' margin	1	1	3	1	1	1
14. 'Leaf' surface resins	2	2	2	2	2	2
15. 'Leaf' glands	2	2	2	2	2	2
16. 'Leaf' consistency	1	1	1	1	1	1
17. 'Leaf' tomentosity	1	1	1	4	4	1
18. 'Leaf' stomata	3	3	4	4	4	4
19. 'Leaf' duration (yrs)	2	2	2	2	2	3
20. 'Leaf' seasonality	1	1	1	1	1	1
21. 'Leaf' area: assim. stem area	1	1	3	1	3	1
22. Organs periodically shed	–	–	–	–	–	–
23. Organs shed – seasonality	–	–	–	–	–	–
24. Shoot growth – seasonality	1/2 Oct–Dec.	1	1	1	1	1
25. Stem – number from base	3	3	1	3	2 (1–3)	3
26. Stem – height (m)	–	–	< 10 cm	–	< 10 cm	–
27. Stem – diameter (cm)	–	–	–	–	–	–
28. Stem lignification	–	–	3	–	3	–
29. Bark consistency	–	–	–	–	–	–
30. Bark thickness (mm)	–	–	–	–	–	–
31. Bark shedding	–	–	–	–	–	–
32. Bark shedding – seasonality	–	–	–	–	–	–
33. Spinescence	4	4	2	4	4	4
34. Underground stems	1	1	abs.	abs.	abs.	1
35. Root morphology	4	4	1	1	1	4
36. Root modification	7	7	7	7	7	7
37. Rootlet modification	8	8	8	8	8	8
38. Root depth (cm)	4/5	4/5	4/5	3/4	3/4	2/3
39. Root spread (cm)	4	4	2	1	1	1
40. Veg. multiplication	2	2	4	4	4	2
41. Veg. regen. after fire	4	4	2	2	2	2
42. Flowering – seasonality	1	1	1	1/4	2/3	1/4
43. Flowering – pyrogenic	2	2	2	2	2	2
44. Fruit dehiscence	1	1	2	2	2	1
45. Seed dissemination	1	1	1	1	1	1
46. Seedling – germ. type	–	–	2	2	2	–
47. Plumule position	2	2	1	1	1	2
48. Number leaves per 10 cm stem	–	–	2	–	10–12	–

Code of Ecomorphological Characters: See Chapter 1.

Table 15. Ecomorphological characters: Western Australia.
Location: Durokoppin Nature Reserve, Western Australia (31° 24′ S, 117° 45′ E, 325–400 m). Kodj Kodjin Nature Reserve, Western Australia (31° 27′ S, 117° 48′ E, 315–360 m).
Vegetation structure: Eucalyptus wandoo 'savanna' woodland.
Age since last fire: >50 years.
Nearest climate station: Kellerberrin, Western Australia (31° 38′ S, 117° 43′ E, 247 m). Annual precipitation – 339 m.
Soil: Sandy loam, which may be gravelly with a clayey subsoil, on flat or slightly undulating country.
Reference: Muir, B.G., Chapman, A., Dell, J. and Kitchener, D.J., 1978. Biological Survey of the Western Australian Wheatbelt. Part 6. Durokoppin and Kodj Kodjin Nature Reserves. Rec. W. Aust. Mus. Suppl. No. 7.
Compiler: L. Atkins, June 1987.

Character	*Eucalyptus wandoo* (Myrtaceae)	*Oxylobium parviflorum* (Fabaceae)	*Acacia acuaria* (Mimosaceae)	*Acacia hemiteles* (Mimosaceae)	*Acacia ixiophylla* (Mimosaceae)	*Olearia muelleri* (Asteraceae)	*Olearia revoluta* (Asteraceae)	*Restio* sp. (Restionaceae)	*Lomandra effusa* (Dasypogonaceae)	*Borya constricta?* (Anthericaceae)	*Podolepis lessonii* (Asteraceae)	*Podolepis capillaris* (Asteraceae)
Foliage Projective Cover (%)												
Upper stratum (>5 m)	86.2	–	–	–	–	–	–	–	–	–	–	–
Mid stratum (30 cm–5 m)	–	0.8	0.5	0.5	0.1	0.4	0.2	–	–	–	–	–
Ground stratum (<30 cm)	–	–	–	–	–	–	–	30	0.1	0.1	0.4	0.5
1. Renewal buds	MM	N	N	N	N	N	N	H	H	H	Th	Th
2. Trophic type	1	2	2	2	2	1	1	1	1	1	1	1
3. Life duration (yrs)	7	6	6	6	6	6	6	6	6	6	1	1
4. Height (m)	8	4	4	4	4	4	4	2	2	1	2	2
5. Crown diameter (m)	7	4	5	5	4	3	3	2	3	2	1	2
6. Canopy density (%)	?	?	?	?	?	?	?	?	?	?	1	1
7. Photosynthetic organs	1	1	2	2	2	1	1	4	1	1	1	1
8. 'Leaf' size	4/5	3	1	4	3	2	1	n/a	3	2	3	1
9. 'Leaf' colour	2	1	1	2	1	1	4	n/a	2	1	2	1
10. 'Leaf' length (cm)	4/5	3	2	4	3	1	2	n/a	6	2	1	1
11. 'Leaf' width (mm)	7	4	1	5	5	4	1	n/a	2	1	5	1
12. 'Leaf' angle (°)	3	2	2	1	2	2	2	n/a	2	1	3 (60°)	2
13. 'Leaf' margin	1	1	1	1	1	1	5	1	2	1	1	5
14. 'Leaf' surface resins	2	2	2	2	1	1	2	2	2	2	2	2
15. 'Leaf' glands	1	2	2	2	2	2	2	2	2	2	2	2
16. 'Leaf' consistency	3	3	3	3	3	2	2	3	3	3	1	1
17. 'Leaf' tomentosity	1	2	1	1	1	1	2	n/a	?	1	4	1
18. 'Leaf' stomata	?	?	?	?	?	?	?	n/a	?	2	?	?
19. 'Leaf' duration (yrs)	3	3	3	3	3	3	3	n/a	?	?	1	2
20. 'Leaf' seasonality	1	1	1	1	1	1	1	n/a	1	2	2	2
21. 'Leaf' area: assim. stem area	1	1	1	1	1	1	1	2	1	1	1	1
22. Organs periodically shed	4	4	4	4	4	4	4	?	?	4	1	1
23. Organs shed – seasonality	?	?	?	?	?	?	?	?	?	?	2	?
24. Shoot growth – seasonality	?	?	?	?	?	?	?	?	?	4/1	4/1	4/1
25. Stem – number from base	1	1	1	1	1	1	1	2	2	2	1	1
26. Stem – height (m)	5	0.2	0.1	0.1	0.2	0.3	0.1	n/a	?	?	0.01	0.02
27. Stem – diameter (cm)	40	2	3	3	3	2	?	n/a	0.3	0.4	<0.1	<0.1
28. Stem lignification	1	1	1	1	1	1	1	n/a	3	3	3	3
29. Bark consistency	1	1	?	?	?	?	?	none	none	none	none	none
30. Bark thickness (mm)	?	?	?	?	?	?	?	n/a	n/a	n/a	n/a	n/a
31. Bark shedding	1	?	?	?	?	?	?	n/a	n/a	n/a	n/a	n/a
32. Bark shedding-seasonality	6/7	?	?	?	?	?	?	n/a	n/a	n/a	n/a	n/a
33. Spinescence	4	4	2	4	4	4	4	4	4	2	4	4
34. Underground stems	7	–	–	–	–	–	–	1	1	1	–	–
35. Root morphology	3	3	3	3	3	3	3	4	4	4	1	1
36. Root modification	7	7	7	7	7	7	7	6	6	6	7	7
37. Rootlet modification	5	6	6	6	6	8	8	4	?	?	8	8
38. Root depth (cm)	?	?	?	?	?	?	?	?	?	?	?	?
39. Root spread (cm)	?	?	?	?	?	?	?	?	?	?	?	?
40. Veg. multiplication	4	4	4	4	4	4	4	2	2	2	4	4
41. Veg. regen. after fire	3	?	?	?	?	?	?	4	4	4	1	1
42. Flowering – seasonality (mths)	10–4	7–10	8–11	9–10	5–7	7–11	6	7–8	6–10	8–9	8–12	1–12
43. Flowering – pyrogenic	2	2	2	2	2	2	2	2	2	2	2	2
44. Fruit dehiscence	2	2	2	2	2	1	1	1	2	2	1	1
45. Seed dissemination	1	1	1	1	1	1	1	1	1	1	1	1
46. Seedling – germ. type	2	2	2	2	2	2	2	–	–	–	2	2
47. Plumule position	1	1	1	1	1	1	1	2	2	2	1	1
48. No. 'leaves' per 10 cm stem	2	8	20	8	11	30	70	–	–	–	6	20

Code of Ecomorphological Characters: See Chapter 1.

Table 16. Ecomorphological characters: Western Australia.
Location: Jurien Bay, Western Australia (30° 18′ S, 150° 00′ E).
Vegetation structure: Banksia sclerophyll low woodland.
Age since last fire: 10 years (?).
Soil: Silicious sands, neutral to acidic; fine textured with low organic contents.
Reference: Pate, J.S. and Beard, J.S. (eds) 1984. Kwongan: Plant Life of the Sandplain. University of Western Australia Press, Nedlands.
Compiler: K. Dixon, October 1986.

Character	Banksia attenuata (Proteaceae)	Banksia menziesii (Proteaceae)	Eucalyptus todtiana (Myrtaceae)	Leucopogon conostephioides (Epacridaceae)	Beaufortia elegans (Myrtaceae)	Eremaea pauciflora (Myrtaceae)	Stirlingia latifolia (Proteaceae)	Ecdeiocolea monostachya (Ecdeiocoleaceae)
Foliage Projective Cover (%)								
Upper stratum (2–10 m)	x	x	x	–	–	–	–	–
Mid stratum (50 cm–2 m)	–	–	–	–	x	x	x	–
Ground stratum (< 50 cm)	–	–	–	x	–	–	–	x
1. Renewal buds	M	M	M	Ch	N	N	N	Hev
2. Trophic type	1	1	1	1	1	1	1	1
3. Life duration (yrs)	5	5	7	4	4	4	4	6
4. Height (m)	7	7	6	3	5	4	4	3
5. Crown diameter (m)	6	6	7	3	4	4	3	2
6. Canopy density (%)	2/3	2/3	2	2	4	3	2	2
7. Photosynthetic organs	1	1	1	1	1	1	1	1
8. 'Leaf' size	4	5	4	2	2	2	4	2
9. 'Leaf' colour	1	1	1	1	1	1	2	1
10. 'Leaf' length (cm)	9 cm	12 cm	10 cm	1 cm	1 cm	1 cm	20 cm	50 cm
11. 'Leaf' width (mm)	8 mm	15 mm	15 mm	2 mm	1 mm	1 mm	5 mm	15 mm
12. 'Leaf' angle (°)	1	3	3	2	3	1	2	2
13. 'Leaf' margin	2	2	1	1	1	1	1	1
14. 'Leaf' surface resins	2	2	2	2	2	2	2	2
15. 'Leaf' glands	2	2	1	2	1	1	2	2
16. 'Leaf' consistency	3	3	3	3	3	3	3	3
17. 'Leaf' tomentosity	2	2	1	1	1	1	1	1
18. 'Leaf' stomata	2	2	4	4	4	4	4	4
19. 'Leaf' duration (yrs)	4	4	3	2	3	2	3	3
20. 'Leaf' seasonality	1	1	1	1	1	1	1	1
21. 'Leaf' area: assim. stem area	1	1	1	1	1	1	1	1
22. Organs periodically shed	3	3	3	4	3	3	4	2
23. Organs shed – seasonality	?	?	?	?	?	?	?	?
24. Shoot growth – seasonality	?	?	?	?	?	?	?	?
25. Stem – number from base	1	1	2	1	1	1	1	?
26. Stem – height (m)	1.5 m	1.5 m	0.5 m	10 cm	10 cm	5 cm	1–5 cm	?
27. Stem – diameter (cm)	25 cm	25 cm	15 cm	–	1.5 cm	1.5 cm	1.5 cm	?
28. Stem lignification	1	1	1	1	1	1	1	2
29. Bark consistency	2	2	2	2	4	4	3	?
30. Bark thickness (mm)	1 mm	1 mm	10 mm	1 mm	2 mm	2 mm	1 mm	?
31. Bark shedding	1	1	1	1	1	1	1	?
32. Bark shedding – seasonality	?	?	?	?	?	?	?	?
33. Spinescence	4	4	4	4	4	4	4	4
34. Underground stems	–	–	7	–	–	–	–	2
35. Root morphology	3	3	3	1	3	3	3	4
36. Root modification	7	7	7	7	7	7	1	7
37. Rootlet modification	3	3	8	8	7	7	–	4
38. Root depth (cm)	?	?	?	?	?	?	?	20
39. Root spread (cm)	?	?	?	?	?	?	?	25
40. Veg. multiplication	4	4	4	4	4	4	4	2
41. Veg. regen. after fire	3	3	2/3	1	1	2	2	2
42. Flowering – seasonality	2	3	2	4	1	1	1	1/2
43. Flowering – pyrogenic	2	2	2	2	2	2	1/2	2
44. Fruit dehiscence	2	2	2	3	2	2	2	2
45. Seed dissemination	1/2	1/2	1/2	1	2	2	1	1
46. Seedling – germ. type	2	2	2	2	2	2	2	–
47. Plumule position	1	1	1	1	1	1	1	2
48. No. leaves per 10 cm stem	15–20	10	8	45	45	45	5	–

Code of Ecomorphological Characters: See Chapter 1.

Table 17. Ecomorphological characters: Western Australia.
Location: Mount Lesueur, Western Australia (30° 11′ S, 115° 12′ E, 313 m).
Vegetation structure: Kwongan (shrub-heathland) dominated by *Banksia tricuspis* and *Dryandra sessilis*.
Age since last fire: 20 years.
Soil: Truncated lateritic podzolic soil developed on Lesueur Sandstone, with a massive duricrust (cemented laterite) covering most of the area.
Reference: Griffin, E.A. and Hopkins, A.J.M., 1985. J.R. Soc. W. Aust. 67: 45–57.
Compiler: A.J.M. Hopkins, May 1987.

Character	*Banksia tricuspis* (Proteaceae)	*Dryandra sessilis* (Proteaceae)	*Calothamnus sanguineus* (Myrtaceae)	*Daviesia* aff. *striata* (Fabaceae)	*Lambertia multiflora* (Proteaceae)	*Kingia australis* (Xanthorrhoeaceae)	*Hakea megalosperma* (Proteaceae)	*Banksia micrantha* (Proteaceae)	*Calothamnus torulosus* (Myrtaceae)	*Dryandra nivea* (Proteaceae)	*Hibbertia acerosa* (Dilleniaceae)	*Petrophile striata* (Proteaceae)	*Hypocalymma xanthopetalum* (Myrtaceae)	*Macropidia fuliginosa* (Haemodoraceae)	*Mesomelaena tetragona* (Cyperaceae)
Foliage Projective Cover (%)															
Upper stratum (1–2 m)	2	5	–	–	–	–	–	–	–	–	–	–	–	–	–
Mid stratum (70 cm)	–	–	8	8	5	2	2	–	–	–	–	–	–	–	–
Ground stratum (< 40 cm)	–	–	–	–	–	–	–	3	3	4	3	8	3	3	1
1. Renewal buds	N	N	N	N	N	N	N	Ch	Ch	Ch	Ch	Ch	Ch	Hev	Hev
2. Trophic type	1	1	1	1	1	1	1	1	1	1	1	1	1	1	1
3. Life duration (yrs)	6	5	5	5	5	5	5	5	5	5	5	5	5	?4	?4
4. Height (m)	5	5	4	4	4	4	4	3	2	2	2	3	2	2	2
5. Crown diameter (m)	6	5	5/4	5/4	4	3	3	3	2	3	2	3	2	3	2
6. Canopy density (%)	?	?	?	?	?	?	?	?	?	?	?	?	?	?	?
7. Photosynthetic organs	1	1	1	4	1	1	1	1	1	1	1	1	1	1	4
8. 'Leaf' size	3	4	1/2	1/2	3	4	4	2	2	3/4	2	3	2	6	3
9. 'Leaf' colour	1	1	1	1	1	1	1	1	1	1	1	1	1	1	1
10. 'Leaf' length (cm)	4	3	2	1	3	6	3	3	2	4	2	3	2	5	6
11. 'Leaf' width (mm)	2	7	3	4	5	3	2	4	2	5	2	7	4	7	2
12. 'Leaf' angle (°)	3	3	3	2	3	1	3	3	3	3	3	1	3	2	2
13. 'Leaf' margin	1	2	1	1	2	1	1	1	1	2	4	3	1	1	1
14. 'Leaf' surface resins	2	2	2	2	2	2	2	2	2	2	2	2	2	2	2
15. 'Leaf' glands	2	2	1	2	2	2	2	2	1	2	2	2	1	2	2
16. 'Leaf' consistency	3	3	3	3	3	3	3	3	3	3	3	3	3	4	3
17. 'Leaf' tomentosity	1	1	4	1	1	1	1	1	1	1	1	1	1	1	1
18. 'Leaf' stomata	?	?	?	?	?	?	?	?	?	?	?	?	?	?	?
19. 'Leaf' duration (yrs)	2/3	2/3	2/3	2/3	2/3	2/3	2/3	2/3	2/3	2/3	2/3	2/3	2/3	2/3	2/3
20. 'Leaf' seasonality	1	1	1	1	1	1	1	1	1	1	1	1	1	1	1
21. 'Leaf' area: assim. stem area	1	1	1	3	1	1	1	1	1	1	1	1	1	1	3
22. Organs periodically shed	4	4	4	3 + 4	4	4	4	4	4	4	4	4	4	2	2
23. Organs shed – seasonality	2	2	2	2	2	2	2	2	2	2	2	2	2	2	2
24. Shoot growth – seasonality	1/2	1/2	1/2	1/2	1/2	1/2	1/2	1	1	1	1	1	1	1	1
25. Stem – number from base	1	1	2 (few)	2 (few)	2 (few)	1	2 (few)	2 (many)	2 (few)	2 (few)	2 (many)	2 (few)	2 (many)	1	2 (many)
26. Stem – height (m)	1.0	0.5	0.0	0.0	0.0	–	0.0	0.0	0.0	0.0	0.0	0.0	0.0	–	–
27. Stem – diameter (cm)	8	2.5	–	–	–	–	–	–	–	–	–	–	–	–	–
28. Stem lignification	1	1	1	2	1	1	1	1	1	1	1	1	1	3	3
29. Bark consistency	1	1	1	1	1	n/a	1	1	1	1	1	1	1	–	–
30. Bark thickness (mm)	?	?	?	?	?	–	?	?	?	?	?	?	?	–	–
31. Bark shedding	–	–	–	–	–	–	–	–	–	–	–	–	–	–	–
32. Bark shedding – seasonality	–	–	–	–	–	–	–	–	–	–	–	–	–	–	–
33. Spinescence	4	2	4	2	4	4	4	4	4	2	2	2	4	4	4
34. Underground stems	–	–	(7)*	–	(7)	8?	(7)	(7)	(7)	(7)	(7)	(7)	(7)	?1	?1
35. Root morphology	?	?	?	?	?	?	?	?	?	?	?	?	?	?	?
36. Root modification	?	?	?	?	?	?	?	?	?	?	?	?	?	?	?
37. Rootlet modification	?	?	?	?	?	?	?	?	?	?	?	?	?	?	?
38. Root depth (cm)	?	?	?	?	?	?	?	?	?	?	?	?	?	?	?
39. Root spread (cm)	?	?	?	?	?	?	?	?	?	?	?	?	?	?	?
40. Veg. multiplication	4	4	4	4	4	4	4	4	4	4	4	4	3	2	2
41. Veg. regen. after fire	3	3	2/4	1	2/4	5	2/4	2/4	2/4	2/4	2/4	2/4	2/4	4	4
42. Flowering – seasonality	3	1/2/3	1	4	1	1	3/4	1	1	1	1	1	1	1	2
43. Flowering – pyrogenic	2	2	2	2	2	1	2	2	2	2	2	2	2	2	2
44. Fruit dehiscence	1	2 (1)	1	2	1	2	1	1	1	1	2	1	2	2	2
45. Seed dissemination	3	1	3	1	2	1	2/3	3	2/3	2/3	1	2/3	1	1	1
46. Seedling – germ. type	2	2	2	2	2	–	2	2	2	2	2	2	2	–	–
47. Plumule position	1	1	1	1	1	2	1	1	1	1	1	1	1	–	2

* (7) = not strictly a lignotuber but has below-ground buds.

Code of Ecomorphological Characters: See Chapter 1.

Table 18. Ecomorphological characters: Western Australia.
Location: Eneabba, Western Australia (29° 49′ S, 115° 16′ E, 249 m).
Vegetation structure: Kwongan (shrub-heathland) dominated by *Banksia hookeriana, B. attenuata* and *B. menziesii.*
Age since last fire: 15 years.
Soil: Deep sandplain soils.
Reference: Pate, J.S. and Beard, J.S. (eds.) 1984. Kwongan: Plant Life of the Sandplain. University of Western Australia Press, Nedlands.
Compiler: B.B. Lamont and R. Struthers, December 1986.

Character	*Banksia hookeriana* (Proteaceae)	*Banksia menziesii* (Proteaceae)	*Banksia attenuata* (Proteaceae)	*Hakea obliqua* (Proteaceae)	*Hakea brachyptera* (Proteaceae)	*Adenanthos cygnorum* (Proteaceae)	*Xylomelum angustifolium* (Proteaceae)	*Petrophile drummondii* (Proteaceae)	*Beaufortia elegans* (Myrtaceae)	*Eucalyptus todtiana* (Myrtaceae)	*Eremaea beaufortioides* (Myrtaceae)	*Phymatocarpus porphyrocephalus* (Myrtaceae)	*Mesomelaena stygia* (Cyperaceae)	*Actinostrobus acuminatus* (Cupressaceae)	*Loxocarya fasciculata* (Restionaceae)
Foliage Projective Cover (%)															
Upper stratum (1–2 m)	13	8	7	<1*	–	<1*	<1*	–	–	<1*	<1	–	–	–	–
Mid stratum (50 cm–1 m)	<1	1	1	–	<1	–	–	2	<1	–	–	<1	–	–	–
Ground stratum (<50 cm)	–	–	–	–	–	–	–	–	–	–	–	–	<1	<1	<1
1. Renewal buds	N	N	N	M	N	M	M	N	N	M	N	N	H	Ch	H
2. Trophic type	1	1	1	1	1	1	1	1	1	1	1	1	1	1	1
3. Life duration (yrs)	5	6	6	5	5	4	6	4	5	7	5	5	5	6	5
4. Height (m)	5	5	5	6	4	6	6	4	4	6	5	4	3	3	2
5. Crown diameter (m)	5	5	5	4	4	6	6	3	4	6	4	4	3	6	4
6. Canopy density (%)	3	2	3	2	5	4	3	3	2	1	1	1	1	1	2
7. Photosynthetic organs	1	1	1	1	1	1	1	1	1	1	1	1	3	1	3
8. 'Leaf' size	4	4	5	3	3	3	5	4	2	5	2	2	2	1	1
9. 'Leaf' colour	1	4	1	1	1	1	1	1	1	1	1	1	1	1	1
10. 'Leaf' length (cm)	4	4	4	3	3	3	4	3	1	4	1	1	5	1	2
11. 'Leaf' width (mm)	5	5	6	2	1	1	5	1	3	7	4	3	2	2	1
12. 'Leaf' angle (°)	3	3	3	3	2	2	3	3	3	3	3	3	2	3	2
13. 'Leaf' margin (T = terete)	2	2	2	T	T	T	1	T	1	1	1	1	4	1	4
14. 'Leaf' surface resins	2	2	2	2	2	2	2	2	2	2	2	2	2	2	2
15. 'Leaf' glands	2	2	2	2	2	1	2	2	1	1	1	1	2	2	2
16. 'Leaf' consistency	2	2	3	3	3	2	2	3	3	3	3	3	3	3	3
17. 'Leaf' tomentosity	1	2	1	1	1	4	1	1	1	1	1	1	1	1	4
18. 'Leaf' stomata	2	2	2	4	4	4	4	4	4	4	4	4	4	4	4
19. 'Leaf' duration (yrs)	5	5	4	4	3	6	4	5	3	4	4	3	–	3	–
20. 'Leaf' seasonality	1	1	1	1	1	1	1	1	1	1	1	1	1	1	1
21. 'Leaf' area: assim. stem area	1	1	1	1	1	1	1	1	1	1	1	1	2	3	2
22. Organs periodically shed	4	4	4	4	4	4	4	4	4	4	4	4	2	4	2
23. Organs shed – seasonality	?	?	?	?	?	?	?	?	?	?	?	?	?	?	?
24. Shoot growth – seasonality	2	2	1	1	1	1	1	1	1	1	2	1	4	1	1
25. Stem – number from base	1	2	2	1	1	2	2	1	1	2	2	1	2	2	2
26. Stem – height (m)	0.4	0.4	0.5	0.2	0.2	0.1	0.2	0.05	0.3	1.0	0.4	0.2	–	0.4	–
27. Stem – diameter (cm)	1.2	0.5	1.5	1	4	6.5	4	2.5	1.2	0.4	2.5	1.8	–	6.5	–
28. Stem lignification	1	1	1	1	1	1	1	1	1	1	1	1	1	1	1
29. Bark consistency	5	5	5	5	5	1	1	1	2	2	3	3	–	3	–
30. Bark thickness (mm)	1	1	1	1	1	1	1	1	1	2	1	1	–	2	–
31. Bark shedding	4	4	4	4	4	4	4	4	4	4	4	4	–	4	–
32. Bark shedding – seasonality	4	4	4	4	4	4	4	4	4	4	4	4	–	4	–
33. Spinescence	2	2	2	2	2	4	2	2	4	4	4	4	1	1	4
34. Underground stems	–	7	7	–	–	–	7	–	–	7	7	7	2	1	1
35. Root morphology	3	3	3	3	–	–	3	1 & 2	1 & 2	3	1 & 2	1 & 2	5	1 & 2	5
36. Root modification	7	7	7	7	7	7	7	7	7	7	7	7	7	7	7
37. Rootlet modification	3	3	3	3	3	3	3	3	–	5	?	?	1	?	1
38. Root depth (cm)	7	7	7	7	6	7	7	7	6	7	6	6	5	7	5
39. Root spread (cm)	7	7	7	6	6	7	7	6	6	7	6	6	3	–	3
40. Veg. multiplication	4	4	4	4	4	4	4	4	4	4	4	4	2	2	2
41. Veg. regen. after fire	1	2	2	1	1	1	2	1	1	3	2	2	4	2	4
42. Flowering – seasonality	4	2	4	1	1	2	1	1	2	2	2	2	–	2	–
43. Flowering – pyrogenic	2	2	2	2	2	2	2	2	2	2	2	2	1	2	1
44. Fruit dehiscence	2	2	2	2	2	1	2	1	2	2	2	2	1	2	1
45. Seed dissemination	3	3	3	3	3	1	3	3	3	3	3	3	?	2	1
46. Seedling – germ. type	2	2	2	2	2	4	2	2	2	2	2	2	–	2	–
47. Plumule position	1	1	1	1	1	1	1	1	1	1	1	1	2	1	2
48. No. 'leaves' per 10 cm stem	81	43	21	17	21	35	12	43	117	8	46	180	0	118	0

* Emergent shrubs 2–5 m.

Code of Ecomorphological Characters: See Chapter 1.

Table 19. Natural vegetation of California, U.S.A. – P.H. Zedler.
(Some of these ecosystems may be found in the summer rainfall zone of Arizona.)

I. Trees (>4 m) dominant (>50% cover)
  1. Coast, Transverse and Peninsular Ranges
     Supra-Mediterranean Zone
     (1). *Sequoia sempervirens* forest (Kuchler 2 and 24)
     Mediterranean Zone
     (2). Cypress and closed-cone pine forest (Kuchler 9) *Cupressus forbsii, Pinus torreyana, P. muricata, P. radiata,* etc.
     (3). Mixed hardwood forest (Kuchler 23) *Arbutus menziesii. Quercus* spp.
     (4). Oak and digger pine forest (Kuchler 25) *Quercus* spp., *Pinus sabiniana*
     (5). *Quercus agrifolia* forest (Kuchler 27)
  2. Sierra Range
     Supra-Mediterranean Zone
     (1). *Pinus jeffreyi, P. ponderosa* forest
     (2). *Sequoiadendron giganteum* forest
  3. Riparian forest (Kuchler 28)

II. Trees present, not dominant – savanna types
  1. Oak savanna (Kuchler 33) – *Quercus agrifolia, Q. douglasii, Q. lobata, Q. wislizenii*

III. Shrubs (woody plants mostly less than 3 m) dominant
  1. Drought deciduous coastal scrub (Kuchler 32)
     (1). Coastal sage scrub (*Salvia* spp.)
     (2). Coastal succulent scrub (succulents – Cactaceae, Crassulaceae and Euphorbiaceae): a variant found in Baja California, Mexico
  2. Evergreen species dominant – chaparral types (Kuchler 29)
     (1). *Adenostoma fasciculatum* (chamise chaparral)
     (2). *Adenostoma sparsifolium* (redshank chaparral)
     (3). *Arctostaphylos* spp. (manzanita chaparral)
     (4). *Quercus dumosa* (scrub oak chaparral)
     (5). *Ceanothus-Adenostoma-Arctostaphylos* chaparral
     (6). Mixed chaparral (dominants vary regionally)

IV. Herbaceous species dominant
  1. Perennial grassland – Native perennial grasses (*Aristida, Poa, Stipa*)
  2. Annual grassland – Introduced annual grasses

Reference

Küchler, A.W., 1977. Natural Vegetation of California. Map. In: Barbour, M.G. and Major, J. (eds) Terrestrial Vegetation of California. John Wiley, New York.

Table 20. Natural vegetation of Arizona, U.S.A.

Needle-leaved evergreen forest formations	
Pinyon pine – juniper forest	(*Pinus edulis* – *Juniperus* spp.)
Yellow pine forest	(*Pinus ponderosa*)
Yellow pine – Douglas fir forest	(*Pinus ponderosa* – *Pseudotsuga menziesii*)
Broad-leaved forest formations	
Oak woodland	(*Quercus arizonica* – *Q. emoryi*)
Shrub formations	
Chaparral	(*Arctostaphylos* spp. – *Ceanothus greggii* – Cercocarpus spp. – *Quercus turbinella*)
Mesquite	(*Prosopis* sp.)
Palo verde	(*Cercidium* spp.)
Low shrub formations	
Sagebrush	(*Artemisia* spp.)
Saltbush	(*Atriplex* spp.)
Creosote bush	(*Larrea divaricata*)
Graminoid formations	
Plains grass	
Desert grass	
Desert formations	
Cacti	

References

Carmichael, R.S., Knipe, O.D., Pase, C.P. and Brady, W.W., 1978. Arizona chaparral: Plant associations and ecology. U.S. Dep. Agric., For. Serv., Rocky Mt For. and Range Exp. Stn, (Fort Collins, Col.), Res. Pap. RM-202, 16 pp.

Humphrey, R.R., 1963. Arizona natural vegetation (map). Univ. Ariz., Tucson, Agric. Exp. Stn Bull. A-45.

Table 21. Ecomorphological characters: California communities.
Community type: Coastal sage scrub (*Artemisia californica*).
Stand age: Approx. 21 yrs.
Location: San Diego County; Rancho Bernardo foothills on Woodson mountain granodiorite.
Compilers: P.H. Zedler, C.A. Zammit and G.A. Scheid, May 1986 (National Science Foundation Grant No. BSR-85076997).

Character	*Artemisia californica* (Asteraceae)	*Salvia apiana* (Lamiaceae)	*Salvia mellifera* (Lamiaceae)	*Keckiella antirrhinoides* (Scrophulariaceae)	*Rhus laurina* (Anacardiaceae)	*Mimulus longiflorus* (Scrophulariaceae)	*Haplopappus squarrosus* (Asteraceae)	*Eriogonum fasciculatum* (Polygonaceae)
Foliage Projective Cover (%)								
Upper stratum (>2 m)	–	–	–	–	50	–	–	–
Mid stratum (30 cm–2 m)	20	10	2	10	–	1	2	5
Ground stratum (<30 cm)	–	–	–	–	–	–	–	–
1. Renewal buds	N	N	N	N	M	N	N	N
2. Trophic type	1	1	1	1	1	1	1	1
3. Life duration (yrs)	5	5	5	4–5	6	4	5	5
4. Height (m)	5	5	5	5	6	4	4	4
5. Crown diameter (m)	5	5	5	5	5–6	3	2	5
6. Canopy density (%)	5	4	5	4	6	2	2	3
7. Photosynthetic organs	1	1	1	1	1	1	1	1
8. 'Leaf' size	1	6	4	3	5–6	4	4	2
9. 'Leaf' colour	1	3	4	1	1	1	1	4
10. 'Leaf' length (cm)	3	4	3	2	3–4	3	3	2–3
11. 'Leaf' width (mm)	1	7	5	3–5	7	5	6	4
12. 'Leaf' angle (°)	3	1	1	3	3	1	1	3
13. 'Leaf' margin	1	3	3	1	1	5	2	5
14. 'Leaf' surface resins	2	2	–	2	2	1	1	–
15. 'Leaf' glands	2	2	2	2	2	1	1	–
16. 'Leaf' consistency	1	1	2	2	3	2	3	2
17. 'Leaf' tomentosity	4	4	2	1	1	2	2	2
18. 'Leaf' stomata	?	?	?	?	?	?	?	?
19. 'Leaf' duration (yrs)	1–2	1–2	1–2	?	?	2	2	2
20. 'Leaf' seasonality	2–3	2–3	2–3	1	1	2–3	1	1
21. 'Leaf' area: assim. stem area	1	1	1	1	1	1	1	1
22. Organs periodically shed	4	4	4	4	–	4	–	–
23. Organs shed – seasonality	2	2	2	–	–	2	–	–
24. Shoot growth – seasonality	1	1	1	1	1	1	1	1
25. Stem – number from base	10–25	10–25	10–25	10–25	10–25	2–15	2–3	10–30
26. Stem – height (m)	0.1	0.1	0.1	0.1	0.05	0.05	0.05	0.01
27. Stem – diameter (cm)	0.1–1	0.1–1	0.1–1	<2	1–10	0.1–0.5	0.1–0.5	0.1–1
28. Stem lignification	1	1	1	1	1	1	1	1
29. Bark consistency	1	1	1	2,4	1	1	1	1
30. Bark thickness (mm)	2	2	2	1–2	1	1	1	2
31. Bark shedding	–	–	–	–	–	–	–	–
32. Bark shedding – seasonality	–	–	–	–	–	–	–	–
33. Spinescence	4	4	4	4	4	4	4	4
34. Underground stems	7	7	7	7	7	7	7	7
35. Root morphology	–	–	–	–	–	–	–	3
36. Root modification	–	–	–	7	–	–	–	–
37. Rootlet modification	–	–	–	–	–	–	–	–
38. Root depth (cm)	–	–	–	–	–	–	–	–
39. Root spread (cm)	–	–	–	–	–	–	–	–
40. Veg. multiplication	4	4	4	4	4	4	4	4
41. Veg. regen. after fire	2	2	2	2	2	–	2	–
42. Flowering – seasonality	1	1	1	1	2	1	2	2
43. Flowering — pyrogenic	2	2	2	1–2	2	2	2	2
44. Fruit dehiscence	1	1	1	2	3	2	1	1
45. Seed dissemination	1	1	1	1	1	1	1	1
46. Seedling – germ. type	2	2	2	2	2	2	2	2
47. Plumule position	1	1	1	1	1	1	1	1
48. No. leaves per 10 cm stem	>100	6	10–20	20–40	<15	30–50	20–30	>80

Code of Ecomorphological Characters: See Chapter 1.

Table 22. Ecomorphological characters: California communities.
Community type: Arctostaphylos chaparral.
Stand age: Approx. 55–60 yrs.
Location: San Diego State University Sky Oaks Biological Field Station (33° 21′ N, 116° 34′ W) at about 1500 m elevation. Site on NE slope of about 45° on Tollhouse soil.
Compilers: P.H. Zedler, C.A. Zammit and G.A. Scheid, May 1986 (National Science Foundation Grant No. BSR-85076997).

Character	*Arctostaphylos pringlei* (Ericaceae)	*Arctostaphylos glandulosa* (Ericaceae)	*Ceanothus greggii* (Rhamnaceae)	*Adenostoma fasciculatum* (Rosaceae)	*Quercus dumosa* (Fagaceae)
Foliage Projective Cover (%)					
Upper stratum (>2 m)	← 70 →		10–15	10	<5
Mid stratum (30 cm–2 m)	–	–	–	–	–
Ground stratum (<30 cm)	–	–	–	–	–
1. Renewal buds	M, N	M, N	N	N	M
2. Trophic type	1	1	2	1	1
3. Life duration (yrs)	7	7	6	7	7
4. Height (m)	5–6	5	5	5–6	5–6
5. Crown diameter (m)	5–6	5–6	5	5	6
6. Canopy density (%)	5	5	4	4	5
7. Photosynthetic organs	1	1	1	1	1
8. 'Leaf' size	4	4	3	1	4
9. 'Leaf' colour	1	1	1	1	1
10. 'Leaf' length (cm)	2–3	2–3	2	1	2
11. 'Leaf' width (mm)	6	6	5–6	1	2
12. 'Leaf' angle (°)	2	2	3	3	3
13. 'Leaf' margin	2	2	1–2	3	2
14. 'Leaf' surface resins	2	2	2	2	2
15. 'Leaf' glands	2	2	2	2	2
16. 'Leaf' consistency	3	3	3	3	3
17. 'Leaf' tomentosity	4	4	2	1	2
18. 'Leaf' stomata	?	?	?	?	?
19. 'Leaf' duration (yrs)	?	?	?	?	2
20. 'Leaf' seasonality	1	1	–	1	1
21. 'Leaf' area: assim. stem area	1	1	1	1	1
22. Organs periodically shed	4	4	4	4	4
23. Organs shed – seasonality	–	–	–	–	1
24. Shoot growth – seasonality	1	1	1	4	1
25. Stem – number from base	2–15	2–15	1	2–50	2–50
26. Stem – height (m)	0	0	<0.5	0	0
27. Stem – diameter (cm)	2–10	2–10	4–12	1–6	3–5
28. Stem lignification	1	1	1	1	1
29. Bark consistency	1	1	1	3	3
30. Bark thickness (mm)	1	1	2	1	2
31. Bark shedding	1	1	–	2	–
32. Bark shedding – seasonality	–	–	–	–	–
33. Spinescence	4	4	4	4	4
34. Underground stems	N	7	N	7	1
35. Root morphology	3	3	3	3	3
36. Root modification	7	7	7	7	7
37. Rootlet modification	–	–	–	–	5
38. Root depth (cm)	5–6	5	5–6	6	5–6
39. Root spread (cm)	5–6	5	4–6	3	5–6
40. Veg. multiplication	4	4	4	2	2
41. Veg. regen. after fire	1	2	1	2	2
42. Flowering – seasonality	1	1	2	1–2	1
43. Flowering — pyrogenic	2	2	2	2	2
44. Fruit dehiscence	3	3	2	1	1
45. Seed dissemination	1	1	1	1	1
46. Seedling – germ. type	2	2	2	2	3
47. Plumule position	1	1	1	1	3
48. No. leaves per 10 cm stem	<20	<20	<20	>100	<25

Code of Ecomorphological Characters: See Chapter 1.

Table 23. Ecomorphological characters: California communities.
Community type: Adenostoma sparsifolium chaparral.
Stand age: Approx. 60 yrs.
Location: San Diego State University Sky Oaks Biological Field Station (33° 21′ N, 116° 34′ W) at about 1500 m elevation on Sheephead soil.
Compilers: P.H. Zedler, C.A. Zammit and G.A. Scheid, May 1986 (National Science Foundation Grant No. BSR-85076997).

Character	*Adenostoma sparsifolium* (Rosaceae)	*Cercocarpus betuloides* (Rosaceae)	*Ceanothus greggii* (Rhamnaceae)	*Adenostoma fasciculatum* (Rosaceae)	*Quercus dumosa* (Fagaceae)	*Tauschia parishii* (Apiaceae)	*Gnaphalium californicum* (Asteraceae)
Foliage Projective Cover (%)							
Upper stratum (>2 m)	75	10	5	5	5	–	–
Mid stratum (30 cm–2 m)	–	–	–	–	–	–	–
Ground stratum (<30 cm)	–	–	–	–	–	<1	<1
1. Renewal buds	M	M	N	N	M	3	3
2. Trophic type	1	2	2	1	1	1	1
3. Life duration (yrs)	7	7	6	7	7	4	2
4. Height (m)	6	6	5–6	5–6	5–6	2	2
5. Crown diameter (m)	6	5–6	5	5	6	3	1
6. Canopy density (%)	3	4	4	4	5	6	5
7. Photosynthetic organs	4	1	1	1	1	1	1
8. 'Leaf' size	1	3	3	1	4	5	4
9. 'Leaf' colour	1	4	1	1	1	1	1
10. 'Leaf' length (cm)	1	2–3	2	1	2	3–4	3–4
11. 'Leaf' width (mm)	1	5	5–6	1	2	8	5
12. 'Leaf' angle (°)	3	3	3	3	3	1	3
13. 'Leaf' margin	1	2	1–2	1	2	2	1
14. 'Leaf' surface resins	2	2	2	2	2	2	2
15. 'Leaf' glands	2	2	2	2	2	2	1
16. 'Leaf' consistency	3	3	3	3	3	1	1
17. 'Leaf' tomentosity	1	2	2	1	2	1	4
18. 'Leaf' stomata	?	?	?	?	?	?	?
19. 'Leaf' duration (yrs)	?	?	?	?	2	1	1
20. 'Leaf' seasonality	1	–	–	1	1	3	3
21. 'Leaf' area: assim. stem area	–	1	1	1	1	3	1
22. Organs periodically shed	4	4	4	4	4	1	1
23. Organs shed – seasonality	1	–	–	–	1	2	2
24. Shoot growth – seasonality	1	1	1	4	1	1	1
25. Stem – number from base	4–20	1–20	1	2–50	2–50	>100	2–5
26. Stem – height (m)	0	<0.5	<0.5	0	0	0	0
27. Stem – diameter (cm)	2–10	1–8	4–12	1–6	3–5	<0.5	<0.5
28. Stem lignification	1	1	1	1	1	3	3
29. Bark consistency	1, 3	1	5	3	3	NA	NA
30. Bark thickness (mm)	1–2	1–2	2	1	2	NA	NA
31. Bark shedding	–	–	–	2	–	NA	NA
32. Bark shedding – seasonality	–	–	–	–	–	NA	NA
33. Spinescence	4	4	4	4	4	4	4
34. Underground stems	7	7	N	7	1	1	N
35. Root morphology	3	3	3	3	3	1	1
36. Root modification	7	7	7	7	7	2	7
37. Rootlet modification	–	–	–	–	5	–	–
38. Root depth (cm)	6–7	6–7	5–6	6	5–6	3–4	2
39. Root spread (cm)	6	6	4–6	3	5–6	2–3	1
40. Veg. multiplication	4	4	4	2	2	1	4
41. Veg. regen. after fire	2	2	1	2	2	4	1
42. Flowering – seasonality	2	2	2	1–2	1	1	1
43. Flowering — pyrogenic	2	2	2	2	2	2	2
44. Fruit dehiscence	1	1	3	1	1	1	1
45. Seed dissemination	1	1	1	1	1	1	1
46. Seedling – germ. type	2	2	2	2	3	2	2
47. Plumule position	1	1	1	1	3	1	1
48. No. leaves per 10 cm stem	<30	<10	<20	>100	<25	3	–

Code of Ecomorphological Characters: See Chapter 1.

Table 24. Ecomorphological characters: California communities.
Community type: Quercus dumosa chaparral.
Stand age: Approx. 60 yrs.
Location: San Diego State University Sky Oaks Biological Field Station (33° 21′ N, 116° 34′ W), north slope on Tollhouse soil.
Compilers: P.H. Zedler, C.A. Zammit and G.A. Scheid, May 1986 (National Science Foundation Grant No. BSR-85076997).

Character	*Quercus dumosa* (Fagaceae)	*Quercus wislizenii* (Fagaceae)	*Ceanothus greggii* (Rhamnaceae)	*Lonicera subspicata* (Caprifoliaceae)	*Cercocarpus betuloides* (Rosaceae)	*Adenostoma sparsifolium* (Rosaceae)	*Garrya flavescens* (Garryaceae)	*Penstemon centranethifolius* (Scrophulariaceae)
Foliage Projective Cover (%)								
Upper stratum (>2 m)	40	20	15–20	–	<5	<5	5–10	–
Mid stratum (30 cm–2 m)	–	–	–	<5	–	–	–	<1
Ground stratum (<30 cm)	–	–	–	–	–	–	–	–
1. Renewal buds	M	M	N	M	M	M	M, N	2
2. Trophic type	1	1	2	1	2	1	1	1
3. Life duration (yrs)	7	7	6	6	7	7	7	3–4
4. Height (m)	5–6	6	5	5	6	6	6	3
5. Crown diameter (m)	6	6	5	5	5–6	6	6	3
6. Canopy density (%)	4	4	4	2	3	2	5	2
7. Photosynthetic organs	1	1	1	1	1	4	1	1
8. 'Leaf' size	4	4	3	3	3	1	4–5	4–5
9. 'Leaf' colour	1	1	1	4	4	1	1	2
10. 'Leaf' length (cm)	2	2	2	2	2–3	1	3	3–4
11. 'Leaf' width (mm)	2	2	5–6	5	5	1	7	7
12. 'Leaf' angle (°)	3	3	3	1	3	3	2	1
13. 'Leaf' margin	2	2	1–2	1	2	1	1	2
14. 'Leaf' surface resins	2	2	2	2	2	2	2	2
15. 'Leaf' glands	2	2	–	2	2	2	2	2
16. 'Leaf' consistency	3	3	3	1	3	3	3	1
17. 'Leaf' tomentosity	1	1	2	4	2	1	4	1
18. 'Leaf' stomata	?	?	?	?	?	?	?	?
19. 'Leaf' duration (yrs)	3	3	?	?	?	?	?	?
20. 'Leaf' seasonality	1	1	–	3	–	1	1	3
21. 'Leaf' area: assim. stem area	1	1	1	1	1	1	1	1
22. Organs periodically shed	4	4	4	4	4	1	1	1
23. Organs shed – seasonality	–	–	–	3	–	1	–	3
24. Shoot growth – seasonality	1	1	1	1	1	1	1	1
25. Stem – number from base	2–25	2–25	1	2	1–20	4–20	5–20	2–10
26. Stem – height (m)	0	0	<0.5	<0.5	<0.5	0	0	0
27. Stem – diameter (cm)	2–10	2–10	4–12	<1	0.5–8	2–10	2–8	<1
28. Stem lignification	1	1	1	1	1	1	1	2
29. Bark consistency	3	3	5	1	1	1, 3	1	1
30. Bark thickness (mm)	2	2	2	1	1	1–2	2	1
31. Bark shedding	–	–	–	4	–	–	–	–
32. Bark shedding – seasonality	–	–	–	–	–	–	–	–
33. Spinescence	4	4	4	4	4	4	4	4
34. Underground stems	7	7	N	–	7	7	7	–
35. Root morphology	3	3	3	3	3	3	3	3
36. Root modification	7	7	7	–	–	7	–	–
37. Rootlet modification	5	5	–	–	–	–	–	–
38. Root depth (cm)	–	–	5–6	–	–	6–7	–	–
39. Root spread (cm)	–	–	4–6	–	–	6	–	–
40. Veg. multiplication	4	4	4	2	4	4	4	–
41. Veg. regen. after fire	2	2	4	2	2	2	2	1
42. Flowering – seasonality	1	1	2	2	2	2	2	2
43. Flowering — pyrogenic	2	2	2	2	2	2	2	2
44. Fruit dehiscence	1	1	2	3	1	1	3	2
45. Seed dissemination	1	1	1	1	1	1	1	1
46. Seedling – germ. type	3	3	2	2	2	2	2	2
47. Plumule position	3	3	1	1	1	1	1	1
48. No. leaves per 10 cm stem	<25	<25	<20	<10	<10	<30	<10	<10

Code of Ecomorphological Characters: See Chapter 1.

Table 25. Ecomorphological characters: California communities.
Community type: Adenostoma fasciculatum chaparral.
Stand age: Approx. 50 yrs.
Location: San Diego County; Del Mar mesa on Redding soil. Site in Caltrans Vernal Pool Reserve.
Compilers: P.H. Zedler, C.A. Zammit and G.A. Scheid, May 1986 (National Science Foundation Grant No. BSR-85076997).

Character	*Adenostoma fasciculatum* (Rosaceae)	*Quercus dumosa* (Fagaceae)	*Rhus integrifolia* (Anacardiaceae)	*Lotus scoparius* (Fabaceae)	*Jepsonia parryi* (Saxifragaceae)
Foliage Projective Cover (%)					
Upper stratum (> 2 m)	85	20	5	<1	–
Mid stratum (30 cm–2 m)	–	–	–	–	–
Ground stratum (<30 cm)	–	–	–	–	<1
1. Renewal buds	N	M	N	N	4
2. Trophic type	1	1	1	2	1
3. Life duration (yrs)	7	7	7	3	4
4. Height (m)	5	5–6	5	3	1
5. Crown diameter (m)	5	6	5	4	1
6. Canopy density (%)	4	5	5	3	6
7. Photosynthetic organs	1	1	1	1	1
8. 'Leaf' size	1	4	4	3	4
9. 'Leaf' colour	1	1	1	1	1
10. 'Leaf' length (cm)	1	2	3	1	3
11. 'Leaf' width (mm)	1	2	2–3	4–5	3–4
12. 'Leaf' angle (°)	3	3	3	1	1
13. 'Leaf' margin	1	2	2	1	2
14. 'Leaf' surface resins	2	2	1	2	2
15. 'Leaf' glands	2	2	2	2	2
16. 'Leaf' consistency	3	3	3	1	1
17. 'Leaf' tomentosity	1	2	1	1	4
18. 'Leaf' stomata	?	?	?	?	?
19. 'Leaf' duration (yrs)	?	2	3	1	1
20. 'Leaf' seasonality	1	1	1	2	2
21. 'Leaf' area: assim. stem area	1	1	1	3	1
22. Organs periodically shed	4	4	4	4	2
23. Organs shed – seasonality	–	1	–	2	1
24. Shoot growth – seasonality	4	1	4–1	1	4
25. Stem – number from base	2–50	2–50	–	1	1
26. Stem – height (m)	0	0	0	0.001	–
27. Stem – diameter (cm)	1–6	3–5	4–6	1	<0.2
28. Stem lignification	1	1	1	1	3
29. Bark consistency	3	3	3	5	NA
30. Bark thickness (mm)	1	2	2	1	NA
31. Bark shedding	2	–	2	–	NA
32. Bark shedding – seasonality	–	–	–	–	NA
33. Spinescence	4	4	4	4	4
34. Underground stems	7	1	7	–	6
35. Root morphology	3	3	3	–	4
36. Root modification	7	–	–	–	–
37. Rootlet modification	–	5	–	–	–
38. Root depth (cm)	6	–	7	–	–
39. Root spread (cm)	3	–	–	–	–
40. Veg. multiplication	2	2	2	4	4
41. Veg. regen. after fire	2	2	2	1	4
42. Flowering – seasonality	1–2	1	4–1	1	3–4
43. Flowering — pyrogenic	2	2	2	2	2
44. Fruit dehiscence	1	1	3	1	2
45. Seed dissemination	1	1	1	1	1
46. Seedling – germ. type	2	3	2	2	2
47. Plumule position	1	3	1	1	1
48. No. leaves per 10 cm stem	>100	<25	–	–	–

Code of Ecomorphological Characters: See Chapter 1.

Table 26. Ecomorphological characters: California communities.
Community type: Quercus agrifolia forest.
Stand age: Approx. 150 yrs.
Location: San Diego State University Sky Oaks Biological Field Station (33° 21' N, 116° 34' W).
Compilers: P.H. Zedler, C.A. Zammit and G.A. Scheid, May 1986 (National Science Foundation Grant No. BSR-85076997).

Character	*Quercus agrifolia* (Fagaceae)	*Lonicera subspicata* (Caprifoliaceae)	*Solidago californica* (Asteraceae)	*Eriogonum wrightii* (Polygonaceae)
Foliage Projective Cover (%)				
Upper stratum (>2 m)	60	–	–	–
Mid stratum (30 cm–2 m)	–	<5	–	–
Ground stratum (<30 cm)	–	–	15	<5
1. Renewal buds	MM	M–N	3	N
2. Trophic type	1	1	1	1
3. Life duration (yrs)	7	6	4	4
4. Height (m)	8	5	3	3
5. Crown diameter (m)	7–8	5	2	3
6. Canopy density (%)	5	2	3	3
7. Photosynthetic organs	1	1	1	1
8. 'Leaf' size	5–6	3	3	2
9. 'Leaf' colour	1	4	1	4
10. 'Leaf' length (cm)	3	2	4	1
11. 'Leaf' width (mm)	7	5	5	2
12. 'Leaf' angle (°)	1	1	3	2
13. 'Leaf' margin	5	1	1	5
14. 'Leaf' surface resins	2	2	2	2
15. 'Leaf' glands	2	2	2	2
16. 'Leaf' consistency	3	1	1	2
17. 'Leaf' tomentosity	4	2	4	4
18. 'Leaf' stomata	?	?	2	2
19. 'Leaf' duration (yrs)	?	?	1	2
20. 'Leaf' seasonality	1	3	3	1
21. 'Leaf' area: assim. stem area	1	1	1	1
22. Organs periodically shed	4	4	2	4
23. Organs shed – seasonality	–	3	3	–
24. Shoot growth – seasonality	–	1	1–2	1–2
25. Stem – number from base	1	2	1–5	4–100
26. Stem – height (m)	2–3	<0.5	<0.4	<0.05
27. Stem – diameter (cm)	>80	<1	<1	<3
28. Stem lignification	1	1	2	1
29. Bark consistency	1	1	NA	3
30. Bark thickness (mm)	3	1	NA	1
31. Bark shedding	4	4	NA	–
32. Bark shedding – seasonality	–	–	NA	–
33. Spinescence	4	4	4	4
34. Underground stems	–	–	1	1
35. Root morphology	3	–	3	3
36. Root modification	7	–	7	7
37. Rootlet modification	5	–	8	8
38. Root depth (cm)	7	–	2	3
39. Root spread (cm)	7	–	2	3
40. Veg. multiplication	4	2	2	2
41. Veg. regen. after fire	2, 3	2	2	1, 2
42. Flowering – seasonality	1	2	2	2
43. Flowering — pyrogenic	2	2	2	2
44. Fruit dehiscence	1	3	1	1
45. Seed dissemination	1	1	1	1
46. Seedling – germ. type	3	2	2	2
47. Plumule position	3	1	1	1
48. No. leaves per 10 cm stem	<10	<10	<15	<50

Code of Ecomorphological Characters: See Chapter 1.

Table 27. Ecomorphological characters: California communities.
Community type: Ceanothus-Adenostoma-Arctostaphylos chaparral.
Stand age: Approx. 40 yrs.
Location: San Diego County; east side of Palomar Mountain on Sheephead soil.
Compilers: P.H. Zedler, C.A. Zammit and G.A. Scheid, May 1986 (National Science Foundation Grant No. BSR-85076997).

Character	*Keckiella antirrhinoides* (Scrophulariaceae)	*Adenostoma fasciculatum* (Rosaceae)	*Arctostaphylos glandulosa* (Ericaceae)	*Rhus ovata* (Anacardiaceae)	*Eriogonum fasciculatum* (Polygonaceae)	*Ceanothus greggii* (Rhamnaceae)	*Eriophyllum confertiflorum* (Asteraceae)	*Stipa pulchra* (Poaceae)	*Arctostaphylos pringlei* (Ericaceae)	*Quercus dumosa* (Fagaceae)
Foliage Projective Cover (%)										
Upper stratum (>2 m)	30	20–30	20	<10	–	20–30	–	–	30	15
Mid stratum (30 cm–2 m)	–	–	–	–	15	–	<5	–	–	–
Ground stratum (<30 cm)	–	–	–	–	–	–	–	<5	–	–
1. Renewal buds	N	N	N	N	N	N	2	5	M	M
2. Trophic type	1	1	1	1	1	2	1	1	1	1
3. Life duration (yrs)	4–5	7	7	7	5	6	3	3	7	7
4. Height (m)	5	5–6	6	6	4	5	3	4	6	5–6
5. Crown diameter (m)	5	5	5–7	6	4	5	2	3	5–6	6
6. Canopy density (%)	4	4	5	4	5	4	2	2	2	5
7. Photosynthetic organs	1	1	1	1	1	1	1	1	1	1
8. 'Leaf' size	3	1	4	5–6	3	3	3	6	1	4
9. 'Leaf' colour	1	1	1	1	4	1	1	1	1	1
10. 'Leaf' length (cm)	2	1	3	3–4	1–2	2	3	7	1	2
11. 'Leaf' width (mm)	3–5	1	6–7	6–7	3–4	5–6	2	5	1	2
12. 'Leaf' angle (°)	3	3	3	3	3	3	1	3	3	3
13. 'Leaf' margin	1	1	1	1	5	1–2	1	2	1	2
14. 'Leaf' surface resins	2	2	2	2	2	2	2	2	2	2
15. 'Leaf' glands	2	2	2	2	2	2	2	2	2	2
16. 'Leaf' consistency	2	3	3	3	3	3	1	1	3	3
17. 'Leaf' tomentosity	1	1	1	1	2	2	2	1	1	2
18. 'Leaf' stomata	?	?	4	?	?	?	?	?	?	?
19. 'Leaf' duration (yrs)	?	?	?	?	?	?	1	2	?	2
20. 'Leaf' seasonality	1	1	1	1	1	1	2	2	1	1
21. 'Leaf' area: assim. stem area	1	1	1	1	1	1	1	1	1	1
22. Organs periodically shed	4	4	4	4	4	4	4	4	4	4
23. Organs shed – seasonality	–	–	–	–	–	–	2	2	1	1
24. Shoot growth – seasonality	1	4	–	–	–	1	1	4	1	1
25. Stem – number from base	10–25	2–50	1	2–15	2–50	1	1	–	4–20	2–50
26. Stem – height (m)	0.1	0	<0.2	0	0	<0.5	<0.1	0	0	0
27. Stem – diameter (cm)	<2	<6	<25	<10	<2	<12	<1	–	<10	<5
28. Stem lignification	1	1	1	1	1	1	1	3	1	1
29. Bark consistency	2, 4	3	1, 4	3	1, 3	5	1	NA	1, 3	3
30. Bark thickness (mm)	1–2	1	1	2	1–2	2	1	NA	1–2	2
31. Bark shedding	–	2	–	–	–	–	–	NA	–	–
32. Bark shedding – seasonality	–	–	–	–	–	–	–	NA	–	–
33. Spinescence	4	4	4	4	4	4	4	4	4	4
34. Underground stems	7	7	7	7	–	–	–	1	7	1
35. Root morphology	3	1	–	–	–	3	–	–	–	3
36. Root modification	–	7	–	–	–	7	–	–	–	–
37. Rootlet modification	–	–	–	–	–	–	–	–	–	5
38. Root depth (cm)	–	6	5–7	–	4–5	5–6	3–4	–	–	–
39. Root spread (cm)	–	3	–	–	4–5	4–6	2–4	–	–	–
40. Veg. multiplication	4	2	4	4	1	4	4	1	–	2
41. Veg. regen. after fire	2	2	2	2	1	2	1	2	1	2
42. Flowering – seasonality	2	1–2	4	1	2	2	1	1	1	1
43. Flowering — pyrogenic	1–2	2	2	2	2	2	2	2	2	2
44. Fruit dehiscence	2	1	3	3	1	2	1	1	–	1
45. Seed dissemination	1	1	1	1	1	1	1	1	1	1
46. Seedling – germ. type	2	2	2	2	2	2	2	–	2	3
47. Plumule position	1	1	1	1	1	1	1	2	1	3
48. No. leaves per 10 cm stem	<40	>100	<20	<10	–	<20	–	–	<30	<25

Code of Ecomorphological Characters: See Chapter 1.

Table 28. Vegetation types of Chile – R. Gajardo.

1. Región del Desierto – Desert Region
   1-A Desierto Absoluto – Absolute Desert
   1-B Desierto Andino – Andean Desert
   1-C Desierto Costero – Coastal Desert
   1-D Desierto Florido – Desert of Ephemerals

2. Región de las Estepas Alto-Andinas – Region of High-Andean Steppes
   2-B Andes Mediterraneos – Mediterranean Andes

3. Región de los Matorrales y Bosques Esclerófilos – Region of Matorrals and Sclerophyllous Forests
   3-A Matorrales Esteparios – Steppe-like Matorrals
   3-B Matorrales y Bosques Espinosos – Matorrals and Thorn Forests
   3-C Bosques Esclerófilos – Sclerophyllous Forests

4. Región de los Bosques Caducifolios – Region of Deciduous Forests
   4-A Bosques Caducifolios Montanos – Montane Deciduous Forests
   4-B Bosques Caducifolios del Llano – Deciduous Forests of the Plain
   4-C Bosques Caducifolios Andinos – Andean Deciduous Forests

6. Región de los Bosques Andino-Patagónicos – Region of Andean-Patagonic Forests
   6-A Cordillera de la Araucanía – Cordillera of the Araucania
   6-B Cordilleras Patagónicas – Patagonic Cordilleras

8. Región de los Matorrales y Estepas Patagónicas – Region of Patagonic Matorrals and Steppes

References

Gajardo R., 1983. Sistema básico de clasificación de la vegetación nativa chilena. Corporación Nacional Forestal. Santiago. 315 pp. + Annex 21 pp. + 15 maps.

Quintanilla V.G., 1981. Carta de las formaciones vegetales de Chile. Contribuciones Científicas y Tecnológicas, Area Geociencias, Universidad de Santiago 11 (47); 5–32. + 1 map.

Table 29. Ecomorphological characters: Chile.
Location: Papudo-Zapallar-Cachagua, Chile (32° 35′ S, 71° 28′ W, 20–80 m).
Vegetation structure: Coastal scrub (coastal matorral) on ocean bluffs and slopes, near the immediate coast.
Age since last fire: 20 years.
Nearest climate station: Zapallar, Chile (32° 33′ S, 71° 30′ W, 30 m).
Soils: Quartz diorite – pH 5.2–6.0, clay 15–23%, organic matter 0.7–1.3%, nitrate nitrogen 2.3–2.6 μg g^{-1} soil, and available phosphorus 1.8–2.1 ppm.
Reference: Orshan, G., Montenegro, G., Avila, G., Aljaro, M.E., Walckowiak, A. and Mujica, A.M., 1985. Plant growth forms of Chilean matorral. A monocharacter growth form analysis along an altitudinal transect from sea level to 2000 m.a.s.l. Bull. Soc. Bot. Fr. 131, Act. Bot. 1984: 411–425.
Compiler: G. Montenegro, April 1984.

Character	*Peumus boldus* (Monimiaceae)	*Flourensia thurifera* (Asteraceae)	*Proustia cuneifolia* (Asteraceae)	*Lobelia tupa* (Lobeliaceae)	*Podanthus mitiqui* (Asteraceae)	*Fuchsia lycioides* (Onagraceae)	*Baccharis concava* (Asteraceae)	*Aster haplopappus* (Asteraceae)	*Bahia ambrosioides* (Asteraceae)	*Haplopappus foliosus* (Asteraceae)	*Puya chilensis* (Bromeliaceae)
* Foliage Projective Cover (%)											
Upper stratum (>2 m)	12.0	6.5	0.9	3.3	0.8	4.6	–	–	–	–	4.4
Mid stratum (25 cm–2 m)	–	–	–	–	–	–	18.6	5.7	1.2	3.1	–
Ground stratum (<25 cm)	–	–	–	–	–	–	–	–	–	–	–
Density (mature plants ha^{-1})	12.0	5.3	1.8	1.6	2.5	1.9	29.8	4.2	5.6	7.6	2.3
Biomass tops (kg ha^{-1})	260,342	14,310	42,192	5,120	22,590	5,890	238,400	2,100	2,810	99,760	
1. Renewal buds	M	N	N	N	N	N	Ch	Ch	Ch	Ch	N
2. Trophic type	1	1	1	1	1	1	1	1	1	1	1
3. Life duration (yrs)	6	5	4	4	5	5	4	4	4	4	5
4. Height (m)	7	5	5	5	5	5	4	3	4	3	5
5. Crown diameter (m)	6	5	4	5	5	5	4	3	4	3	6
6. Canopy density (%)	4	3	3	3	4	4	4	3	3	3	6
7. Photosynthetic organs	1	1	1	1	4	1	1	1	1	1	1
8. 'Leaf' size	4	4	4	4	4	4	4	3	1	4	7
9. 'Leaf' colour	1	1	1	2	1	1	1	1	1	1	2
10. 'Leaf' length (cm)	3	3	3	6	3	2	2	3	1	3	7
11. 'Leaf' width (mm)	7	6	6	5	6	5	2	5	1	6	7
12. 'Leaf' angle (°)	2	3	3	3	3	3	3	3	2	3	1
13. 'Leaf' margin	4	2	1	1	1	1	2	2	3	2	2
14. 'Leaf' surface resins	2	1	1	2	2	2	2	2	2	1	2
15. 'Leaf' glands	1	2	2	2	2	2	2	1	1	1	2
16. 'Leaf' consistency	3	1	1	1	1	1	2	1	1	2	3
17. 'Leaf' tomentosity	4	1	1	2	3	1	1	4	1	4	2
18. 'Leaf' stomata	2	4	4	2	2	2	2	4	4	4	2
19. 'Leaf' duration (yrs)	3	2	1	3	2	2	2	2	2	2	5
20. 'Leaf' seasonality	1	2	2	2	1	2	1	2	1	1	1
21. 'Leaf' area: assim. stem area	1	1	1	1	3	1	1	1	1	1	1
22. Organs periodically shed	3	4	4	3	3	4	3	3	4	3	4
23. Organs shed – seasonality	April	Jan.–Feb.	Jan.–Feb.	March	April	April	April	March	March	April	March
24. Shoot growth – seasonality	Sept.	Aug–Sept.	Jun–Jul.	Sept.	Aug–Sept.	Sept.	Sept.	Sept.	Sept.	Oct.	Sept.
25. Stem – number from base	1	1	2	1	1	1	2	2	2	2	2
26. Stem – height (m)	1.5	0.3	0.1	0.1	0.5	0.8	0.2	0.2	0.2	0.1	none
27. Stem – diameter (cm)	30	4	5	2	5	10	3	2	2	1	none
28. Stem lignification	1	1	1	1	1	1	1	1	1	2	none
29. Bark consistency	1	3	1	6	4	3	1	none	2	2	none
30. Bark thickness (mm)	3	1	1	1	1	1	1	none	1	1	none
31. Bark shedding	3	3	2	3	3	2	3	4	3	2	none
32. Bark shedding – seasonality	Mar.	Jun	May–Jun.	Apr.	Apr.	May–Jun.	Mar–Jun.	none	Apr.	Apr.	none
33. Spinescence	4	4	1	4	4	4	4	4	4	4	2
34. Underground stems	7	none	none	3	7	3	7	none	none	none	1
35. Root morphology	1	3	3	3	1	3	3	3	3	3	2
36. Root modification	7	7	7	7	7	7	7	7	7	7	7
37. Rootlet modification	8	8	8	8	8	8	8	8	8	8	8
38. Root depth (cm)	6	4	4	3	4	4	4	4	4	3	3
39. Root spread (cm)	6	5	4	5	4	5	4	4	4	3	5
40. Veg. multiplication	4	4	4	4	4	4	4	4	1	4	2
41. Veg. regen. after fire	2	2	3	1	2	1	2	1	3	3	2
42. Flowering – seasonality	Dec.	Nov–Dec.	Dec.	Dec.	Nov.	Oct.	Dec.	Jan–Feb.	Jan.	Jan.	Dec.
43. Flowering — pyrogenic	2	2	2	2	2	2	2	2	2	2	2
44. Fruit dehiscence	3	1	1	2	1	3	1	1	1	1	1
45. Seed dissemination	1	1	1	1	1	1	1	1	1	1	2
46. Seedling – germ. type	2	2	2	2	2	2	2	2	2	2	–
47. Plumule position	1	1	1	1	1	1	1	1	1	1	2

* F.P.C. of Bare ground = 43.82%.

Code of Ecomorphological Characters: See Chapter 1.

Table 30. Ecomorphological characters: Chile.
Location: Fundo Santa Laura, Chile (33°04′ S, 71°00′ W, 1000 m) – Primary IBP Site.
Vegetation structure: Typical matorral (mid elevation). Sclerophyll scrub, evergreen woodland.
Age since last fire: 25 years.
Nearest climate station: Til-Til, Chile (33°04′ S, 71°00′ W, 950 m).
Soil: Quartz diorite and andesite – pH 6.3–6.5, clay 15–32%, organic matter 0.8–1.7%, nitrate nitrogen 1–2.6 µg g^{-1} soil, available phosphorus 0.4–5.2 ppm. Coarse textured.
Reference: Orshan, G., Montenegro, G., Avila, G., Aljaro, M.E., Walckowiak, A. and Mujica, A.M., 1985. Plant growth forms of Chilean matorral. A monocharacter growth form analysis along an altitudinal transect from sea level to 2000 m.a.s.l. Bull. Soc. Bot. Fr. 131, Act. Bot. 1984: 411–425.
Compiler: G. Montenegro, April 1984.

Character	Quillaja saponaria (Rosaceae)	Cryptocarya alba (Lauraceae)	Lithraea caustica (Anacardiaceae)	Kageneckia oblonga (Rosaceae)	Trevoa trinervis (Rhamnaceae)	Talguenea quinquenervia (Rhamnaceae)	Trichocereus chilensis (Cactaceae)	Colliguaya odorifera (Euphorbiaceae)	Satureja gilliesii (Lamiaceae)	Muehlenbeckia hastulata (Polygonaceae)	Puya berteroniana (Bromeliaceae)
*Foliage Projective Cover (%)											
Upper stratum (>2 m)	10.42	14.92	21.73	2.43	12.8	7.3	2.0	18.52	–	–	2.4
Mid stratum (30 cm–2 m)	–	–	–	–	–	–	–	–	3.64	2.0	–
Ground stratum (<30 cm)	–	–	–	–	–	–	–	–	–	–	–
Density (mature plants ha^{-1})	9.8	10.3	25.3	12.0	11.2	7.6	2.8	44.1	11.8	4.2	3.1
Biomass tops (kg ha^{-1})	360,504	594,328	490,318	174,108	384,092	211,783	54,190	502,230	84,920	73,660	104,830
1. Renewal buds	MM	MM	M	M	M	M	MM	N	N	Ch	H
2. Trophic type	1	1	1	1	2	2	1	1	1	1	1
3. Life duration (yrs)	6	6	7	5	7	5	5	6	5	6	5
4. Height (m)	8	7	7	6	6	5	7	5	4	4	6
5. Crown diameter (m)	7	6	6	5	5	6	4	5	4	6	6
6. Canopy density (%)	3	4	3	4	3	2	1	4	3	4	6
7. Photosynthetic organs	1	1	1	1	4	1	3	1	1	1	1
8. 'Leaf' size	4	5	4	5	4	5	none	4	3	3	7
9. 'Leaf' colour	1	1	2	1	1	1	none	1	1	1	none
10. 'Leaf' length (cm)	3	4	4	3	2	2	none	3	1	2	7
11. 'Leaf' width (mm)	6	7	7	8	2	6	none	6	1	5	8
12. 'Leaf' angle (°)	3	3	3	3	3	3	none	3	3	3	1
13. 'Leaf' margin	1	1	1	2	1	1	none	2	1	1	2
14. 'Leaf' surface resins	2	2	2	2	2	2	2	2	2	2	2
15. 'Leaf' glands	2	2	2	2	2	2	2	2	1	2	2
16. 'Leaf' consistency	3	3	3	3	1	1	none	1	1	1	3
17. 'Leaf' tomentosity	1	1	2	1	1	4	1	1	4	1	2
18. 'Leaf' stomata	4	2	2	2	2	2	1	4	4	4	2
19. 'Leaf' duration (yrs)	2	5	5	2	1	2	none	2	2	2	5
20. 'Leaf' seasonality	1	1	1	1	2	2	none	1	2	1	1
21. 'Leaf' area: assim. stem area	1	1	1	1	3	1	2	1	1	1	1
22. Organs periodically shed	4	4	4	4	3	3	none	4	3	3	4
23. Organs shed – seasonality	May	May	May–Jun.	May–Jun.	Dec.	Dec.	none	April	April	May	May
24. Shoot growth – seasonality	Oct.	Oct–Nov.	Nov–Dec.	Nov–Dec.	Jul–Aug.	Jul–Aug.	Oct–Nov.	Jul–Aug.	Jul.	Nov–Dec.	Sept.
25. Stem – number from base	1	1	2	2	2	1	1	2	2	2	2
26. Stem – height (m)	2.0	2.0	0.8	0.6	0.2	1.0	0.8	0.5	0.2	0.1	none
27. Stem – diameter (cm)	60	80	30	6	10	15	25	5	3	6	none
28. Stem lignification	1	1	1	1	2	1	2	1	1	1	none
29. Bark consistency	6	1	3	1	1	2	none	2	3	1	none
30. Bark thickness (mm)	2	1	2	1	1	1	none	1	2	1	none
31. Bark shedding	3	3	3	3	3	3	none	2	2	3	none
32. Bark shedding – seasonality	March	March	April	March	March	Mar–Jul.	none	Mar–Jul.	April	May–Jul.	none
33. Spinescence	4	4	4	4	1	1	1	4	4	4	2
34. Underground stems	7	7	7	7	7	7	none	7	7	7	1
35. Root morphology	3	3	3	3	3	3	2	3	3	3	2
36. Root modification	7	7	7	7	7	7	3	7	7	7	7
37. Rootlet modification	8	8	8	8	6	6	8	8	8	8	8
38. Root depth (cm)	6	6	7	5	5	5	3	6	4	5	3
39. Root spread (cm)	7	6	7	5	5	5	4	6	4	6	5
40. Veg. multiplication	4	4	4	4	4	2	1	4	3	1	2
41. Veg. regen. after fire	2	2	2	2	2	2	1	2	3	3	2
42. Flowering – seasonality	Dec–Jan.	Nov–Dec.	Nov.	Nov.	Sept.	Sept.	Dec.	Jul–Aug.	Sept.	Sept–Oct.	Dec.
43. Flowering — pyrogenic	2	2	2	2	1	2	2	2	2	2	2
44. Fruit dehiscence	2	3	3	2	1	1	3	2	2	3	1
45. Seed dissemination	1	1	1	1	1	1	1	1	1	1	2
46. Seedling – germ. type	2	2	4	2	2	2	2	2	2	2	–
47. Plumule position	1	1	2	1	1	1	1	1	1	1	2

* F.P.C. of Bare ground = 40.71%.

Code of Ecomorphological Characters: See Chapter 1.

Table 31. Ecomorphological characters: Chile.
Location: Paso Marchant, at Andes Mountain, Chile (1500–1800 m)
Vegetation structure: Montane matorral – Low evergreen community with a few dominant taller shrubs (*Kageneckia angustifolia*, etc.).
Age since last fire: 25 years.
Nearest climate station: Juncal, Chile (32° 52′ S, 70° 10′ W, 2250 m).
Soil: No data available.
Reference: Orshan, G., Montenegro, G., Avila, G., Aljaro, M.E., Walckowiak, A. and Mujica, A.M., 1985. Plant growth forms of Chilean matorral. A monocharacter growth form analysis along an altitudinal transect from sea level to 2000 m.a.s.l. Bull. Soc. Bot. Fr. 131, Act. Bot. 1984: 411–425.
Compiler: G. Montenegro, April 1984.

Character	*Kageneckia angustifolia* (Rosaceae)	*Schinus montanus* (Anacardiaceae)	*Escallonia illinita* (Escalloniaceae)	*Colliguaya salicifolia* (Euphorbiaceae)	*Adesmia arborea* (Fabaceae)	*Valenzuelia trinervis* (Sapindaceae)	*Berberis chilensis* (Berberidaceae)	*Chuquiragua oppositifolia* (Asteraceae)	*Haplopappus glutinosus* (Asteraceae)	*Mulinum spinosum* (Hydrocotylaceae)	*Oxalis parvifolia* (Oxalidaceae)
* Foliage Projective Cover (%)											
Upper stratum (>2 m)	8.1	18.14	9.0	–	–	–	–	–	–	–	–
Mid stratum (30 cm–2 m)	–	–	–	8.3	3.6	11.5	0.9	2.7	0.75	8.5	–
Ground stratum (<30 cm)	–	–	–	–	–	–	–	–	–	–	6.3
Density (mature plants ha^{-1})	6.3	5.4	3.8	18.1	2.4	16.7	2.6	10.5	8.6	5.4	12.3
Biomass tops (kg ha^{-1})	102,211	72,576	97,432	223,852	223,852	215,760	6.094	5,710	3,182	2,365	1,010
1. Renewal buds	M	M	M	N	N	N	N	Ch	Ch	Ch	H
2. Trophic type	1	1	1	1	2	1	1	1	1	1	1
3. Life duration (yrs)	6	5	5	6	5	5	5	5	4	5	3
4. Height (m)	6	6	6	5	5	5	5	4	4	4	1
5. Crown diameter (m)	6	5	6	5	5	5	4	5	5	4	2
6. Canopy density (%)	4	3	4	2	3	1	3	3	3	3	2
7. Photosynthetic organs	1	1	1	1	1	1	1	1	1	1	1
8. 'Leaf' size	3	4	4	4	1	3	3	3	5	3	2
9. 'Leaf' colour	1	1	1	1	1	1	1	2	1	1	1
10. 'Leaf' length (cm)	4	3	3	4	1	2	3	1	4	2	3
11. 'Leaf' width (mm)	5	7	6	5	2	5	5	1	7	5	4
12. 'Leaf' angle (°)	3	3	2	3	3	3	3	3	3	3	3
13. 'Leaf' margin	1	1	2	1	1	1	2	2	6	2	3
14. 'Leaf' surface resins	2	2	1	2	2	2	2	2	1	2	2
15. 'Leaf' glands	2	2	1	2	2	2	2	2	1	2	2
16. 'Leaf' consistency	2	2	1	2	1	2	3	2	2	2	4
17. 'Leaf' tomentosity	1	1	4	1	4	1	1	4	1	1	2
18. 'Leaf' stomata	4	4	2	4	4	4	2	4	4	4	4
19. 'Leaf' duration (yrs)	3	3	3	2	2	2	2	2	2	2	1
20. 'Leaf' seasonality	1	1	1	1	2	3	1	1	1	1	3
21. 'Leaf' area: assim. stem area	1	1	1	1	1	1	1	1	1	1	1
22. Organs periodically shed	4	4	4	4	4	4	3	4	3	4	2
23. Organs shed – seasonality	Aug.	Jul.	Jul–Aug.	Jun.	March	Aug.	Aug.	Jul–Aug.	Jul–Aug.	Aug.	Jul–Aug.
24. Shoot growth – seasonality	Nov–Dec.	Nov–Dec.	Jan.	Sept.	Aug.	Oct–Nov.	Sept–Oct.	Oct–Nov.	Sept–Oct.	Oct.	Nov–Dec.
25. Stem – number from base	1	2	1	2	1	2	2	2	2	2	1
26. Stem – height (m)	1.5	1.5	1.5	1	0.8	0.8	0.5	0.5	0.5	0.5	0.1
27. Stem – diameter (cm)	10	4	10	4	4	3	2	2	1.5	2	2
28. Stem lignification	1	1	1	1	1	1	1	1	1	1	3
29. Bark consistency	3	1	1	3	2	3	1	2	2	6	none
30. Bark thickness (mm)	1	1	1	2	1	1	1	1	1	1	none
31. Bark shedding	2	3	3	3	3	3	3	2	2	3	none
32. Bark shedding – seasonality	Apr.	Jan.	Dec.	Jan.	Mar.	Jan.	Jan–Feb.	Apr.	Apr.	Jan.	none
33. Spinescence	4	4	4	4	1	4	2	2	2	2	4
34. Underground stems	7	none	none	2	none	none	none	none	none	none	none
35. Root morphology	1	1	3	3	3	1	3	3	1	1	1
36. Root modification	8	8	8	8	6	8	8	8	8	8	6
37. Rootlet modification	8	8	8	8	8	8	8	8	8	2	2
38. Root depth (cm)	6	5	5	4	5	5	5	4	4	4	3
39. Root spread (cm)	6	5	6	5	6	4	4	5	4	3	1
40. Veg. multiplication	4	4	4	2	4	4	4	3	4	4	1
41. Veg. regen. after fire	2	3	3	2	3	1	1	3	1	3	5
42. Flowering – seasonality	Jan–Feb.	Dec–Jan.	Feb.	Nov.	Aug–Sept.	Nov–Dec.	Nov.	Dec–Jan.	Dec.–Jan.	Jan.	Jan.
43. Flowering — pyrogenic	2	2	2	2	2	2	2	2	2	2	2
44. Fruit dehiscence	2	3	2	2	2	2	3	1	1	1	2
45. Seed dissemination	1	2	1	1	1	1	1	1	1	1	1
46. Seedling – germ. type	2	2	2	2	2	4	2	2	2	2	2
47. Plumule position	1	1	1	1	1	3	1	1	1	1	1

* F.P.C. of Bare ground = 58.8%.

Code of Ecomorphological Characters: See Chapter 1.

Table 32. Ecomorphological characters: Chile.
Location: Polpaico, Central Valley, Chile (33° 14′ S, 70° 40′ W).
Vegetation structure: Steppe-like formation dominated by *Acacia caven* (espinal).
Nearest climate station: Colina, Chile (33° 12′ S, 70° 40′ W, 542 m).
Soils: No data available.
Reference: Montenegro, G., Gatti, R. and Riveros, F., 1976. Modificaciones foliares en *Acacia caven* (Mol.) Hook. et Arn. por efecto de la contaminación ambiental. Medio Ambiente 2: 29–34.
Compiler: G. Montenegro, August 1984.

Character	*Acacia caven* (Mimosaceae)	*Sisyrinchium junceum* (Iridaceae)	*Solenomelus pedunculatus* (Iridaceae)	*Helenium aromaticum* (Asteraceae)	*Clarkia tenella* (Onagraceae)	*Geranium core-core* (Geraniaceae)	*Stellaria cuspidata* (Caryophyllaceae)	*Briza minor* (Poaceae)
Foliage Projective Cover (%)								
Upper stratum (>2 m)	18.7	–	–	–	–	–	–	–
Ground stratum (<80 cm)	–	2.6	2.9	3.6	10.3	3.8	2.3	1.9
Density (mature plants m^{-2})	21.2*	14.8	4.6	16.3	5.6	2.1	1.1	6.8
Biomass tops (kg m^{-2})	358.72*	3.16	2.83	8.5	18.9	1.3	2.6	2.1
1. Renewal buds	M	G	G	Th	Th	Th	Th	Th
2. Trophic type	2	1	1	1	1	1	1	1
3. Life duration (yrs)	7	3	3	1	1	1	1	1
4. Height (m)	6	3	4	2	2	3	4	2
5. Crown diameter (m)	6	2	3	1	1	2	3	1
6. Canopy density (%)	3	1	2	1	1	1	1	1
7. Photosynthetic organs	1	4	4	4	4	4	4	1
8. 'Leaf' size	3	3	3	3	3	3	3	3
9. 'Leaf' colour	1	1	1	1	1	1	1	1
10. 'Leaf' length (cm)	3	5	4	2	2	3	3	4
11. 'Leaf' width (mm)	7	1	5	6	6	3	4	4
12. 'Leaf' angle (°)	3	2	2	2	3	1	1	1
13. 'Leaf' margin	1	1	1	1	1	3	1	1
14. 'Leaf' surface resins	2	2	2	2	2	2	2	2
15. 'Leaf' glands	2	2	2	2	2	2	2	2
16. 'Leaf' consistency	1	1	1	1	1	1	1	1
17. 'Leaf' tomentosity	1	1	1	1	1	1	1	1
18. 'Leaf' stomata	2	4	4	4	2	2	2	3
19. 'Leaf' duration (yrs)	2	1	1	1	·1	1	1	1
20. 'Leaf' seasonality	1	2	2	2	2	2	2	2
21. 'Leaf' area: assim. stem area	1	3	3	3	3	3	3	3
22. Organs periodically shed	4	2	2	1	1	1	1	1
23. Organs shed – seasonality	May	Dec.	Dec.	Dec.	Dec.	Dec.	Dec.	Dec.
24. Shoot growth – seasonality	1	Sept.	Sept.	Jul.	Jul.	Jul.	Jul.	Jul.
25. Stem – number from base	2	2	2	1	1	1	1	1
26. Stem – height (m)	1	–	–	–	–	–	–	–
27. Stem – diameter (cm)	50	–	–	–	–	–	–	–
28. Stem lignification	1	3	3	3	3	3	3	3
29. Bark consistency	2	none	none	none	none	none	none	none
30. Bark thickness (mm)	2	none	none	none	none	none	none	none
31. Bark shedding	3	none	none	none	none	none	none	none
32. Bark shedding – seasonality	May	none	none	none	none	none	none	none
33. Spinescence	1	4	4	4	4	4	4	4
34. Underground stems	7	1	1	none	none	none	none	none
35. Root morphology	1	3	3	1	1	2	1	2
36. Root modification	7	6	6	none	none	none	none	none
37. Rootlet modification	6	none	none	none	none	none	none	none
38. Root depth (cm)	6	2	2	2	2	2	3	1
39. Root spread (cm)	5	2	2	2	2	2	2	1
40. Veg. multiplication	4	2	2	4	4	4	4	4
41. Veg. regen. after fire	2	4	4	1	1	1	1	1
42. Flowering – seasonality	Aug–Sept.	Nov.	Nov.	Oct–Nov.	Oct–Nov.	Sept.	Sept.	Nov.
43. Flowering – pyrogenic	2	2	2	2	2	2	2	2
44. Fruit dehiscence	1	2	2	1	2	2	2	1
45. Seed dissemination	1	1	1	1	1	1	1	1
46. Seedling – germ. type	2	–	–	2	2	2	2	–
47. Plumule position	1	2	2	1	1	1	1	2

* Value represents mature plants ha^{-1}.

Code of Ecomorphological Characters: See Chapter 1.

Table 33. Natural vegetation of the Mediterranean Basin – P. Quézel.

Mediterranean Zone
  Arborescent Matorral
    Western Mediterranean
      Calcareous Rocks
        1. *Quercus ilex*
        2. *Olea europaea – Pistacia lentiscus (Ceratonia siliqua)*
        3. *Juniperus* spp.
        4. *Pinus halepensis (P. mesogeensis, P. pinea)*
        5. *Tetraclinis articulata*
        6. *Argania, Maytenus, Ziziphus*
      Non-Calcareous Rocks
        7. *Quercus suber* (with various combinations of the above dominants)
    Adriatic area
      8. *Quercus ilex (Q. coccifera)*
      9. *Pinus halepensis*
      10. *Pinus* ssp. *dalmatica*
    Eastern Mediterranean
      11. *Olea-Pistacia (Ceratonia)*
      12. *Quercus ilex*
      13. *Q. calliprinos*
      14. *Q. infectoria*
      15. *Q. boissieri*
      16. *Q. aegilops* (sensu lato)
      17. *Pinus halepensis (P. brutia)*
      18. *Cupressus sempervirens*
      19. *Pinus nigra*
  Maquis, Garrigue and Phrygana
    20. *Quercus coccifera*
    21. *Rosmarinus* spp.
    22. *Thymus capitatus*
    23. *Lavandula* spp.
    24. *Phillyrea* spp.
    25. *Erica arborea* (+ spp.)
    26. *Cistus* spp.
    27. *Centaurea spinosa*
    28. *Calicotome villosa*
    29. *Sarcopoterium spinosum*
    30. *Genista acanthoclada*
Supra-Mediterranean Zone
  Deciduous Oak Forests
    31. *Quercus cerris* (Italy to Syria)
    32. *Quercus faginea* (southern and western Mediterranean)
    33. *Quercus infectoria* etc. (eastern Mediterranean)
    34. *Quercus ithaburensis* (Israel and Jordan)
    35. *Quercus pubescens* (northern Mediterranean)
Montane Mediterranean Zone
  Cedar Forests
    36. *Cedrus atlantica* (Atlas Mountains)
    37. *Cedrus brevifolia* (Cyprus)
    38. *Cedrus libani* (Turkey)
  Mediterranean Fir Forests
    39. *Abies cephalonica* (Greece)
    40. *Abies cilicica* (Turkey)
    41. *Abies maroccana* (Algeria and Morocco)
    42. *Abies nebrodensis* (Calabria and Sicily)
    43. *Abies pinsapo* (Spain)
  Mountain Pine Forests
    44. *Pinus calabra* (Calabria and Sicily)
    45. *Pinus clusiana* (Spain)
    46. *Pinus italica* (Appenines)
    47. *Pinus laricio* (Corsica)
    48. *Pinus mauretanica* (Algeria and Morocco)
    49. *Pinus nigra* (disjunct areas throughout Mediterranean)
    50. *Pinus pallasiana* (Greece and Turkey)
    51. *Pinus salzmanni* (Spain)

Reference

Quézel, P., 1981. The study of plant groupings in the countries surrounding the Mediterranean: Some methodological aspects. In: di Castri, F., Goodall, D.W. and Specht, R.L. (eds) Ecosystems of the World. Vol. 11. Mediterranean-Type Shrublands. pp. 87–93. Elsevier, Amsterdam.

Table 34. Ecomorphological characters: Corsica.
Location: Cauro, Corsica (41° 55′ N, 8° 55′ E).
Vegetation structure: Maquis arboré (*Erica arborea – Arbutus unedo – Quercus ilex* community).
Age since last fire: 15 years.
Annual precipitation: c. 750 mm.
Soil: Siliceous soil developed on acid granite.
Reference: Brun, B. & L., Conrad, M. and Gamisans, J., 1975. La Nature en France: Corse. Horizons de France, France, 223 pp.
Compilers: R.L. Specht, L. Amandier and C. Martinez, May 1983.

Character	*Arbutus unedo* (Ericaceae)	*Erica arborea* (Ericaceae)	*Phillyrea latifolia* (Oleaceae)	*Pistacia lentiscus* (Pistaciaceae)	*Rhamnus alaternus* (Rhamnaceae)	*Quercus ilex* (Fagaceae)	*Cistus monspeliensis* (Cistaceae)	*Cistus salviifolius* (Cistaceae)	*Cistus incanus* (Cistaceae)	Ephemeral spp.	Total FPC%
Foliage Projective Cover (%)											
Upper stratum (2–5 m)	10	27	16	5	6	16	–	–	–	–	80
Mid stratum (25 cm–2 m)	–	–	–	–	–	–	3	10	9	–	27
Ground stratum (<25 cm)	–	–	–	–	–	–	–	–	–	27	27
1. Renewal buds	M	M	M	M	M	M	N	N	N	Th	
2. Trophic type	1	1	1	1	1	1	1	1	1	?	
3. Life duration (yrs)	5	5	7	7	6	7	4	4	4	1	
4. Height (m)	6	6	6	5	6	6	5	3	4	2	
5. Crown diameter (m)	6	6	6	5	5	6	4	5	4	?	
6. Canopy density (%)	5	4	6	5	5	5	4	4	4	?	
7. Photosynthetic organs	1	1	1	1	1	1	1	1	1	?	
8. 'Leaf' size	4	2	4	3	3/4	4	3	3/4	3/4	?	
9. 'Leaf' colour	1	4	1	1	1	4	4	1	1	?	
10. 'Leaf' length (cm)	4	2	3	3	3	3	3	2/3	3	?	
11. 'Leaf' width (mm)	7	1	6	6	5–7	7	5	5/6	6/7	?	
12. 'Leaf' angle (°)	3	3	3	3	3	3	3	3	3	?	
13. 'Leaf' margin	1/2	5	2	1	2	2	5	1	1	?	
14. 'Leaf' surface resins	2	2	2	2	2	2	1	1	1	?	
15. 'Leaf' glands	2	2	2	1	2	2	2	2	2	?	
16. 'Leaf' consistency	3	2	2	2	3	3	2	2	2	1	
17. 'Leaf' tomentosity	1	2	1	1	1	2	4	4	4	?	
18. 'Leaf' stomata	?	?	?	?	?	?	?	?	?	?	
19. 'Leaf' duration (yrs)	2	3	2	4	3	2	2	2	2	1	
20. 'Leaf' seasonality	1	1	1	1	1	1	1	1	1	3	
21. 'Leaf' area: assim. stem area	1	1	1	1	1	1	1	1	1	?	
22. Organs periodically shed	4	4	4	4	4	4	4	4	4	?	
23. Organs shed – seasonality	2	2	1	1	2	1+3	1+3	1+3	1+3	?	
24. Shoot growth – seasonality	1	1	1 & 2	1	1	1–2	1	1	1	?	
25. Stem – number from base	1	1	2 (1–3)	1	1	2 (1–8)	1	1	1	?	
26. Stem – height (m)	1–2	?	0.5–2	?	?	0.5–2	0.3	?	?	?	
27. Stem – diameter (cm)	10–20	?	1–4	?	?	1–8	2	?	?	?	
28. Stem lignification	1	1	1	1	1	1	1	1	1	?	
29. Bark consistency	3	1	1	1	1	1	1	1	1	?	
30. Bark thickness (mm)	2	1	1	1	1	2	1	1	1	?	
31. Bark shedding	4	4	3	3	3	3	4	4	4	?	
32. Bark shedding – seasonality	abs	abs	abs	abs	abs	abs	abs	abs	abs	?	
33. Spinescence	4	4	4	4	4	4/(2)	4	4	4	?	
34. Underground stems	7	7	nil	nil	nil	nil	nil	nil	nil	?	
35. Root morphology	3	1	?	3	3	3	3	3	3	?	
36. Root modification	7	7	7	7	7	7	7	7	7	?	
37. Rootlet modification	5	5	8	?	8	8	8	8	8	?	
38. Root depth (cm)	6	5	?	7	5	7	5	5	5	?	
39. Root spread (cm)	6 (?)	5 (?)	7	6	5	7	4	4	4	?	
40. Veg. multiplication	4	4	4	2	4	4	4	4	4	?	
41. Veg. regen. after fire	2	2	2 (?)	2	2 (?)	4	1	1	1	1	
42. Flowering – seasonality	3	1 (early)	1 & 2	1	1	1	1	1	1	1	
43. Flowering — pyrogenic	2	2	2	2	2	2	2	2	2	?	
44. Fruit dehiscence	3	2	3	3	3	1	2	2	2	?	
45. Seed dissemination	1	1	1	1	1	1	1	1	1	?	
46. Seedling – germ. type	2	2	2	3	2	3	2	2	2	?	
47. Plumule position	1	1	1	3	1	3	1	1	1	?	

Code of Ecomorphological Characters: See Chapter 1.

Table 35. Ecomorphological characters: Corsica.
Location: Désert des Agriates, Corsica (42° 40′ N, 9° 10′ E).
Vegetation structure: Maquis (*Erica arborea – Arbutus unedo* community).
Age since last fire: 12 years.
Annual precipitation: less than 500 mm.
Soil: Siliceous soil developed on acid granite.
Reference: Brun, B. & L., Conrad, M. and Gamisans, J., 1975. La Nature en France: Corse. Horizons de France, France, 223 pp.
Compilers: R.L. Specht, L. Amandier and C. Martinez, May 1983.

Character	*Arbutus unedo* (Ericaceae)	*Erica arborea* (Ericaceae)	*Phillyrea angustifolia* (Oleaceae)	*Myrtus communis* (Myrtaceae)	*Cistus monspeliensis* (Cistaceae)	*Rosmarinus officinalis* (Lamiaceae)	Ephemeral spp.	Total FPC%
Foliage Projective Cover (%)								
Upper stratum (2–5 m)	8	12	7	–	–	–	–	27
Mid stratum (25 cm–2 m)	–	–	–	1	25	10	–	36
Ground stratum (< 25 cm)	–	–	–	–	–	–	85	85
1. Renewal buds	M	M	M	M	N	N	Th	
2. Trophic type	1	1	1	1	1	1	?	
3. Life duration (yrs)	5	5	6	6	4	5	1	
4. Height (m)	6	6	6	5	5	4	?	
5. Crown diameter (m)	6	6	5	6	4	4	?	
6. Canopy density (%)	5	4	5	5	4	3	?	
7. Photosynthetic organs	1	1	1	1	1	1	?	
8. 'Leaf' size	4	2	3	4	3	2	?	
9. 'Leaf' colour	1	4	1	1	4	4	?	
10. 'Leaf' length (cm)	4	2	3	3	3	2	?	
11. 'Leaf' width (mm)	7	1	5	6	5	1	?	
12. 'Leaf' angle (°)	3	3	3	3	3	2	?	
13. 'Leaf' margin	1/2	5	2	1	5	5	?	
14. 'Leaf' surface resins	2	2	2	2	1	1	?	
15. 'Leaf' glands	2	2	2	1	2	2	?	
16. 'Leaf' consistency	3	2	3	3	2	1	1	
17. 'Leaf' tomentosity	1	2	1	1	4	3	?	
18. 'Leaf' stomata	?	?	?	?	?	?	?	
19. 'Leaf' duration (yrs)	2	3	3	2	2	2	1	
20. 'Leaf' seasonality	1	1	1	1	1	1	3	
21. 'Leaf' area: assim. stem area	1	1	1	1	1	1	?	
22. Organs periodically shed	4	4	4	4	4	4	?	
23. Organs shed – seasonality	2	2	1 + 3	2	1 + 3	1 + 3	?	
24. Shoot growth – seasonality	1	1	1	1	1	1	?	
25. Stem – number from base	1	1	2	1	1	1–2	?	
26. Stem – height (m)	1–2	?	0.2	?	0.3	?	?	
27. Stem – diameter (cm)	10–20	?	5	?	2	?	?	
28. Stem lignification	1	1	1	1	1	1	?	
29. Bark consistency	3	1	1	1	1	3	?	
30. Bark thickness (mm)	2	1	2	1	1	1	?	
31. Bark shedding	4	4	4	4	4	4	?	
32. Bark shedding – seasonality	abs	abs	abs	abs	abs	abs	?	
33. Spinescence	4	4	4	4	4	4	?	
34. Underground stems	7	7	nil	7	nil	nil	?	
35. Root morphology	3	1	3	1	3	3 (?)	?	
36. Root modification	7	7	7	7	7	7	?	
37. Rootlet modification	5	5	8	5 (?)	8	8	?	
38. Root depth (cm)	6	5	6	5 (?)	5	4	?	
39. Root spread (cm)	6 (?)	5 (?)	5/6	5 (?)	4	3/4	?	
40. Veg. multiplication	4	4	2	4	4	4	?	
41. Veg. regen. after fire	2	2	4	2	–	1	1	
42. Flowering – seasonality	3	1 (early)	1	1 + 3	1	4	1	
43. Flowering – pyrogenic	2	2	2	2	2	2	?	
44. Fruit dehiscence	3	2	3	3	2	1	?	
45. Seed dissemination	1	1	1	1	1	1	1	
46. Seedling – germ. type	2	2	2	2	2	2	?	
47. Plumule position	1	1	1	1	1	1	?	

Code of Ecomorphological Characters: See Chapter 1.

Table 36. Ecomorphological characters: France.
Location: Puech-du-Mas-du-Juge, near St Gély-du-Fesc (Montpellier), France (43° 41′ N, 3° 48′ E, 130 m).
Vegetation structure: dense garrigue of *Quercus coccifera – Brachypodium ramosum*.
Age since last fire: 30 years.
Annual precipitation: 1071 mm.
Soil: Calcareous brown soil.
References: Long, G., Visona, L. and Rami, J., 1961. Boll. Inst. Bot. Univ. Catania 3: 5–52.
Poissonet, P., 1966. Thèse 3ème cycle Ecologie, Fac. Sci. Univ. Montpellier 107 p.
Trabaud, L., 1962. Thèse 3ème cycle Ecologie, Fac. Sci. Univ. Montpellier 131 p.
Trabaud, L., 1980. Thèse Dr. Etat, Univ. Sci. Tech. Languedoc, Montpellier 288 p.
Compiler: L. Trabaud, February 1983.

Character	*Quercus coccifera* (Fagaceae)	*Dorycnium pentaphyllum* (Fabaceae)	*Rubia peregrina* (Rubiaceae)	*Brachypodium retusum* (Poaceae)	*Phillyrea angustifolia* (Oleaceae)	*Asparagus acutifolius* (Liliaceae)	*Aphyllanthes monspeliensis* (Liliaceae)	*Teucrium chamaedrys* (Lamiaceae)	*Cistus monspeliensis* (Cistaceae)	*Rubus ulmifolius* (Rosaceae)	*Genista scorpius* (Fabaceae)	Total
Foliage Projective Cover (%)												
Upper stratum (1–2 m)	80.3	–	–	–	<5	–	–	–	5	tr.	5	
Mid stratum (25 cm–1 m)	–	17.2	tr.	–	–	tr.	–	–	–	–	–	
Ground stratum (<25 cm)	–	–	–	26.0	–	–	<1	tr.	–	–	–	
Biomass tops (kg ha^{-1})	23,822			32.6								27,894
1. Renewal buds	1 N	1 N	2 Ch	2 Ch/4 G	1 M	2 Ch/4 G	3 H	2 Ch	1 N	1 N	1 N	
2. Trophic type	1	2	1	1	1	1	1	1	1	1	2	
3. Life duration (yrs)	6	4	4	5	6	4	4	4	4	5	4	
4. Height (m)	5	3	4	2	6	3	2	2	5	5	5	
5. Crown diameter (m)	8	3	1	8	5	1	2	1	4	2	3	
6. Canopy density (%)	6	3	1	4	5	1	4	2	4	3	3	
7. Photosynthetic organs	1	1	4	1	1	3	4	1	1	1	4	
8. 'Leaf' size	4	3	4	2	3	1	2	2	3	4	1	
9. 'Leaf' colour	1	1	1	1	1	1	1	1	1	1	1	
10. 'Leaf' length (cm)	3	1	2	4	3	1	?	1	3	3	1	
11. 'Leaf' width (mm)	6	3	5	1	5	1	1	4	5	6	4	
12. 'Leaf' angle (°)	3	3	3	3	3	3	3	3	3	3	3	
13. 'Leaf' margin	2	3	1	4	2	1	1	3	5	2	3	
14. 'Leaf' surface resins	2	2	2	2	2	2	2	2	1	2	2	
15. 'Leaf' glands	2	2	2	2	2	2	2	2	1	2	2	
16. 'Leaf' consistency	3	1	2	1	3	3	2	2	2	1	2	
17. 'Leaf' tomentosity	1	1	1	1	1	1	1	1	4	2	1	
18. 'Leaf' stomata	?	?	?	?	?	?	?	?	?	?	?	
19. 'Leaf' duration (yrs)	4	?	3	2	3	?	–	2	2	?	1	
20. 'Leaf' seasonality	1	1	1	1	1	1	1	1	1	1	1	
21. 'Leaf' area: assim. stem area	1	1	3	1	1	1	3	1	1	1	3	
22. Organs periodically shed	4	4	4	4	4	2	2	4	4	4	4	
23. Organs shed – seasonality	1	3 + 4	2	1	1 + 3	?	?	?	1 + 3	?	?	
24. Shoot growth – seasonality	1	1	1	1	1	1	1	1	1	1	1	
25. Stem – number from base	2	1	2	2	2	2	2	2	1	2	1	
26. Stem – height (m)	0.2	0.1	–	–	0.2	0.1	–	0.1	0.3	0.1	0.2	
27. Stem – diameter (cm)	3	0.5	–	–	5	–	–	–	2	0.5	1	
28. Stem lignification	1	1	3	3	1	2	3	2	1	1	1	
29. Bark consistency	1	1	–	–	1	1	–	–	1	1	1	
30. Bark thickness (mm)	2	1	–	–	2	–	–	–	1	1	1	
31. Bark shedding	4	4	–	–	4	–	–	–	4	4	4	
32. Bark shedding – seasonality	–	–	–	–	–	–	–	–	–	–	–	
33. Spinescence	2	4	4	4	4	3	4	4	4	1	3	
34. Underground stems	–	–	1	3	–	1	1	2	–	7	7	
35. Root morphology	3	3	4	5	3	4	4	3	3	3	3	
36. Root modification	7	7	7	7	7	6	7	7	7	7	7	
37. Rootlet modification	8	8	8	8	8	8	8	8	8	8	8	
38. Root depth (cm)	7	5	3	3	6	3	4	2	5	5	5	
39. Root spread (cm)	7	4	3	4	5/6	3	3	2	4	?	3	
40. Veg. multiplication	2	2	2	2	2	2	2	2	4	2	2	
41. Veg. regen. after fire	4	4	4	4	4	4	4	4	–	4	4	
42. Flowering – seasonality	1	1	1	1	1	1	1	1	1	1	1	
43. Flowering – pyrogenic	2	2	2	2	2	2	2	2	2	2	2	
44. Fruit dehiscence	1	2	3	1	3	3	1	2	2	3	2	
45. Seed dissemination	1	1	1	1	1	1	1	1	1	1	1	
46. Seedling – germ. type	3	2	2	–	2	–	–	2	2	2	2	
47. Plumule position	3	1	1	2	1	2	2	1	1	1	1	
48. No. 'leaves' per 10 cm stem	15–22	20	8	–	8–12	300	–	16–24	12–20	3–5	–	

Code of Ecomorphological Characters: See Chapter 1.

Table 37. Ecomorphological characters: France.
Location: Puechabon, France (43° 44′ N, 3° 35′ E, 300 m).
Vegetation structure: Coppice of *Quercus ilex* forest *(Quercetum ilicis gallo provinciale)*.
Age since last fire: 40 years.
Soil: Calcareous fersialitic soil (Mediterranean red soil).
References: Grillas, P., 1980. Structure et phytomasse de taillis de chêne vert (*Quercus ilex* L.). Etude de trois stations du Montpelliérais. Diplôme d'Etudes Approfondies. Université des Sciences et Techniques du Languedoc. Montpellier, 33 pp.
Piron, C., 1983. Etude historique et structurale d'un taillis de chêne vert (*Quercus ilex* L.) du Montpelliérais. Diplôme d'Etudes Approfondies. Faculté des Sciences et Techniques, St Jérôme, Marseille, 30 pp.
Rhanem, M., 1983. Contribution à l'étude de la structure d'un taillis de chêne vert (*Quercus ilex* L.) à Puechabon (Hérault). Diplôme d'Etudes Approfondies. Université de Sciences et Techniques du Languedoc. Montpellier, 43 pp.
Compilers: C. Floret and F. Romane, May 1984.

Character	*Quercus ilex* (Fagaceae)	*Phillyrea latifolia* (Oleaceae)	*Buxus sempervirens* (Buxaceae)	*Juniperus oxycedrus* (Cupressaceae)	*Pistacia terebinthus* (Pistaciaceae)	*Lonicera implexa* (Caprifoliaceae)	*Smilax aspera* (Smilacaceae)	*Brachypodium retusum* (Poaceae)	*Ruscus aculeatus* (Ruscaceae)	*Teucrium chamaedrys* (Lamiaceae)	*Carex hallerana* (Cyperaceae)
Foliage Projective Cover (%)											
Upper stratum (>2 m)	72	2	–	–	1	–	–	–	–	–	–
Mid stratum (30 cm–2 m)	10	2	10	2	–	–	–	–	1	–	–
Ground stratum (<30 cm)	2	–	1	–	–	–	–	1	–	1	1
Biomass tops (kg ha^{-1})	74000										
1. Renewal buds	M	M	N	N	M	AMPH	AMPH	AMPH	H	AMPH	H
2. Trophic type	1	1	1	1	1	1	1	2 (?)	1	1	1
3. Life duration (yrs)	7	7	5 (?)	7	6	4	4	3	5	4	3
4. Height (m)	6	6	5	6	6	6	6	3	3	2	2
5. Crown diameter (m)	7	6	5	6	6	3	1	3	3	2	2
6. Canopy density (%)	4	5	4	4	3	1	1	2	2	1	4
7. Photosynthetic organs	1	1	1	1	1	1	4	1	3	1	1
8. 'Leaf' size	4	4	4	1	4	3	3	2	3	3	2
9. 'Leaf' colour	4	1	1	4	1	1	1	1	1	1	1
10. 'Leaf' length (cm)	3	3	2	2	3	3	3	3	2	2	4
11. 'Leaf' width (mm)	7	6	5	2	7	7	7	2	6	5	3
12. 'Leaf' angle (°)	3	3	3	3	3	3	3	3	3	3	3
13. 'Leaf' margin	2	2	1	1	1	1	2	1	1	3	1
14. 'Leaf' surface resins	2	2	2	2	2	2	2	2	2	1	2
15. 'Leaf' glands	2	2	2	1	1	2	2	2	2	2	2
16. 'Leaf' consistency	3	2	3	3	1	2	3	1	3	1	1
17. 'Leaf' tomentosity	4	1	1	1	1	1	1	1	1	4	1
18. 'Leaf' stomata	?	?	?	?	?	?	?	?	?	?	?
19. 'Leaf' duration (yrs)	4	2	2	4	2	2	?	2	4	2	2
20. 'Leaf' seasonality	1	1	1	1	3	1	1	1	1	1	1
21. 'Leaf' area: assim. stem area	1	1	1	1	1	1	3	1	1	1	1
22. Organs periodically shed	4	4	4	4	4	4	2	2	2	3	2
23. Organs shed – seasonality	1 & 3	?	?	?	4	?	?	?	?	?	?
24. Shoot growth – seasonality	1–3	1 & 2	1	1	1	1	1	1 & 3	1	1	1
25. Stem – number from base	1–8	1–3	1–10	1–2	1	1	1	–	–	–	–
26. Stem – height (m)	0.5–2	0.5–2	0–0.5	0–1	1–2	–	–	–	–	–	–
27. Stem – diameter (cm)	1–8	1–4	1–3	1–2	1–5	0.5	0.25	–	–	–	–
28. Stem lignification	1	1	1	1	1	1	3	3	3	3	3
29. Bark consistency	6	1	1	2	3	2	–	–	–	–	–
30. Bark thickness (mm)	2	1	1	1	1	1	–	–	–	–	–
31. Bark shedding	3	3	1	2	3	2	–	–	–	–	–
32. Bark shedding – seasonality	abs	abs	abs	abs	abs	abs	–	–	–	–	–
33. Spinescence	(2)	4	4	2	4	4	3	4	2	4	4
34. Underground stems	Nil	Nil	1	Nil	Nil	?	2	3	1	2	1
35. Root morphology	3	?	3	?	?	?	4	5	4	3	4
36. Root modification	7	7	7	7	7	?	6	7	6	7	7
37. Rootlet modification	8	8	8	8	8	?	8	8	8	8	8
38. Root depth (cm)	7	?	?	?	?	3	3	3	3	2	4
39. Root spread (cm)	7	7	6	?	?	?	4	4	3	2	3
40. Veg. multiplication	4	4	2	4	4	2	2	2	2	2	2
41. Veg. regen. after fire	4	2 (?)	3 (?)	3	?	?	?	4	?	?	4
42. Flowering – seasonality	1	1 & 2	1	3	1	1	?	1	?	?	1
43. Flowering — pyrogenic	2	2	2 (?)	2	2	2	?	?	?	?	?
44. Fruit dehiscence	1	3	2	3	3	3	3	1	3	1	1
45. Seed dissemination	1	1	1	1	1	1	1	1	1	1	1
46. Seedling – germ. type	3	2	2	2	3	2	–	–	–	2	–
47. Plumule position	3	1	1	1	3	1	2	2	2	1	2

Code of Ecomorphological Characters: See Chapter 1.

Table 38. Ecomorphological characters: Greece.
Location: Hymettus, Athens, Central Greece (37° 57′ N, 23° 49′ E, 400 m).
Vegetation structure: Phrygana (low elevation), with seasonal dimorphic shrubs (*Sarcopoterium*, etc.) (Cover 43%) (Number of woody spp. 21 m^{-2}).
Age since last fire: c. 25 years.
Nearest climate station: Hellinikon, Greece.
Soil: pH alkaline 7–8, on limestone.
References: Arianoutsou-Faraggitaki, M. and Diamantopoulos, J., 1985. Comparative phenology of five dominant plant species in maquis and phrygana ecosystems in Greece. Phyton (Austria) 25: 77–85.
Margaris, N., 1976. Structure and function in a phryganic ecosystem. J. Biogeogr. 3: 249–259.
Compiler: M. Arianoutsou and S. Kokkini, July 1986.

Character	*Anthyllis hermanniae* (Fabaceae)	*Ballota acetabulosa* (Lamiaceae)	*Cistus incanus* (Cistaceae)	*Euphorbia acanthothamnos* (Euphorbiaceae)	*Genista acanthoclada* (Fabaceae)	*Helianthemum nummularum* (Cistaceae)	*Phagnalon graecum* (Asteraceae)	*Phlomis fruticosa* (Lamiaceae)	*Sarcopoterium spinosum* (Rosaceae)	*Teucrium polium* (Lamiaceae)	*Teucrium divaricatum* (Lamiaceae)	*Thymus capitatus* (Lamiaceae)	*Hypericum empetrifolium* (Hypericaceae)
Foliage Projective Cover (%)	0.01	0.32	4.41	12.86	0.01	0.45	0.32	13.31	6.22	0.13	0.09	1.90	0.01
Density (mature plants ha^{-1})		400	15600	14800		12400	800	8400	16000	400	2800	3200	400
Biomass (kg ha^{-1})			835	2700		803		3665	1435			360	
1. Renewal buds	Ch	Ch	N	N	N	Ch	Ch	N	N	Ch	Ch	Ch	H
2. Trophic type	2	1	1	1	2	1	1	1	1	1	1	1	1
3. Life duration (yrs)	4	4	4	5	5	4	4	6	5	4	4	4	1
4. Height (m)	3	4	4	4	3	3	3	4	4	2	2	3	2
5. Crown diameter (m)	3	3	4	3	3	2	3	3	3	2	3	3	3
6. Canopy density (%)	3	2	4	3	3	2	3	3	2	2	4	4	5
7. Photosynthetic organs	1	1	1	1	4	1	1	1	1	4	1 (?)	1	1
8. 'Leaf' size	3	4	4	3	3	3	3	6	3	3	3	1	3
9. 'Leaf' colour	2	2	1	1	1	1	4	5	1	2	1	1	1
10. 'Leaf' length (cm)	1.5–2	3–5	1–7	1–1	0.5–1*	0.5–5	1.5–2	6–9	1.5–4*	0.7–2.7	0.6–2.5	0.5–1	0.2–1.2
11. 'Leaf' width (mm)	5–10	10–30	8–30	3–5	1–3+	2–5	2–4	20–30	2–4	2–4	4–15	1–1.2	1–2
12. 'Leaf' angle (°)	?	?	?	?	?	?	?	?	?	?	?	?	?
13. 'Leaf' margin	1 (or 3), 5	2	none	1	1	1	2	1	3	2,5	2	1	1
14. 'Leaf' surface resins	2	2	2	2	2	2	2	2	2	2	2	2	2
15. 'Leaf' glands	2	2	1	2	2	1	1	2	1	2	2	2	1
16. 'Leaf' consistency	2	2	2	2	2	2	2	2	2	2	2	2	2
17. 'Leaf' tomentosity	4	2	4	1		4	4	4	3	4	1	2	1
18. 'Leaf' stomata	2	2	2	2	2	2	2	2	2	2	2	2	2
19. 'Leaf' duration (yrs)	2	2	2	2 wl / 1 sl	2 wl / 1 sl	2	2	2	2	2	2	2	2
20. 'Leaf' seasonality	4	4	4	2	2	4	4	4	4	4	4	4	
21. 'Leaf' area: assim. stem area	1:0	1:0	1:0	1:0	1:0	1:0	1:0	1:0	1:0	1:1	1:0	1:0	
22. Organs periodically shed	4	4	4 (?)	4	4	3	3	4	4	3	4	4	
23. Organs shed – seasonality	Ap–lS	Ap–D	Mr–lS	Mr–Jn	Jn	e Mr–Jl	Ma	Ma–S	Ma–S	Ap–S		Ma–S	
24. Shoot growth – seasonality	Mr–Ap	Mr–Ap	J–Ap	Mr–Ap–Ma	Ap–Ma–Jn	J–Jn	Mr–Ap	F–Ma	Ma–Ap	Mr–Ap–Ma		Ma	
25. Stem – number from base	2 (4–6)	1	2 (6–10)	2 (6–10)	2 (5–8)	2 (6–10)	2 (5–10)	2 (6–8)	2 (6–10)	2 (10–15)	2 (5–10)	2 (5–10)	
26. Stem – height (m)	none	?	none	none				none	none	none	none	none	
Stem – diameter (cm)													
– at breast height (cm)	none	none	none	none				none	none	none	none	none	
– near ground level	?	?	?	?				?	?	?	?	?	
28. Stem lignification	1	1	1	1	1	1	1	1	1	1	1	1	
29. Bark consistency	2	2	1	1 or 2	1	2	1	1	2	1	1	2	
30. Bark thickness (mm)	1	1	1	1	1	1	1	1	1	1	1	1	
31. Bark shedding	?	?	?	?	?	?	?	?	?	?	?	?	
32. Bark shedding – seasonality	?	?	?	?	?	?	?	?	?	?	?	?	
33. Spinescence	1	4	4	1	1	4	4	4	1	4	4	4	
34. Underground stems	?	?	?	?	?	?	?	?	?	?	?	?	
35. Root morphology	?	?	?	?	?	?	?	?	?	?	?	?	
36. Root modification	?	?	?	?	?	?	?	?	?	?	?	?	
37. Rootlet modification	?	?	?	?	?	?	?	?	?	?	?	?	
38. Root depth (cm)	?	?	?	?	?	?	?	?	?	?	?	?	
39. Root spread (cm)	?	?	?	?	?	?	?	?	?	?	?	?	
40. Veg. multiplication	1	1	4	1	1	1	1	1	1	1	1	4	
41. Veg. regen. after fire	3	3	1	3	3	3	3	3	3	3	3	1	
42. Flowering – seasonality	Ap–Ma	Ma	Ap–Ma	Mr–Ap	Jn	Ap–Ma–Jn	Ap–Ma	Ma	F	Jn	?	Jn–Jl	
43. Flowering — pyrogenic	2	2	1	2	2	2	1	2	2	2	2	2	2
44. Fruit dehiscence	2	1	2	2	2	2	1	1	3	1	1	1	1
45. Seed dissemination	1	1	1	1	1	1	1	1	1	1	1	1	1
46. Seedling – germ. type	2	2	2	2	2	2	2	2	2	2	2	2	2
47. Plumule position	1	1	1	1	1	1	1	1	1	1	1	1	1
48. No. leaves per 10 cm stem	40	25	40	10	–	127	25	22	12	65	50	50	31

* There are leaves only in the 5 top cms.
+ Leaflets.
Abbreviations: J – January; F – February; Mr – March; Ap – April; Ma – May; Jn – June; Jl – July; Aug – August; S – September; O – October; N – November; D – December / lS – late September; e Mr – early March.

Code of Ecomorphological Characters: See Chapter 1.

Table 39. Ecomorphological characters: Greece.
Location: Stavros, Halkidiki, Northern Greece (40° 39′ N, 23° 43′ E, 20 m).
Vegetation structure: Maquis (low elevation), with evergreen sclerophyll tall shrubs.
Age since last fire: 25–30 years.
Nearest climate station: Vassilika, Thermi, Greece.
Soil: pH slightly acid 5.0–5.5, on metamorphic rocks of biotite gneiss and amphiboles.
Reference: Arianoutsou, M. and Mardiris, T.A., 1987. Observations on the phenology of two dominant plants of the Greek maquis. In: Tenhunen, J.D., Catarino, F.M., Lange, O.L. and Oechel, W.C. (eds) Plant Response to Stress – Functional Analysis in Mediterranean Ecosystems. Springer-Verlag, Berlin.
Compiler: M. Arianoutsou, July 1986.

Character	*Arbutus unedo* (Ericaceae)	*Quercus coccifera* (Fagaceae)	*Quercus ilex* (Fagaceae)	*Phillyrea media* (Oleaceae)	*Erica arborea* (Ericaceae)	*Erica verticillata* (Ericaceae)	*Olea europaea* var. *oleaster* (Oleaceae)	*Calicotome villosa* (Fabaceae)
Foliage Projective Cover (%)	25	20	5	28	10	8	2	2
Density (mature plants ha^{-1})	3000	4000	1500	9000	5000	4000	1000	2000
Biomass (kg ha^{-1})	6270	13830		14210	3500	3000	2860	2500
1. Renewal buds	M	M	MM	MM	M	M	MM	M
2. Trophic type	1	1	1	1	1	1	1	2
3. Life duration (yrs)	5	5	5	4	4	4	6	4
4. Height (m)	5–6	5–6	8–9	8	5–6	5–6	8	6
5. Crown diameter (m)	5	5	5	4–5	5	5	5	5
6. Canopy density (%)	2	3	3	3–4	3	3	3	3
7. Photosynthetic organs	1	1	1	1	1	1	1	1
8. 'Leaf' size	5	3	4	3	1	1	3	3
9. 'Leaf' colour	1	1	1	1	1	1	2	1
10. 'Leaf' length (cm)	4	3	4	3	1	1	3	0.15–1.5
11. 'Leaf' width (mm)	7	6	7	6	3	2	5	4–6
12. 'Leaf' angle (°)	?	?	?	?	?	?	?	?
13. 'Leaf' margin	2	2	2	2	5	5	1	3
14. 'Leaf' surface resins	2	2	2	2	2	2	2	2
15. 'Leaf' glands	2	2	2	2	2	2	2	2
16. 'Leaf' consistency	2	3	3	3	3	3	3	2
17. 'Leaf' tomentosity	1	2	2	2	1	1	2	2
18. 'Leaf' stomata	2	2	2	2	2	2	2	2
19. 'Leaf' duration (yrs)	3	3	3	3	3	3	3	2 wl / 1 sl
20. 'Leaf' seasonality	1	1	1	1	1	1	1	2
21. 'Leaf' area: assim. stem area	1:0	1:0	1:0	1:0	1:0	1:0	1:0	1:1
22. Organs periodically shed	4	4	4	4	4	4	4	4
23. Organs shed – seasonality	------------------------------------ all the year: peak season June–July ------------------------------------ Jn							
24. Shoot growth – seasonality	Ap–Ma–e Jn ------------------------------------ April – May ------------------------------------ F–Ma							
25. Stem – number from base	2 (5–10)	2 (4–5)	2 (4–5)	2 (4–5)	2 (8–10)	2 (8–10)	2 (1–3)	2 (1–2)
26. Stem – height (m)	–	–	–	–	–	–	–	–
27. Stem – diameter (cm)	5	2.5	2.5	3.5	4	4	5	1.5
28. Stem lignification	1	1	1	1	1	1	1	1
29. Bark consistency	1↔2	1↔2	1	1	2	2	1↔2	1↔2
30. Bark thickness (mm)	?	?	?	?	?	?	?	1
31. Bark shedding	?	?	?	?	?	?	?	?
32. Bark shedding – seasonality	?	?	?	?	?	?	?	?
33. Spinescence	4	2	2	4	4	4	4	1
34. Underground stems	1 (?)	1 (?)	1 (?)	1 (?)	1 (?)	1 (?)	1 (?)	?
35. Root morphology	?	?	?	?	?	?	?	?
36. Root modification	?	?	?	?	?	?	?	?
37. Rootlet modification	?	?	?	?	?	?	?	?
38. Root depth (cm)	?	?	?	?	?	?	?	?
39. Root spread (cm)	?	?	?	?	?	?	?	?
40. Veg. multiplication	1	1	1	1	1	1	1	1
41. Veg. regen. after fire	3	3	3	3	3	3	3	3
42. Flowering – seasonality	Oc–D	Mr–Ma	Ap–Ma	Ap–Ma	Mr–Ma	Aug–D	Ma–Jn	F–Ap
43. Flowering – pyrogenic	2	2	2	2	2	2	2	2
44. Fruit dehiscence	3	1	1	3	2	2	3	2
45. Seed dissemination	1	1	1	1	1	1	1	1
46. Seedling – germ. type	2	3	3	2	2	2	2	2
47. Plumule position	1	3	3	1	1	1	1	1
48. No. leaves per 10 cm stem	11	8	6	9	> 300	> 300	30	18

Abbreviations: J – January; F – February; Mr – March; Ap – April; Ma – May; Jn – June; Jl – July; Aug – August; S – September; Oc – October; N – November; D – December / e Jn – early June.

Code of Ecomorphological Characters: See Chapter 1.

Table 40. Ecomorphological characters: Israel.
Location: Upper Galilee, Israel (32° 55′ N, 35° 30′ E).
Vegetation structure: Quercus calliprinos – Pistacia palaestina association (Number of records: 23, listing 41 perennial spp. and 250 annual spp.).
Age since last fire: Unknown.
Reference: Rabinovitz-Vyn, A., 1979. Ph. D. Thesis, Hebrew University of Jerusalem.
Compilers: G. Orshan and A. Rabinovitz-Vyn, November 1985.

Character	*Quercus calliprinos* (Fagaceae)	*Quercus boissieri* (Fagaceae)	*Pistacia palaestina* (Pistaciaceae)	*Rhamnus punctata* (Rhamnaceae)	*Crataegus aronia* (Rosaceae)	*Calicotome villosa* (Fabaceae)	*Sarcopoterium spinosum* (Rosaceae)	*Cistus creticus* (Cistaceae)	*Cistus salviifolius* (Cistaceae)	*Majorana syriaca* (Lamiaceae)	Ephemeral spp.
Foliage Projective Cover (%)											
Upper stratum (2–5 m)	23.1	8.1*	5.5	2.3	2.3	–	–	–	–	–	–
Mid stratum (1–2 m)	–	–	–	–	–	3.9	–	–	–	–	–
Ground stratum (<1 m)	–	–	–	–	–	–	6.3	3.9	4.7	1.6	(250 spp.)
Frequency (Presence %)	91	37	83	74	48	52	65	69	57	43	?
1. Renewal buds	M	M	M	M	M	N	Ch	Ch	Ch	Ch	Th
2. Trophic type	1	1	1	1	1	2	1	1	1	1	1
3. Life duration (yrs)	7	7	7	?	7	6	4	5	5	4	1
4. Height (m)	6	7	6	6	6	5	3	4	4	4	1–3
5. Crown diameter (m)	6	7	6	5	5	5	4	4	4	4	1/2
6. Canopy density (%)	5	4	2	3	3	2	4	2	2	2	1/2
7. Photosynthetic organs	1	1	1	1	1	4	1	1	1	1	1
8. 'Leaf' size	4	5	4	3	4	3	4 & 2	4 & 3	4 & 3	3 & 2	?
9. 'Leaf' colour	1	1	1	1	1	1	1	1	1	2	?
10. 'Leaf' length (cm)	3	4	3	2	3	1	3 & 2	3 & 2	3 & 2	2 & 1	?
11. 'Leaf' width (mm)	7	7	7	5	7	5	5 & 2	6 & 5	6 & 5	6 & 4	?
12. 'Leaf' angle (°)	3	3	3	3	3	3	3	3	3	3	?
13. 'Leaf' margin	2/1	2	1	1	1/3	1	1	1	1	1	?
14. 'Leaf' surface resins	2	2	2	2	2	2	2	1	1	1	?
15. 'Leaf' glands	2	2	1	2	2	2	2	2	2	2	?
16. 'Leaf' consistency	3	3	1	3	3	1	1	1	1	1	1
17. 'Leaf' tomentosity	1	2	1	4	2	1	1	4	4	4	?
18. 'Leaf' stomata	?	?	?	?	?	?	?	?	?	?	?
19. 'Leaf' duration (yrs)	2	2	2	2	2	1	1 & 2	1 & 2	1 & 2	2 & 2	?
20. 'Leaf' seasonality	1	2	2	2	2	1	1	1	1	1	3
21. 'Leaf' area: assim. stem area	1	1	1	1	1	2	1	2	1	1	?
22. Organs periodically shed	4	4	4	4	4	4	3	3	nil	3	?
23. Organs shed – seasonality											
24. Shoot growth – seasonality	1 + 2	1 + 2	1	1	1	1	1	1	1	1	1
25. Stem – number from base	1	1									–
26. Stem – height (m)											–
27. Stem – diameter (cm)											
28. Stem lignification	1	1	1	1	1	1	2	1	1	2	3
29. Bark consistency	1	6	1	1	3	1	2	1	1	4	nil
30. Bark thickness (mm)	?	2	1	?	1	1	1	1	1	1	nil
31. Bark shedding	?	3	?	3	3	3	2	3	3	2	nil
32. Bark shedding – seasonality											nil
33. Spinescence	2	4	4	2	1	1	1	4	4	4	?
34. Underground stems	nil	nil	?	nil	nil	nil	3	nil	nil	nil	?
35. Root morphology	3	3	3	3	3	3	3	3	3	?	?
36. Root modification	7	7	?	?	7	7	7	7	7	7	?
37. Rootlet modification	8	8	?	8	8	6	8	5	5	?	?
38. Root depth (cm)	7	7	6	?	7	?	5	5	5	?	?
39. Root spread (cm)	6	7	6	?	6	?	4	4	4	?	?
40. Veg. multiplication	4	4	?	?	?	4	4	4	4	4	4
41. Veg. regen. after fire	2	2	2	?	2	2	2	2	2	2	1
42. Flowering – seasonality	1	1	1	1	1	1	1	1	1	1	1
43. Flowering – pyrogenic	2	2	2	2	2	2	2	2	2	2	2
44. Fruit dehiscence	1	1	3	3	3	2	3	2	2	1	?
45. Seed dissemination	1	1	1	1	1	1	1	1	1	1	?
46. Seedling – germ. type	3	3	3	2	2	2	2	2	2	2	?
47. Plumule position	3	3	3	1	1	1	1	1	1	1	?

* Emergent 5–10 m.

Code of Ecomorphological Characters: See Chapter 1.

Table 41. Ecomorphological characters: Israel.
Location: Upper Galilee, Israel (32° 55' N, 35° 30' E).
Vegetation structure: Ceratonia siliqua – Pistacia lentiscus association (Number of records: 10, listing 31 perennial spp. and 91 annual spp.).
Age since last fire: Unknown.
Reference: Rabinovitz-Vyn, A., 1979. Ph. D. Thesis, Hebrew University of Jerusalem.
Compilers: G. Orshan and A. Rabinovitz-Vyn, November 1985.

Character	*Ceratonia siliqua* (Caesalpiniaceae)	*Quercus calliprinos* (Fagaceae)	*Styrax officinalis* (Styracaceae)	*Pistacia lentiscus* (Pistaciaceae)	*Rhamnus palaestina* (Rhamnaceae)	*Calicotome villosa* (Fabaceae)	*Sarcopoterium spinosum* (Rosaceae)	*Salvia triloba* (Lamiaceae)	*Majorana syriaca* (Lamiaceae)	*Cistus salviifolius* (Cistaceae)	Ephemeral spp.
Foliage Projective Cover (%)											
Upper stratum (2–5 m)	6.8*	13.4	2.9	–	–	–	–	–	–	–	
Mid stratum (1–2 m)	–	–	–	15.3	5.6	9.6	–	–	–	–	
Ground stratum (<1 mm)	–	–	–	–	–	–	20.1	5.8	4.8	2.9	(91 spp.)
Frequency (presence %)	100	60	20	100	90	30	100	40	40	10	?
1. Renewal buds	M	M	M	N	N	N	Ch	Ch	Ch	Ch	Th
2. Trophic type	2	1	1	1	1	2	1	1	1	1	1
3. Life duration (yrs)	7	7	7	7	6	6	4	5	4	5	1
4. Height (m)	7	6	6	5	5	5	3	4	4	4	1–3
5. Crown diameter (m)	6/7	6	5	5	5	5	4	4	4	4	1/2
6. Canopy density (%)	6	5	4	5	3	2	4	2	2	2	1/2
7. Photosynthetic organs	1	1	1	1	1	4	1	1	1	1	1
8. 'Leaf' size	4	4	4	3	3	3	4 & 2	4	3 & 2	4 & 3	?
9. 'Leaf' colour	1	1	5	1	1	1	1	2	2	1	?
10. 'Leaf' length (cm)	3	3	4	3	2	1	3 & 2	3	2 & 1	3 & 2	?
11. 'Leaf' width (mm)	7	7	6	6	5	5	5 & 2	6	6 & 4	6 & 5	?
12. 'Leaf' angle (°)	3	3	3	3	3	3	3	3	3	3	?
13. 'Leaf' margin	1	2/1	1	1	1/(2)	1	1	3	1	1	?
14. 'Leaf' surface resins	2	2	2	2	2	2	2	1	1	1	?
15. 'Leaf' glands	2	2	2	2	2	2	2	2	2	2	?
16. 'Leaf' consistency	3	3	2	2	2	1	1	1	1	1	1
17. 'Leaf' tomentosity	1	1	2	1	1	1	1	4	4	4	?
18. 'Leaf' stomata	?	?	?	?	?	?	?	?	?	?	?
19. 'Leaf' duration (yrs)	?	2	3	4	2	1	1 & 2	2	2 & 2	1 & 2	?
20. 'Leaf' seasonality	1	1	1	1	2	1	1	1	1	1	3
21. 'Leaf' area: assim. stem area	1	1	1	1	1	2	1	1	1	1	?
22. Organs periodically shed	4	4	4	4	4	4	3	3	3	Nil	?
23. Organs shed – seasonality											
24. Shoot growth – seasonality	1	1	1	1	1	1	1	1	1	1	1
25. Stem – number from base	1	1									–
26. Stem – height (m)											–
27. Stem – diameter (cm)											–
28. Stem lignification	1	1	1	1	1	1	2	2	2	1	3
29. Bark consistency	?	1	6	1	1	1	2	2	4	1	Nil
30. Bark thickness (mm)	?	?	1	1	1	1	1	1	1	1	Nil
31. Bark shedding	?	?	3	3	3	3	2	3	2	3	Nil
32. Bark shedding – seasonality											Nil
33. Spinescence	4	2	4	4	1	1	1	4	4	4	?
34. Underground stems	Nil	Nil	7	Nil	Nil	Nil	3	Nil	Nil	Nil	?
35. Root morphology	3	3	?	3	3	3	3	?	?	3	?
36. Root modification	7	7	?	7	7	7	7	?	7	7	?
37. Rootlet modification	8	8	?	?	8	6	8	?	?	5	?
38. Root depth (cm)	7	7	6	7	6	?	5	?	?	5	?
39. Root spread (cm)	7	6	6	6	6	?	4	?	?	4	?
40. Veg. multiplication	4	4	4	2	4	4	4	4	4	4	4
41. Veg. regen. after fire	2/3	2	2	2	?	2	2	2	2	2	1
42. Flowering – seasonality	1	1	1	1	1	1	1	1	1	1	1
43. Flowering – pyrogenic	2	2	2	2	2	2	2	2	2	2	2
44. Fruit dehiscence	1	1	3	3	3	2	3	1	1	2	?
45. Seed dissemination	1	1	1	1	1	1	1	1	1	1	?
46. Seedling – germ. type	2	3	2	3	2	2	2	2	2	2	?
47. Plumule position	1	3	1	3	1	1	1	1	1	1	?

* Emergent (5–8 m).

Code of Ecomorphological Characters: See Chapter 1.

Table 42. Ecomorphological characters: Israel.
Location: Upper Galilee, Israel (32° 55′ N, 35° 30′ E).
Vegetation structure: Pinus halepensis association. (Number of records: 4, listing 26 perennial spp. and many annual spp.).
Age since last fire: Unknown.
Reference: Rabinovitz-Vyn, A., 1979. Ph. D. Thesis, Hebrew University of Jerusalem.
Compilers: G. Orshan and A. Rabinovitz-Vyn, November 1985.

Character	*Pinus halepensis* (Pinaceae)	*Quercus calliprinos* (Fagaceae)	*Pistacia palaestina* (Pistaciaceae)	*Pistacia lentiscus* (Pistaciaceae)	*Rhamnus palaestina* (Rhamnaceae)	*Cistus creticus* (Cistaceae)	*Cistus salviifolius* (Cistaceae)	*Salvia triloba* (Lamiaceae)	*Sarcopoterium spinosum* (Rosaceae)	*Satureja thymbra* (Lamiaceae)	Ephemeral spp.
Foliage Projective Cover (%)											
Upper stratum (10–20 m)	37.4	–	–	–	–	–	–	–	–	–	–
Mid stratum (2–5 m)	–	8.3	3.3	–	–	–	–	–	–	–	–
Mid stratum (1–2 m)	–	–	–	7.5	2.5	–	–	–	–	–	–
Ground stratum (< 1 mm)	–	–	–	–	–	8.3	13.3	4.8	6.7	3.3	?
Frequency (Presence %)	75	100	75	75	50	100	100	75	50	75	?
1. Renewal buds	MM	M	M	N	N	Ch	Ch	Ch	Ch	Ch	Th
2. Trophic type	1	1	1	1	1	1	1	1	1	1	1
3. Life duration (yrs)	6	7	7	7	6	5	5	5	4	4	1
4. Height (m)	8	6	6	5	5	4	4	4	4	3	1–3
5. Crown diameter (m)	7	6	6	5–6	5	4	4	4	4	4	1/2
6. Canopy density (%)	4	5	2	5	3	2	2	2	4	3	1/2
7. Photosynthetic organs	1	1	1	1	1	1	1	1	1	1	1
8. 'Leaf' size	3	4	4	3	3	4 & 3	4 & 3	4	4 & 2	3 & 2	?
9. 'Leaf' colour	1	1	1	1	1	1	1	2	1	1	?
10. 'Leaf' length (cm)	4	3	3	3	2	3 & 2	3 & 2	3	3 & 2	1 & 1	?
11. 'Leaf' width (mm)	2	7	7	6	5	6 & 5	6 & 5	6	5 & 2	4 & 3	?
12. 'Leaf' angle (°)	3	3	3	3	3	3	3	3	3	3	?
13. 'Leaf' margin	1	2/(1)	1	1	1	1	1	3	1	1	?
14. 'Leaf' surface resins	2	2	2	2	2	1	1	1	2	1	?
15. 'Leaf' glands	1	2	1	1	2	2	2	2	2	2	?
16. 'Leaf' consistency	3	3	1	2	2	1	1	1	1	2	1
17. 'Leaf' tomentosity	1	1	1	1	1	4	4	4	1	4	?
18. 'Leaf' stomata	4	?	?	?	?	?	?	?	?	?	?
19. 'Leaf' duration (yrs)	5	2	2	4	2	1 & 2	1 & 2	1 & 2	1 & 2	1 & 2	?
20. 'Leaf' seasonality	1	1	2	1	2	1	1	1	1	1	3
21. 'Leaf' area: assim. stem area	1	1	1	1	1	1	1	1	1	1	?
22. Organs periodically shed	4	4	4	4	4	3	3	3	3	3	?
23. Organs shed – seasonality	–	–	–	–	–	–	–	–	–	–	–
24. Shoot growth – seasonality	1	1 & 2	1	1 & 2	1	1*	1*	1*	1*	1*	1
25. Stem – number from base	1	1	1–2	2	2	2	2	2	2	2	?
26. Stem – height (m)	?	?	?	?	?	?	?	?	?	?	?
27. Stem – diameter (cm)	?	?	?	?	?	?	?	?	?	?	?
28. Stem lignification	1	1	1	1	1	1	1	2	2	2	3
29. Bark consistency	6	1	1	1	1	1	1	2	2	1	Nil
30. Bark thickness (mm)	4	?	1	1	1	1	1	1	1	1	Nil
31. Bark shedding	3	?	?	3	3	3	3	3	2	2	Nil
32. Bark shedding – seasonality											Nil
33. Spinescence	4	2	4	4	1	4	4	4	1	4	?
34. Underground stems	Nil	Nil	Nil	Nil	Nil	Nil	Nil	Nil	3	Nil	?
35. Root morphology	2	3	3	3	3	3	3	3	3	3	?
36. Root modification	7	7	?	7	7	7	7	?	7	7	?
37. Rootlet modification	5	8	?	?	8	5	5	?	8	?	?
38. Root depth (cm)	5	7	6	7	6	5	5	?	5	4	?
39. Root spread (cm)	6	6	6	6	6	4	4	?	4	4	?
40. Veg. multiplication	4	4	?	2	4	4	4	4	4	4	4
41. Veg. regen. after fire	1	2	2	2	?	2	2	2	2	?	1
42. Flowering – seasonality	1	1	1	1	1	1	1	1	1	1	1
43. Flowering – pyrogenic	2	2	2	2	2	2	2	2	2	2	2
44. Fruit dehiscence	2	1	3	3	3	2	2	1	3	1	?
45. Seed dissemination	2	1	1	1	1	1	1	1	1	1	?
46. Seedling – germ. type	2	3	3	3	2	2	2	2	2	2	?
47. Plumule position	1	3	3	3	1	1	1	1	1	1	?

* Almost the whole year round – either brachyblasts or dolichoblasts.

Code of Ecomorphological Characters: See Chapter 1.

Table 43. Ecomorphological characters: Israel.
Location: Upper Galilee, Israel (32° 55′ N, 35° 30′ E).
Vegetation structure: Quercus ithaburensis association. (Number of records: 18, listing 16 perennial spp. and 231 annual spp.).
Age since last fire: Unknown.
Reference: Rabinovitz-Vyn, A., 1979. Ph. D. Thesis, Hebrew University of Jerusalem.
Compilers: G. Orshan and A. Rabinovitz-Vyn, November 1985.

Character	*Quercus ithaburensis* (Fagaceae)	*Pistacia palaestina* (Pistaciaceae)	*Styrax officinalis* (Styracaceae)	*Calicotome villosa* (Fabaceae)	*Phlomis viscosa* (Lamiaceae)	*Majorana syriaca* (Lamiaceae)	*Phagnalon rupestre* (Asteraceae)	*Sarcopoterium spinosum* (Rosaceae)	*Chiliadenus iphionoides* (Asteraceae)	Ephemeral spp.
Foliage Projective Cover (%)										
Upper stratum (2–5 m)	2.8+	3.0	2.0	–	–	–	–	–	–	–
Mid stratum (1–2 m)	–	–	–	2.0	–	–	–	–	–	–
Ground stratum (<1 m)	–	–	–	–	8.0	30.0	3.0	13.0	8.0	(231 spp.)
Frequency (Presence %)	100	77	77	61	11	100	66	11	44	?
1. Renewal buds	M	M	M	N	Ch	Ch	Ch	Ch	Ch	Th
2. Trophic type	1	1	1	2	1	1	1	1	1	1
3. Life duration (yrs)	7	7	7	6	5	4	4	4	4	1
4. Height (m)	7	6	6	5	4	4	3	4	3	1–3
5. Crown diameter (m)	6	6	5	5	4	4	3	4	3	1/2
6. Canopy density (%)	4	2	4	2	2	2	2	4	2	1/2
7. Photosynthetic organs	1	1	1	4	1	1	1	1	1	1
8. 'Leaf' size	5	4	4	3	4	3 & 2	2 & 1	4 & 2	3 & 2	?
9. 'Leaf' colour	1	1	5	1	1	2	1	1	1	?
10. 'Leaf' length (cm)	5	3	4	1	3 & 2	2 & 1	3 & 2	3 & 2	3 & 1	?
11. 'Leaf' width (mm)	7	7	6	5	6	6 & 4	3 & 2	5 & 2	5 & 3	?
12. 'Leaf' angle (°)	3	3	3	3	3	3	3	3	3	?
13. 'Leaf' margin	2	1	1	1	1	1	1	1	1	?
14. 'Leaf' surface resins	2	2	2	2	1	2	2	2	1	?
15. 'Leaf' glands	2	1	2	2	2	2	2	2	2	?
16. 'Leaf' consistency	1	1	2	1	1	1	1	1	1	1
17. 'Leaf' tomentosity	2	1	2	1	4	4	2	1	4	?
18. 'Leaf' stomata	?	?	?	?	?	?	?	?	?	?
19. 'Leaf' duration (yrs)	2	2	3	1	2 & 2	2 & 2	2	1 & 2	2 & 2	?
20. 'Leaf' seasonality	2	2	1	1	1	1	1	1	1	3
21. 'Leaf' area: assim. stem area	1	1	1	3	1	1	1	1	1	?
22. Organs periodically shed	4	4	4	4	3	3	3	3	3	?
23. Organs shed – seasonality										
24. Shoot growth – seasonality	1	1	1	1	1* & 2	1* & 2	1* & 2	1* & 2	1 & 3	1
25. Stem – number from base	1	1	2	2	2	2	2	2	2	–
26. Stem – height (m)										
27. Stem – diameter (cm)										
28. Stem lignification	1	1	1	1	2	2	2	2	2	3
29. Bark consistency	6	1	6	1	2	4	1	2	Nil	Nil
30. Bark thickness (mm)	3	1	1	1	1	1	1	1	Nil	Nil
31. Bark shedding	3	?	3	3	2	2	3	2	Nil	Nil
32. Bark shedding – seasonality										Nil
33. Spinescence	4	4	4	1	4	4	4	1	4	?
34. Underground stems	Nil	Nil	7	Nil	Nil	Nil	Nil	3	Nil	?
35. Root morphology	3	3	3	3	3	3	3	3	3	?
36. Root modification	7	7	7	7	7	7	7	7	7	?
37. Rootlet modification	8	?	?	6	5	?	?	8	8	?
38. Root depth (cm)	6	6	6	?	4	?	3	5	2	?
39. Root spread (cm)	6	6	6	5 (?)	4	3	3	4	2	?
40. Veg. multiplication	4	?	4	4	4	4	?	4	4	4
41. Veg. regen. after fire	?	2	2	2	2	2	1	2	2	1
42. Flowering – seasonality	1	1	1	1	1	1	1	1	3	1
43. Flowering – pyrogenic	2	2	2	2	2	2	2	2	2	2
44. Fruit dehiscence	1	3	3	2	1	1	1	3	1	?
45. Seed dissemination	1	1	1	1	1	1	1	1	1	?
46. Seedling – germ. type	3	3	2	2	2	2	2	2	2	?
47. Plumule position	3	3	1	1	1	1	1	1	1	?

+ Emergents (5–8 m).
* Almost the whole year round.

Code of Ecomorphological Characters: See Chapter 1.

Table 44. Ecomorphological characters: Portugal.
Location: Jaspe, Serra da Arrábida, Central Portugal (38° 27' N, 9° 00' W, 260 m).
Vegetation structure: Quercus coccifera carrascal. (Total Foliage Projective Cover: 70%).
Age since last fire: Unknown.
Soil: Calcic rhodo-chromic luvisols + calcareous chromic cambisols (FAO System).
Reference: Catarino, F.M., Correia, O.C.A. and Correia, A.I.V.D., 1982. Structure and dynamics of Serra da Arrabida mediterranean vegetation. Ecologia Mediterranea 8 (1/2): 203–222.
Compilers: F.M. Catarino and A.I.V.D. Correia, May 1986.

Character	*Quercus coccifera* (Fagaceae)	*Phillyrea latifolia* (Oleaceae)	*Pistacia lentiscus* (Pistaciaceae)	*Phillyrea angustifolia* (Oleaceae)	*Arbutus unedo* (Ericaceae)	*Juniperus phoenicea* (Cupressaceae)	*Rosmarinus officinalis* (Lamiaceae)	*Daphne gnidium* (Thymelaeaceae)	*Cistus monspeliensis* (Cistaceae)	*Cistus ladaniferus* (Cistaceae)	*Myrtus communis* (Myrtaceae)	*Olea europaea* var. *sylvestris* (Oleaceae)
1. Renewal buds	N, M	M, MM	M	N	N	M	N	N	N	N	N, M	MM
2. Trophic type	1	1	1	1	1	1	1	1	1	1	1	1
3. Life duration (yrs)	7	7	–*	–	7	7	–	–	–	–	–	7
4. Height (m)	5	5	5	5	5	6	4/5	4	4	4/5	4/5	6
5. Crown diameter (m)	–	–	–	–	–	–	–	–	–	–	–	–
6. Canopy density (%)	–	–	–	–	–	–	–	–	–	–	–	–
7. Photosynthetic organs	1	1	1	1	1	1	1	1	1	1	1	1
8. 'Leaf' size	4	4	4	3	4	1	2	3	3	4	4	3
9. 'Leaf' colour	1	1	1	1	1	1	4	1	4	4	1	5
10. 'Leaf' length (cm)	3	3	4	3	4	1	3	3	3	4	3	3
11. 'Leaf' width (mm)	6	6	7/8	5	7	2	2	5	4	6	6	5
12. 'Leaf' angle (°)	3	3	3	3	3	2	3	3	3	3	3	3
13. 'Leaf' margin	1/2	1/2	3	1	1/2	1	1	1	1	1	1	1
14. 'Leaf' surface resins	2	2	2	2	2	2	2	2	1	1	2	2
15. 'Leaf' glands	2	2	1	2	2	2	1	2	1	1	1	2
16. 'Leaf' consistency	3	2	2	2	2	3	2	2	2	2	2	3
17. 'Leaf' tomentosity	1	1	1	1	1	1	2	1	4	2	2	2
18. 'Leaf' stomata	2	2	2	2	2	–	2	–	2	2	2	2
19. 'Leaf' duration (yrs)	3	3	3	–	–	–	–	–	2	2	–	–
20. 'Leaf' seasonality	1	1	1	1	1	1	1	1	1	1	1	1
21. 'Leaf' area: assim. stem area	1	1	1	1	1	1	1	1	1	1	1	1
22. Organs periodically shed	4	4	4	4	4	4	4	4	4	4	4	4
23. Organs shed – seasonality	Sept–Nov.	Sept–Nov.	Sept–Nov.	–	–	–	–	–	Jun–Aug.	Jun–Aug.	–	–
24. Shoot growth – seasonality	Mar–May	Mar–May	Mar–May	–	–	–	–	–	Mar–May	Mar–May	–	–
25. Stem – number from base	2	2	2	2	2	2	2	2	2	2	2	2
26. Stem – height (m)	–	–	–	–	–	–	–	–	–	–	–	–
27. Stem – diameter (cm)	3	1	1–3	0.5–1	1–5	20	1	1	0.5–1	up to 2	1	–
28. Stem lignification	1	1	1	1	1	1	1	1	1	1	1	1
29. Bark consistency	1	1	1	1	3	1	1	1	1	1	1	1
30. Bark thickness (mm)	–	–	–	–	–	–	–	–	–	–	–	–
31. Bark shedding	–	–	–	–	–	–	–	–	–	–	–	–
32. Bark shedding – seasonality	–	–	–	–	–	–	–	–	–	–	–	–
33. Spinescence	2	4	4	4	4	4	4	4	4	4	4	4
34. Underground stems	–	–	–	–	–	–	–	–	–	–	–	–
35. Root morphology	–	–	–	–	–	–	–	–	–	–	–	–
36. Root modification	–	–	–	–	–	–	–	–	–	–	–	–
37. Rootlet modification	–	–	–	–	–	–	–	–	–	–	–	–
38. Root depth (cm)	–	–	–	–	–	–	–	–	–	–	–	–
39. Root spread (cm)	–	–	–	–	–	–	–	–	–	–	–	–
40. Veg. multiplication	–	–	–	–	–	–	–	–	–	–	–	–
41. Veg. regen. after fire	–	–	–	–	–	–	–	–	–	–	–	–
42. Flowering – seasonality	Apr–May	Apr–May	Apr–May	Mar–May	Oct–Jan.	Feb–Mar.	Jan–Dec.	Mar–Oct.	Mar–May	Mar–May	May–Jul.	May–Jul.
43. Flowering – pyrogenic	2	2	2	2	2	2	2	2	2	2	2	2
44. Fruit dehiscence	1	3	3	3	3	4	1	3	2	2	3	3
45. Seed dissemination	1	1	1	1	1	1	1	1	1	1	1	1
46. Seedling – germ. type	3	2	3	2	2	2	2	2	2	2	2	2
47. Plumule position	3	1	3	1	1	1	1	1	1	1	1	1
48. No. leaves per 10 cm stem	30	50	10	10	20	–	80	50	30	15	–	20

* In this Table, – indicates that information is unavailable.

Code of Ecomorphological Characters: See Chapter 1.

Table 45. Ecomorphological characters: Spain.
Location: Garraf, Catalonia, Spain (41° 18′ N, 1° 49′ E, 450 m).
Vegetation structure: Garrigue under *Pinus halepensis* (over 60 years old).
Age since last fire: 13 years.
Soil: Calcareous karst.
Compiler: C.A. Gracia, May 1986.

Character	*Quercus coccifera* (Fagaceae)	*Pistacia lentiscus* (Pistaciaceae)	*Pinus halepensis* (Pinaceae)
Foliage Projective Cover (%)			
Upper stratum (>2 m)	–	–	50–75
Mid stratum (30 cm–2 m)	57	9	–
Ground stratum (<30 cm)	–	–	–
Biomass (kg ha^{-1})	9170	960	–
1. Renewal buds	N	N(M)	MM
2. Trophic type	1	1	1
3. Life duration (yrs)	5	5	6–7
4. Height (m)	4–5	4–5	8–9
5. Crown diameter (m)	3	4	6
6. Canopy density (%)	6	6	4
7. Photosynthetic organs	1	1	1
8. 'Leaf' size	3	3[b]	2
9. 'Leaf' colour	1	1	1
10. 'Leaf' length (cm)	2	2[b]	4
11. 'Leaf' width (mm)	6	6[b]	2
12. 'Leaf' angle (°)	3 (35°)	1	3
13. 'Leaf' margin	1	1	1
14. 'Leaf' surface resins	2	2	2
15. 'Leaf' glands	2	2	2
16. 'Leaf' consistency	3	2	3
17. 'Leaf' tomentosity	1	1	1
18. 'Leaf' stomata	2	2	4
19. 'Leaf' duration (yrs)	3	3	4
20. 'Leaf' seasonality	1	1	1
21. 'Leaf' area: assim. stem area	1:0	1:0	1:0
22. Organs periodically shed	4	4	4
23. Organs shed – seasonality	1 (Apr–Jun.)–3 (Sept–Nov.)	1 (Apr–Jun.)–3 (Sept–Nov.)	1 (Apr–Jun.)–3 (Sept–Nov.)
24. Shoot growth – seasonality	1–3	1–3	1–3
25. Stem – number from base	2 (30–130)[a]	2 (10–60)[a]	1
26. Stem – height (m)	0	0.1–0.5	1–2
27. Stem – diameter (cm)	2	2	25–35
28. Stem lignification	1	1	1
29. Bark consistency	1	1	3
30. Bark thickness (mm)	1	1	4
31. Bark shedding	4	4	3
32. Bark shedding – seasonality	–	–	–
33. Spinescence	2	4	4
34. Underground stems	–	–	–
35. Root morphology	3	3	1
36. Root modification	7	7	7
37. Rootlet modification	5	5	5
38. Root depth (cm)	7	7	7
39. Root spread (cm)	7	7	7
40. Veg. multiplication	2	2	4
41. Veg. regen. after fire	2	2	1
42. Flowering – seasonality	1 (Mar–Jun.)	1 (Mar–Jun.)	1 (Mar–Jun.)
43. Flowering – pyrogenic	2	2	2
44. Fruit dehiscence	1	3	2
45. Seed dissemination	1	1	2
46. Seedling – germ. type	3	3	2
47. Plumule position	3	3	1

[a] Stem number/m^2.
[b] Leaflets.

Code of Ecomorphological Characters: See Chapter 1.

Table 46. Ecomorphological characters: Spain.
Location: Doñana National Park, SW Spain (36° 59' N, 6° 25' W).
Vegetation structure: Open-scrub (*Cistus – Halimium – Lavandula* community).
Age since last fire: c. 80 years.
Annual precipitation: 578 mm.
Soil: Regosol (sand dune).
Compiler: J. Merino, June 1986.

Character	*Halimium halimifolium* (Cistaceae)	*Cistus libanotis* (Cistaceae)	*Halimium commutatum* (Cistaceae)	*Rosmarinus officinalis* (Lamiaceae)	*Lavandula stoechas* (Lamiaceae)	*Helichrysum picardii* (Asteraceae)	*Stauracanthus genistioides* (Fabaceae)	
Foliage Projective Cover (%)								
Upper stratum (1–2 m)	**	–	–	x	–	–	–	Total 50%
Mid stratum (50–100 cm)	–	**	**	–	**	1.5	1.5	
Ground stratum (< 50 cm)	–	–	–	–	–	–	–	
1. Renewal buds	1 N	1 N	1 N	1 N	1 N	1 N	1 N	
2. Trophic type	1	1	1	1	1	1	2	
3. Life duration (yrs)	4	4	4	4	4	4	4	
4. Height (m)	5	4	4	5	4	4	4	
5. Crown diameter (m)	4	4	3	4	3	3	4	
6. Canopy density (%)	?	?	?	?	?	?	?	
7. Photosynthetic organs	1	1	1	1	1	1	4–5	
8. 'Leaf' size	3	3	2	2	1	2	3	
9. 'Leaf' colour	1	4	4	4	3	3	1	
10. 'Leaf' length (cm)	2	3	2	2	2	3	2	
11. 'Leaf' width (mm)	5	4	3	2	2	2	4	
12. 'Leaf' angle (°)	3	3	3	3	2	2	?	
13. 'Leaf' margin	1	5	5	5	5	5	1	
14. 'Leaf' surface resins	2	1	2	2	2	2	2	
15. 'Leaf' glands	2	2	2	2	2	2	2	
16. 'Leaf' consistency	2	2	2	2	2	2	1	
17. 'Leaf' tomentosity	4	2	2	2	4	4	1	
18. 'Leaf' stomata	2	2	2	2	2	2	?	
19. 'Leaf' duration (yrs)	2	2	2	2	2	2	1	
20. 'Leaf' seasonality	1	1	1	1	1	1	?	
21. 'Leaf' area: assim. stem area	1	1	1	1	1	1	3	
22. Organs periodically shed	4	4	4	4	4	4	4	
23. Organs shed – seasonality	2 (early)	2 (early)	2 (early)	2 (early)	2 (early)	2 (early)	?	
24. Shoot growth – seasonality	1/2	1/2	1/2	1/2	4/1	?	?	
25. Stem – number from base	2	2	2	2	1	1	2	
26. Stem – height (m)	?	?	?	?	?	?	?	
27. Stem – diameter (cm)	3	3	<3	3	<3	<3	3	
28. Stem lignification	1	1	1	1	1	1	1	
29. Bark consistency	1	1	1	1	1	1	1	
30. Bark thickness (mm)	1	1	1	1	1	1	1	
31. Bark shedding	?	?	?	?	?	?	?	
32. Bark shedding – seasonality	?	?	?	?	?	?	?	
33. Spinescence	4	4	4	4	4	4	1	
34. Underground stems	abs.	abs.	abs.	abs.	abs.	abs.	abs.	
35. Root morphology	3	2	2	2	3	3	?	
36. Root modification	7	7	7	7	7	7	7	
37. Rootlet modification	?	?	?	?	?	?	?	
38. Root depth (cm)	3	2	2	2	3	3	?	
39. Root spread (cm)	6	6	6	6	4	4	?	
40. Veg. multiplication	4	4	4	4	4	4	4	
41. Veg. regen. after fire	1	1	1	1	1	2 or 4	2 or 4	
42. Flowering – seasonality	1/2	1	4/1	4/1	1/2	2	4	
43. Flowering – pyrogenic	?	?	?	?	?	?	?	
44. Fruit dehiscence	2	2	2	2	2	2	2	
45. Seed dissemination	1	1	1	1	1	1	1	
46. Seedling – germ. type	2	2	2	2	2	2	2	
47. Plumule position	1	1	1	1	1	1	1	
48. No. leaves per 10 cm stem	20	15	14	7	90	40	–	

Code of Ecomorphological Characters: See Chapter 1.

Table 47. Ecomorphological characters: Spain.
Location: Doñana National Park, SW Spain (36° 59′ N, 6° 25′ W).
Vegetation structure: Heathland (*Erica-Calluna* community). Mean number of species (5 × 5 m) = 8.3.
Age since last fire: c. 80 years.
Annual precipitation: 578 mm.
Soil: Hydromorphic (sand dune).
Compiler: J. Merino, June 1986.

Character	*Erica scoparia* (Ericaceae)	*Erica ciliaris* (Ericaceae)	*Calluna vulgaris* (Ericaceae)	Total
Foliage Projective Cover (%)				
Upper stratum (1>2 m)	← c. 150 →		c. 10	} 160
Ground stratum (<1 m)	–	–	–	
1. Renewal buds	1 M	1 N	1 N	
2. Trophic type	1	1	1	
3. Life duration (yrs)	4–5	4–5	4	
4. Height (m)	5	5	5	
5. Crown diameter (m)	6	5	5	
6. Canopy density (%)	?	?	?	
7. Photosynthetic organs	1	1	1	
8. 'Leaf' size	1	1	1	
9. 'Leaf' colour	1	1	1	
10. 'Leaf' length (cm)	1	1	1	
11. 'Leaf' width (mm)	1	1	1	
12. 'Leaf' angle (°)	3	3	3	
13. 'Leaf' margin	1	1	1	
14. 'Leaf' surface resins	2	2	2	
15. 'Leaf' glands	2	2	2	
16. 'Leaf' consistency	2	2	2	
17. 'Leaf' tomentosity	1	1	1	
18. 'Leaf' stomata	2	2	2	
19. 'Leaf' duration (yrs)	?	?	?	
20. 'Leaf' seasonality	1	1	1	
21. 'Leaf' area: assim. stem area	1	1	1	
22. Organs periodically shed	4	4	4	
23. Organs shed – seasonality	3 (early)	3 (early)	3 (early)	
24. Shoot growth – seasonality	2	?	?	
25. Stem – number from base	2	2	2	
26. Stem – height (m)	2	1–2	1–2	
27. Stem – diameter (cm)	10	?	5	
28. Stem lignification	1	1	1	
29. Bark consistency	2	2	2	
30. Bark thickness (mm)	2	2	2	
31. Bark shedding	?	?	?	
32. Bark shedding – seasonality	?	?	?	
33. Spinescence	4	4	4	
34. Underground stems	7	7	abs.	
35. Root morphology	2	2	2	
36. Root modification	7	7	7	
37. Rootlet modification	?	?	?	
38. Root depth (cm)	0–25	0–25	?	
39. Root spread (cm)	?	?	?	
40. Veg. multiplication	4	4	4	
41. Veg. regen. after fire	4	4	1	
42. Flowering – seasonality	1/2	2/3	3/4	
43. Flowering – pyrogenic	2	2	2	
44. Fruit dehiscence	1	1	1	
45. Seed dissemination	1	1	1	
46. Seedling – germ. type	2	2	2	
47. Plumule position	1	1	1	
48. No. leaves per 10 cm stem	79	120	?	

Code of Ecomorphological Characters: See Chapter 1.

Table 48. Natural vegetation of the Western Cape, South Africa, including some ecosystems which extend into the summer rainfall zone of the Eastern Cape. – E.J. Moll.

Formation	Nutrient-poor soil		Nutrient-rich soil		High pH-Ca rich soil	
Open-scrub	(1)	Mountain fynbos (± small trees) Western Cape Central Cape	(5)	*Olea-Sideroxylon* Western Cape Central Cape (?)		
Heathland	(2)	Mountain fynbos Western Cape Central Cape	(6)	Renosterveld Western Cape Central Cape		
	(3)	Coastal fynbos Western Cape Central Cape Eastern Cape			(7)	Calcareous coastal fynbos Western Cape
Arid heathland	(4)	Arid fynbos			(8)	Strandveld

References

Acocks, J.P.H., 1975. Veld Types of South Africa. Mem. Bot. Surv. S. Afr. No. 40, 128 pp.

University of Cape Town, Botany Dept, 1983. Vegetation of the Fynbos Biome. (General Map 1:1,000,000; Worcester Map 1:250,000). Chief Director of Surveys and Mapping, Mowbray, S. Afr.

Vegetation, nutrition and climate – data-tables

(2) Foliar analyses

Co-ordinator: P.W. Rundel

Contributors

P.W. Rundel, R.L. Specht (Southeastern and South Australia)
A.J.M. Hopkins, P.W. Rundel (Southwestern Australia)
P.W. Rundel, R.L. Specht (California)
G. Montenegro, P.W. Rundel (Chile)
P.W. Rundel, R.L. Specht, L. Trabaud (France)
N.S. Margaris (Greece)
P.W. Rundel (South Africa)

Contents

Australia – Southeastern and South (Tables 1–10)	63
Australia – Southwestern (Table 11)	71
California (Tables 12–14)	72
Chile – Central (Tables 15–18)	74
France (Tables 19–21)	75
Greece (Table 22)	78
South Africa – Cape Province (Tables 23–24)	80

Table 1. Foliar analyses: Australia (Prepared by R.L. Specht and P.W. Rundel).
Vegetation structure: heathy open-forest (dry sclerophyll forest).
Alliance: Eucalyptus obliqua – E. baxteri.
Location: Mount Lofty, South Australia. *Latitude:* 34° 59′ S *Longitude:* 138° 40′ E *Elevation:* 725 m.
Reference: Specht, R.L. and Perry, R.A., 1948. The plant ecology of part of the Mount Lofty Ranges. Trans. R. Soc. S. Aust. 72: 91–132.

Species	Thickness of lamina (mm)	Dry weight of lamina (mg cm^{-2})	Lignin (%)	Cellulose (%)	Crude protein (%) (= N × 6.25)	Sclerophyll Index*	Mineral nutrient content											
							% oven dry weight								p.p.m. oven dry weight			
							N	P	S	K	Ca	Mg	Na	Cl	Fe	Cu	Mn	Zn
Upper stratum (20 m)																		
Eucalyptus obliqua	0.42	23	13.0	17.3	5.9	513	0.94	0.04	0.09	0.39	0.78	0.28	0.27	0.39	66	10	248	12
E. baxteri	0.44	25	10.3	18.0	5.1	555	0.82	0.04	0.06	0.41	0.78	0.20	0.31	0.39	66	12	170	14
Mid stratum (1–2 m)																		
Banksia marginata	0.35	15	21.0	30.0	5.3	962	0.85	0.04	0.11	0.53	0.70	0.18	0.35	0.47	122	12	204	6
Leptospermum myrsinoides	0.21	15	18.1	29.1	6.8	694	1.08	0.04	0.13	0.40	0.68	0.30	0.18	0.13	74	14	240	14
Ground stratum																		
Lepidosperma semiteres	0.74	46	14.0	34.7	3.4	1,432	0.55	0.03	0.12	0.67	0.23	0.04	0.05	0.43	32	1	64	17

* Sclerophyll Index = [(lignin + cellulose)/crude protein].

R.L. Specht (ed.) Mediterranean-type Ecosystems. ISBN 90-6193-652-7.
© Kluwer Academic Publishers.

Table 2. Foliar analyses: Australia (Prepared by R.L. Specht and P.W. Rundel).
Vegetation structure: open-heathland (dry heathland).
Alliance: Banksia ornata – Xanthorrhoea australis – Allocasuarina pusilla.
Age: 28 years after fire (leaves collected Jan. 1982).
Location: Dark Island Soak, near Keith, South Australia.
Latitude: 36° 06′ S *Longitude:* 140° 31′ E *Elevation:* c. 40 m.
Reference: Specht, R.L., Rayson, R. and Jackman, M.E., 1958. Dark Island heath (Ninety-Mile Plain, South Australia). VI. Pyric succession: Changes in composition, coverage, dry weight and mineral nutrient status. Aust. J. Bot. 6: 59–88.

Species	Thickness of lamina (mm)	Dry weight of lamina (mg cm^{-2})	Lignin (%)	Cellulose (%)	Crude protein (%) (= N × 6.25)	Sclerophyll Index*	Mineral nutrient content											
							% oven dry weight								p.p.m. oven dry weight			
							N	P	S	K	Ca	Mg	Na	Cl	Fe	Cu	Mn	Zn
Upper stratum																		
Banksia ornata	0.50	26	15.2	41.7	4.8	1,185	0.76	0.04	0.13	0.25	0.65	0.10	0.10	0.08	189	8	380	10
Banksia marginata	0.62	22	14.4	41.6	4.5	1,244	0.72	0.03	0.09	0.17	0.46	0.15	0.08	0.13	140	3	190	4
Xanthorrhoea australis	1.96	56△	17.8	37.3	2.9	1,900	0.47	0.03	0.10	0.80	0.55	0.11	0.05	0.24	36	2	12	13
Mid stratum																		
Allocasuarina pusilla	0.68*	38*	14.8	39.5	6.6	823	1.06	0.03	0.17	0.50	0.64	0.14	0.16	0.28	54	2	18	7
Leptospermum myrsinoides	0.30	40	–	–	5.6	–	0.90	0.03	0.15	0.43	0.69	0.20	0.12	0.04	50	5	108	6
Spyridium subochreatum	0.50	55†	–	–	8.1	–	1.30	0.04	0.15	0.45	2.21	0.18	0.01	0.22	63	5	20	13
Phyllota remota	0.38	18	14.7	50.0	8.9	727	1.43	0.04	0.13	0.40	0.53	0.11	0.14	0.29	77	3	20	12
Ground stratum																		
Dicotyledons																		
Calytrix alpestris	0.44*	47*	9.7	40.0	6.5	765	1.04	0.04	0.15	0.40	0.93	0.18	0.08	0.10	311	8	36	15
Hibbertia riparia	0.45	19	18.4	21.0	6.8	579	1.08	0.04	0.44	0.40	0.74	0.23	0.17	0.61	59	2	20	10
Hibbertia sericea	0.97†	47†	12.6	20.6	6.6	503	1.05	0.04	0.34	0.50	1.41	0.27	0.06	0.43	167	7	24	11
Monocotyledons																		
Hypolanea fastigiata	0.71*	48*	10.0	32.7	3.4	1,256	0.55	0.02	0.30	0.50	0.28	0.04	0.04	0.24	180	3	48	4
Lepidosperma carphoides	0.65*	44*	14.8	31.9	3.6	1,297	0.57	0.02	0.26	0.30	0.41	0.04	0.02	0.12	68	1	26	4
Lepidosperma laterale	0.70	44	11.9	28.6	4.6	880	0.74	0.03	0.19	0.50	0.47	0.04	0.02	0.26	149	1	26	9

* Cylindrical.
† Inrolled.
△ Triangular.

Table 3. Foliar analyses: Australia (Prepared by R.L. Specht and P.W. Rundel).
Vegetation structure: wet-heathland + scattered small trees of *Banksia aemula*.
Alliances: Banksia aemula and *Banksia oblongifolia – Melaleuca sieberi*.
Location: Beerwah, Queensland *Latitude:* 26° 51′ S *Longitude:* 153° 00′ E *Elevation:* c. 15 m.
Reference: Bolton, M.P., 1986. Community Dynamics and Productivity in a Subtropical Wet Heatland. Ph.D. Thesis, University of Queensland.

Species	Thickness of lamina (mm)	Dry weight of lamina (mg cm^{-2})	Lignin (%)	Cellulose (%)	Crude protein (%) (= N × 6.25)	Sclerophyll Index*	Mineral nutrient content											
							% oven dry weight								p.p.m. oven dry weight			
							N	P	S	K	Ca	Mg	Na	Cl	Fe	Cu	Mn	Zn
Emergent stratum																		
Banksia aemula																		
new leaves	0.48	19	18.4	48.1	3.3	2,015	0.63	0.03	0.08	0.23	0.26	0.14	0.35	0.31	78	10	44	4
old leaves	0.46	22	17.5	46.0	3.8	1,671	0.60	0.02	0.11	0.15	0.42	0.16	0.29	0.20	128	12	74	4
Upper stratum																		
Banksia oblongifolia																		
new leaves	0.41	20	16.6	45.6	4.3	1,447	0.69	0.03	0.06	0.20	0.18	0.08	0.27	0.26	58	12	36	2
old leaves (1)	0.41	21	16.2	44.1	3.5	1,723	0.56	0.02	0.08	0.15	0.36	0.12	0.18	0.23	82	10	82	1
old leaves (2)	0.41	21	19.3	47.0	2.3	2,883	0.45	0.02	0.08	0.17	0.40	0.14	0.10	0.16	76	4	82	2
Banksia robur	0.97	32	18.2	45.3	2.3	2,761	0.45	0.02	0.09	0.18	0.42	0.14	0.22	0.44	74	6	84	2
Hakea gibbosa	1.24+	44	18.5	38.2	5.1	1,118	0.81	0.02	0.12	0.20	0.32	0.16	0.28	0.25	96	4	36	4
Xanthorrhoea resinosa	1.65	44	–	–	3.5	–	0.56	0.01	0.19	0.53	0.24	0.10	0.10	0.27	104	2	14	2
Melaleuca sieberi	0.36	36	–	–	5.6	–	0.91	0.03	0.10	0.20	0.20	0.36	0.10	0.15	80	10	64	10
Melaleuca nodosa	0.63+	57	–	–	4.8	–	0.76	0.04	0.27	0.30	0.86	0.26	0.30	0.12	146	6	44	8
Mid stratum																		
Pultenaea myrtoides	0.26	11	13.9	25.0	10.3	378	1.73	0.03	0.19	0.32	0.52	0.30	0.32	0.55	94	14	94	14
Ground stratum																		
Empodisma minus	0.83+	46	9.5	31.0	2.9	1,397	0.46	0.01	0.14	0.31	0.14	0.12	0.14	0.34	192	10	246	6
Schoenus spp.	0.70+	32	10.1	35.2	2.3	1,970	0.47	0.01	0.11	0.25	0.10	0.10	0.23	0.43	138	10	38	1

* Sclerophyll Index = [(lignin + cellulose)/crude protein].
+ Cylindrical leaves.

Table 4. Foliar analyses: Australia (Prepared by R.L. Specht and P.W. Rundel).
Vegetation structure: savannah woodland.
Alliances: Eucalyptus microcarpa and *Eucalyptus leucoxylon – E. viminalis*.
Location: Green Hill, Mount Lofty Ranges, South Australia (*Latitude:* 34° 57′ S *Longitude:* 138° 38′ E *Elevation:* 350 m).
References: Specht, R.L. and Perry, R.A., 1948. The plant ecology of part of the Mount Lofty Ranges. Trans. R. Soc. S. Aust. 72: 91–132.
Davies, J.G. and Sim, A.H., 1931. The influence of frequency of cutting on the productivity, botanical and chemical composition, and the nutritive value of 'natural' pastures in southern Australia. Counc. Sci. Industr. Res. Aust. Pamphlet No. 18.

Species	Thickness of lamina (mm)	Dry weight of lamina (mg cm^{-2})	Lignin (%)	Cellulose (%)	Crude protein (%) (= N × 6.25)	Sclerophyll Index*	Mineral nutrient content											
							% oven dry weight								p.p.m. oven dry weight			
							N	P	S	K	Ca	Mg	Na	Cl	Fe	Cu	Mn	Zn
Upper stratum																		
Eucalyptus microcarpa	0.32	23	11.1	17.6	6.1	470	0.97	0.08	0.09	1.55	1.32	0.26	0.10	0.78	68	12	442	16
E. leucoxylon	0.42	26	15.0	20.3	7.3	484	1.17	0.10	0.11	0.80	1.06	0.26	0.39	0.85	66	12	160	42
E. viminalis	0.28	15	15.1	20.4	11.8	301	1.89	0.13	0.11	1.10	0.82	0.26	0.15	0.56	70	12	280	24
Ground stratum																		
Danthonia sp. } *Erodium botrys* }	–	–	–	–	11.3	–	1.81	0.21	–	1.44	1.50	–	0.31	0.71	–	–	–	–

* Sclerophyll Index = [(lignin + cellulose)/crude protein].

Table 5. Foliar analyses: Australia (Prepared by R.L. Specht and P.W. Rundel).
Vegetation structure: mallee open-scrub.
Alliance: Eucalyptus behriana.
Reference: Myers, B.A., Ashton, D.H. and Osborne, J.A., 1986. The ecology of the mallee outlier of *Eucalyptus behriana* F. Muell. near Melton, Victoria. Aust. J. Bot. 34: 15–39.
Whittaker, R.H., Niering, W.A. and Crisp, M.D. 1979. Structure, pattern and diversity of a mallee community in New South Wales. Vegetatio 39: 65–76.

Species	Thickness of lamina (mm)	Dry weight of lamina (mg cm^{-2})	Lignin (%)	Cellulose (%)	Crude protein (%) (= N × 6.25)	Sclerophyll Index*	Mineral nutrient content											
							% oven dry weight								p.p.m. oven dry weight			
							N	P	S	K	Ca	Mg	Na	Cl	Fe	Cu	Mn	Zn
West Wyalong, N.S.W. (33°55′S, 147°12′E)																		
Upper stratum																		
Eucalyptus behriana	0.47	28	14.0	20.4	8.1	425	1.29	0.09	0.17	0.53	0.72	0.24	0.42	0.60	100	8	800	10
Mid stratum																		
Melaleuca lanceolata	0.45	56	15.1	26.7	8.1	516	1.30	0.06	0.33	0.60	0.66	0.42	0.48	0.66	180	2	14	4
Cassinia laevis	0.41	31	–	–	6.3	–	1.00	0.09	0.12	1.56	0.20	0.23	0.01	0.22	306	5	270	33
Melton, Vic. (37°41′S, 114°35′E)																		
Upper stratum																		
Eucalyptus behriana			11.4	15.2	6.8	391	1.08	0.07	0.17	0.48	0.84	0.20	0.42	0.48	116	6	116	16

* Sclerophyll Index = [(lignin + cellulose)/crude protein].

Table 6. Foliar analyses: Australia (Prepared by R.L. Specht and P.W. Rundel).
Vegetation structure: mallee open-scrub.
Alliance: Eucalyptus socialis – E. dumosa.
Reference: Specht, R.L., 1981. Mallee ecosystems in southern Australia. pp. 203–231. In: di Castri F., Goodall, D.W. and Specht, R.L. (eds). Ecosystems of the World. Vol. 11. Mediterranean-type Shrublands. Elsevier, Amsterdam.

Species	Thickness of lamina (mm)	Dry weight of lamina (mg cm^{-2})	Lignin (%)	Cellulose (%)	Crude protein (%) (= N × 6.25)	Sclerophyll Index*	Mineral nutrient content											
							% oven dry weight								p.p.m. oven dry weight			
							N	P	S	K	Ca	Mg	Na	Cl	Fe	Cu	Mn	Zn
Balranald-Euston, N.S.W. (34° 33′ S, 143° 00′ E)																		
Upper stratum																		
Eucalyptus socialis			9.7	12.9	5.2	435	0.83	0.05	0.09	0.72	0.94	0.16	0.38	0.66	132	8	38	10
Ground stratum																		
Atriplex stipitata			5.0	10.1	15.3	99	2.44	0.09	0.34	2.50	0.76	0.48	6.34	8.4	222	6	72	14
Mildura, Vic. (90 km W) (34° 17′ S, 141° 10′ E)																		
Upper stratum																		
Eucalyptus socialis	0.60	36	9.6	12.3	6.5	337	1.04	0.07	0.09	0.72	1.08	0.24	0.22	0.56	252	8	160	6
Ground stratum																		
Atriplex-Maireana			4.0	8.3	20.5	60	3.28	0.13	0.33	2.10	0.70	0.48	7.76	3.35	444	8	132	12
Blanchetown – Truro, S. Aust. (34° 23′ S, 139° 20′ E)																		
Upper stratum																		
Eucalyptus socialis	0.63	38	12.2	15.2	6.5	422	1.04	0.04	0.12	0.48	0.80	0.20	0.36	0.53	100	6	76	6
Ground stratum																		
Atriplex stipitata			4.5	9.0	12.4	108	1.98	0.10	0.30	2.70	0.96	0.48	6.00	11.10	194	6	76	12
Moorlands, S. Aust. (35° 18′ S, 139° 39′ E)																		
Upper stratum																		
Eucalyptus socialis			12.4	15.4	5.4	515	0.86	0.04	0.10	0.42	1.40	0.24	0.28	0.55	74	6	24	10
Mid stratum																		
Melaleuca lanceolata			13.7	30.9	7.8	572	1.24	0.06	0.39	0.42	1.00	0.40	0.66	0.90	182	4	8	8
Ground stratum Introd. grass/herbs																		

* Sclerophyll Index = [(lignin + cellulose)/crude protein].

Table 7. Foliar analyses: Australia (Prepared by R.L. Specht and P.W. Rundel).
Vegetation structure: mallee open-scrub.
Alliance: Eucalyptus diversifolia.
References: Specht, R.L., 1981. Mallee ecosystems in southern Australia. pp. 203–231. In: di Castri, F., Goodall, D.W. and Specht, R.L. (eds). Ecosystems of the World. Vol. 11. Mediterranean-type Shrublands. Elsevier, Amsterdam.

Species	Thickness of lamina (mm)	Dry weight of lamina (mg cm^{-2})	Lignin (%)	Cellulose (%)	Crude protein (%) (= N × 6.25)	Sclerophyll Index*	Mineral nutrient content											
							% oven dry weight								p.p.m. oven dry weight			
							N	P	S	K	Ca	Mg	Na	Cl	Fe	Cu	Mn	Zn
Ki Ki, S. Aust. (35°35′ S, 139°46′ E)																		
Upper stratum																		
Eucalyptus diversifolia	0.52	31	17.3	21.2	4.3	895	0.68	0.03	0.11	0.38	0.80	0.18	0.30	0.36	80	10	46	12
Ground stratum																		
Gahnia deusta			10.0	29.2	2.4	1,633	0.38	0.01	0.10	0.33	0.42	0.04	0.02	0.17	206	4	58	6
Keith-Bordertown, S. Aust. (36°10′ S, 140°30′ E)																		
Upper stratum																		
Eucalyptus diversifolia			18.0	22.0	3.6	1,111	0.59	0.04	0.10	0.40	0.64	0.16	0.28	0.27	60	8	58	6
Ground stratum																		
Gahnia deusta			10.1	31.5	2.9	1,434	0.46	0.02	0.11	0.33	0.54	0.06	0.02	0.13	250	2	62	4
Lasiopetalum behrii			13.1	26.8	6.4	623	1.02	0.05	0.23	0.90	1.56	0.14	0.08	0.25	306	4	14	30

* Sclerophyll Index = [(lignin + cellulose)/crude protein].

Table 8. Foliar analyses: Australia (Prepared by R.L. Specht and P.W. Rundel).
Vegetation structure: mallee open-scrub (mallee-broombush).
Alliance: Eucalyptus incrassata – Melaleuca uncinata.
Age: 28 years after fire (leaves collected Jan. 1982).
Location: Dark Island Soak, near Keith, South Australia.
Latitude: 36° 06′ S *Longitude:* 140° 31′ E *Elevation: c. 40 m.*
Reference: Specht, R.L., 1966. The growth and distribution of mallee-broombush (*Eucalyptus incrassata-Melaleuca uncinata* association) and heath vegetation near Dark Island Soak, Ninety-Mile Plain, South Australia. Aust. J. Bot. 14: 361–371.

Species	Thickness of lamina (mm)	Dry weight of lamina (mg cm^{-2})	Lignin (%)	Cellulose (%)	Crude protein (%) (= N × 6.25)	Sclerophyll Index*	Mineral nutrient content											
							% oven dry weight								p.p.m. oven dry weight			
							N	P	S	K	Ca	Mg	Na	Cl	Fe	Cu	Mn	Zn
Upper stratum																		
Eucalyptus incrassata	0.62	44	10.5	18.1	4.9	583	0.78	0.05	0.04	0.45	0.80	0.11	0.28	0.34	45	1	52	7
Eucalyptus foecunda	0.50	33	14.3	16.7	6.2	500	0.99	0.05	0.05	0.59	0.76	0.17	0.17	0.40	59	3	248	9
Mid stratum																		
Melaleuca uncinata	0.74*	46*	10.0	27.7	7.8	483	1.25	0.06	0.21	0.59	0.80	0.24	0.35	0.75	45	1	12	10
Hakea muellerana	1.07*	31*	12.1	46.3	3.4	1,718	0.54	0.04	0.23	0.50	0.72	0.13	0.14	0.15	32	1	48	4
Ground stratum																		
Dicotyledon																		
Pultenaea tenuifolia	0.32*	27*	–	–	8.4	–	1.34	0.04	0.13	0.45	1.68	0.15	0.08	0.15	229	1	12	22
Calytrix tetragona	0.36*	24*	–	–	5.9	–	0.95	0.04	0.15	0.50	0.84	0.14	0.07	0.10	203	2	28	10
Baeckea crassifolia	0.49*	35*	14.7	19.0	5.9	571	0.94	0.04	0.13	0.50	0.68	0.15	0.08	0.03	450	2	30	9
Brachyloma ericoides	0.21	27	–	–	3.9	–	0.63	0.04	0.09	0.29	0.54	0.08	0.04	0.05	59	1	26	8
Monocotyledons																		
Lepidosperma laterale	0.67	38	–	–	2.8	–	0.45	0.03	0.11	0.50	0.26	0.02	0.02	0.23	45	1	24	5

* Cylindrical.

Table 9. Foliar analyses: Australia (Prepared by R.L. Specht and P.W. Rundel).
Vegetation structure: mallee open-scrub.
Alliance: Eucalyptus incrassata – Melaleuca uncinata.
Reference: Specht, R.L., 1981. Mallee ecosystems in Southern Australia. pp. 203–231. In: di Castri, F., Goodall, D.W. and Specht, R.L. (eds). Ecosystems of the World. Vol. 11. Mediterranean-type Shrublands. Elsevier, Amsterdam.

Species	Thickness of lamina (mm)	Dry weight of lamina (mg cm^{-2})	Lignin (%)	Cellulose (%)	Crude protein (%) (= N × 6.25)	Sclerophyll Index*	Mineral nutrient content											
							% oven dry weight								p.p.m. oven dry weight			
							N	P	S	K	Ca	Mg	Na	Cl	Fe	Cu	Mn	Zn
Goolgowi, N.S.W. (33°59′ S, 145°43′ E)																		
Upper stratum																		
Eucalyptus incrassata	0.81	45	13.3	18.0	6.8	460	1.08	0.04	0.10	0.52	0.68	0.18	0.38	0.52	164	6	212	10
Mid stratum																		
Melaleuca uncinata			12.6	25.8	7.5	512	1.20	0.06	0.22	0.60	0.92	0.24	0.32	0.60	130	4	66	12
Ferries-McDonald Conservation Park, S. Aust. (35°13′ S, 139°09′ E)																		
Upper stratum																		
Eucalyptus incrassata			9.5	13.5	4.1	561	0.66	0.03	0.08	0.45	1.14	0.14	0.30	0.47	72	6	66	8
Mid stratum																		
Melaleuca uncinata			13.1	25.9	7.3	534	1.16	0.04	0.27	0.51	0.90	0.24	0.44	0.81	60	2	16	12
Little Desert National Park, Vic. (36°30′ S, 141°45′ E)																		
Upper stratum																		
Eucalyptus incrassata			12.6	19.5	5.1	629	0.81	0.03	0.07	0.37	1.04	0.18	0.42	0.44	60	6	108	6
Mid stratum																		
Melaleuca uncinata			12.8	24.7	7.6	493	1.22	0.04	0.24	0.62	0.66	0.24	0.68	1.24	114	4	24	8

* Sclerophyll = [(lignin + cellulose)/crude protein].

Table 10. Foliar analyses: Australia (Prepared by R.L. Specht and P.W. Rundel).
Vegetation structure: mallee open-scrub.
Alliance: Eucalyptus incrassata – Triodia irritans.
Reference: Specht, R.L., 1981. Mallee ecosystems in southern Australia. pp. 203–231. In: di Castri, F., Goodall, D.W. and Specht, R.L. (eds). Ecosystems of the World. Vol. 11. Mediterranean-type Shrublands. Elsevier, Amsterdam.

Species	Thickness of lamina (mm)	Dry weight of lamina (mg cm^{-2})	Lignin (%)	Cellulose (%)	Crude protein (%) (= N × 6.25)	Sclerophyll Index*	Mineral nutrient content											
							% oven dry weight								p.p.m. oven dry weight			
							N	P	S	K	Ca	Mg	Na	Cl	Fe	Cu	Mn	Zn
Waikeri, S. Aust. (34° 11′ S, 139° 59′ E) Upper stratum																		
Eucalyptus incrassata	0.81	45	11.6	18.3	4.5	664	0.72	0.03	0.09	0.48	1.28	0.18	0.38	0.37	106	6	36	6
Eucalyptus socialis	0.60	36	12.5	17.4	6.4	467	1.02	0.04	0.10	0.64	1.20	0.18	0.38	0.52	110	6	108	10
Ground stratum																		
Triodia irritans	–	–	11.7	37.3	3.5	1,400	0.56	0.02	0.08	0.23	0.28	0.04	0.03	0.12	204	2	26	2

* Sclerophyll Index = [(lignin + cellulose)/crude protein].

Table 11. Foliar analyses: Western Australia.
Vegetation structure: Kwongan.
Location: Tutanning Nature Reserve, Western Australia (collected by A.J.M. Hopkins).
Reference: Brown, J.M. and Hopkins, A.J.M., 1983. The kwongan (sclerophyllous shrublands) of Tutanning Nature Reserve, Western Australia. Aust. J. Ecol. 8: 63–73.

Species	% oven dry weight					Lignin	Cellulose	Ether extractive	Sclerophyll index
	N	P	K	Ca	Mg				
Calytrix flavescens (Myrtaceae)	0.56	0.03	0.74	1.00	0.13	7.1	36.7	ND	1251
Baeckea priessiana (Myrtaceae)	0.69	0.03	0.72	0.78	0.21	15.2	23.8	ND	904
Eucalyptus drummondii (Myrtaceae)	0.71	0.03	0.74	1.48	0.28	14.0	18.0	6.95	721
Eremaea pauciflora (Myrtaceae)	0.71	0.03	0.60	1.88	0.33	11.9	22.2	6.90	768
Melaleuca trichophylla (Myrtaceae)	0.79	0.03	0.63	1.13	0.19	10.1	28.8	ND	808
Leptospermum erubescens (Myrtaceae)	0.84	0.03	0.61	1.21	0.16	7.0	22.3	ND	558
Stirlingia latifolia (Proteaceae)	0.65	0.02	0.97	1.59	0.35	10.1	22.3	2.08	795
Dryandra sessilis (Proteaceae)	0.72	0.03	0.58	0.90	0.19	15.4	32.9	ND	1073
Dryandra cynaroides (Proteaceae)	0.51	0.02	0.52	0.26	0.13	14.4	40.3	1.80	1716
Banksia attenuata (Proteaceae)	0.64	0.03	0.35	0.93	0.29	13.0	41.0	ND	1350
Hakea ryscifolia (Proteaceae)	0.60	0.03	0.55	1.54	0.22	13.5	36.6	ND	1336
Conospermum stoechadis (Proteaceae)	0.64	0.03	0.52	1.30	0.58	13.6	22.6	ND	905
Banksia sphaerocarpa (Proteaceae)	0.77	0.04	0.48	0.35	0.13	15.7	38.0	ND	1116
Petrophile media (Proteaceae)	0.52	0.02	0.75	0.81	0.08	11.9	32.8	1.04	1375
Leucopogon conostephioides (Epacridaceae)	0.64	0.03	0.43	0.92	0.08	13.7	28.7	ND	1085
Daviesia incrassata (Fabaceae)	0.85	0.03	0.80	0.56	0.18	15.8	32.9	0.58	917
Allocasuarina humilis (Casuarinaceae)	0.52	0.03	0.59	1.56	0.19	ND	ND	ND	ND
Mean	0.67	0.03	0.62	1.07	0.22	12.6	30.0		1042

Table 12. Foliar analyses: California.
Vegetation structure: Maritime chaparral.
Location: Monterey County (collected by P.W. Rundel).
Reference: Griffin, J.R., 1978. Maritime chaparral and endemic shrubs of the Monterey Bay Region, California. Madroño 25: 65–81.

Species	% oven dry weight					Sclerophyll Index
	N	P	Lignin	Cellulose	Ether extractive	
Arctostaphylos montereyensis (Ericaceae)	0.86	0.06	7.5	9.4	7.7	314
Arctostaphylos tomentosa (Ericaceae)	0.76	0.07	7.9	15.6	9.6	495
Arctostaphylos glandulosa (Ericaceae)	0.80	0.10	8.6	14.3	12.0	458
Arctostaphylos hookeri (Ericaceae)	0.78	0.09	5.2	12.0	7.0	353
Arctostaphylos pumila (Ericaceae)	0.78	0.07	9.0	12.2	8.6	435
Castanopsis chrysophylla (Fagaceae)	0.86	0.06	12.0	25.0	4.5	688
Vaccinium ovatum (Ericaceae)	0.72	0.06	9.9	16.0	4.6	575
Gaultheria shallon (Ericaceae)	0.80	0.07	10.9	16.7	5.0	552
Ceanothus dentatus (Rhamnaceae)	1.23	0.06	6.1	12.0	5.0	235
Mean	0.84	0.07	8.6	14.8	7.1	456

Table 13. Foliar analyses: California.
Vegetation structure: Chaparral.
Location: Echo Valley.
Reference: Miller, P.C. (ed.) 1981. Resource Use by Chaparral and Matorral. Springer-Verlag, New York.

Species	% oven dry weight								Sclerophyll index	Leaf specific weight (mg cm^{-2})
	N	P	K	Ca	Mg	Lignin	Cellulose	Ether extractives		
Adenostoma fasciculatum (Rosaceae)	0.90	0.17	0.61	0.75	0.27	5.7	19.7	7.76	452	8.0
Ceanothus greggii (Rhamnaceae)	1.30	0.22	0.71	1.95	0.29	6.6	15.0	5.86	266	37.0
Rhus ovata (Anacardiaceae)	1.00	0.23	0.58	0.75	0.22	2.2	14.9	3.68	274	21.0
Arctostaphylos glauca (Ericaceae)	1.00	0.19	0.68	0.66	0.27	9.4	16.8	7.70	419	27.0

Table 14. Foliar analyses: California.
Vegetation structure: Mediterranean-type shrublands.
Location: California, U.S.A. (collected by R.L. Specht, November-December 1979).
Reference: Specht, R.L. and Moll, E.J., 1983. In: Mediterranean-type heathlands and sclerophyllous shrublands of the world: an overview. pp. 41–65. In: Mediterranean-type Ecosystems: The Role of Nutrients. Springer-Verlag, Berlin.

Species	Mineral nutrient content											
	% oven dry weight								p.p.m. oven dry weight			
	N	P	S	K	Ca	Mg	Na	Cl	Fe	Cu	Mn	Zn
Flicker Ridge, near Berkeley, Cal. (Moraga Ridge 90–100 years old)												
Adenostoma fasciculatum	1.04	0.20	0.09	0.61	0.44	0.24	0.01	0.04	128	10	126	60
Arctostaphylos crustacea	0.64	0.11	0.06	0.37	0.44	0.18	0.01	0.01	66	6	48	32
Vaccinium obovalis	0.65	0.05	0.11	0.87	0.44	0.16	0.02	0.12	46	8	60	24
Yosemite Nat. Park, Cal. (Site B4, granite)												
Arctostaphylos viscida	0.80	0.11	0.06	0.63	1.16	0.14	0.01	0.01	150	6	62	56
San Dimas, Cal. (Lower Munroe Canyon Truck Trail, 19 years after burn, on San Gabriel metamorphics)												
Adenostoma fasciculatum	1.08	0.16	0.07	0.73	1.18	0.18	0.01	0.02	94	6	322	50
Arctostaphylos pungens	0.80	0.07	0.06	0.43	0.66	0.12	0.01	0.01	176	8	26	18
Ceanothus crassifolius	1.10	0.10	0.07	0.53	1.44	0.18	0.01	0.02	120	4	144	8
San Dimas, Cal. (Bell Canyon Saddle, 4 years after burn, on San Gabriel metamorphics)												
Adenostoma fasciculatum	1.04	0.11	0.13	0.60	0.72	0.22	0.01	0.02	194	4	102	48
Arctostaphylos pungens	0.66	0.06	0.07	0.46	0.72	0.10	0.01	0.01	116	2	38	26
Ceanothus crassifolius	1.64	0.10	0.18	0.56	0.78	0.28	0.01	0.11	572	6	280	34
Quercus dumosa	1.18	0.09	0.09	0.69	0.82	0.12	0.01	0.04	148	2	430	12
Malibu Canyon, Los Angeles, Cal. (Miocene coarse sandstone)												
Adenostoma fasciculatum	1.12	0.08	0.07	0.40	0.96	0.18	0.02	0.12	200	4	78	34
Ceanothus spinosus	1.76	0.11	0.13	0.72	0.88	0.24	0.01	0.24	166	4	74	32
Cercocarpus betuloides	1.62	0.11	0.09	0.35	2.64	0.26	0.02	0.10	140	6	78	24
Salvia mellifera	2.18	0.28	0.24	2.90	1.30	0.14	0.36	0.74	450	12	54	40
Laguna Beach, Cal. (San Joaquin Hills sandstone)												
Adenostoma fasciculatum var. *obtusifolium*	1.04	0.06	0.08	0.39	0.82	0.22	0.03	0.17	240	6	184	40

Table 15. Foliar analyses: Chile.
Vegetation structure: Coastal matorral.
Location: Pichidangui (collected by P.W. Rundel).
References: Rundel, P.W. 1981. The matorral zone of central Chile. pp. 175–201. In: di Castri, F., Goodall, D.W. and Specht R.L. (eds). Mediterranean-type Shrublands. Elsevier, Amsterdam..
Weisser, P. and Rundel, P.W., 1981. Estudio comparativo del matorral de Pichidangui con el de Los Molles. Ann. Mus. Hist. Nat. Valparaiso 13: 47–57.

Species	% oven dry weight								Sclerophyll index
	N	P	K	Ca	Mg	Lignin	Cellulose	Ether extractive	
Flourensia thurifera (Asteraceae)	1.41	0.19	0.79	0.83	0.50	8.2	17.7	7.1	294
Haplopappus glutinosus (Asteraceae)	1.59	0.17	1.46	0.75	0.18	6.4	11.8	12.4	183
Gochnatia fascicularis (Asteraceae)	1.56	0.08	1.32	1.28	0.53	6.3	14.2	8.9	210
Haplopappus foliosus (Asteraceae)	1.05	0.09	1.32	0.92	0.36	8.3	13.8	9.9	337
Bahia ambrosioides (Asteraceae)	1.08	0.18	0.41	1.43	1.17	10.3	11.5	5.9	323
Mean	1.34	0.14	1.06	1.04	0.55	7.9	13.8	8.8	269

Table 16. Foliar analyses: Chile.
Vegetation: Matorral.
Location: Fundo Santa Laura.
References: Miller, P.C. (ed.) 1981. Resource Use by Chaparral and Matorral. Springer-Verlag, New York.
Rundel, P.W. 1981. The matorral zone of central Chile. pp. 175–201. In: di Castri, F., Goodall D.W. and Specht R.L. (eds). Mediterranean-type Shrublands. Elsevier, Amsterdam.

Species	% oven dry weight								Sclerophyll index	Leaf specific weight ($mg\,cm^{-2}$)
	N	P	K	Ca	Mg	Lignin	Cellulose	Ether extractives		
Quillaja saponaria (Rosaceae)	1.05	0.19	0.68	1.60	0.59	4.9	15.3	1.02	308	21.0
Trevoa trinervis (Rhamnaceae)	2.73	0.20	0.65	3.65	0.26	1.5	5.5	0.22	41	8.0
Colliguaya odorifera (Euphorbiaceae)	1.16	0.12	0.50	1.15	0.31	5.8	8.8	1.96	201	27.0
Satureja gilliesii (Lamiaceae)	1.72	0.26	1.18	1.45	0.34	8.3	1.05	2.02	175	8.0
Cryptocarya alba (Lauraceae)	1.05	0.16	0.80	1.50	0.19	9.3	1.68	2.38	398	20.0
Kageneckia oblonga (Rosaceae)	1.72	0.26	1.00	2.10	0.35	5.4	1.62	3.70	201	21.0
Lithraea caustica (Anacardiaceae)	1.00	0.10	0.76	0.80	0.19	1.00	2.48	1.80	413	26.0

Table 17. Foliar analyses: Chile.
Vegetation structure: Montane matorral.
Location: Cordillera de los Andes (collected by G. Montenegro).
Reference: Rundel, P.W., 1981. The matorral zone of central Chile. pp. 175–201. In: di Castri, F., Goodall, D.W. and Specht, R.L. (eds). Mediterranean-type Shrublands. Elsevier, Amsterdam.

Species	% oven dry weight								Sclerophyll index
	N	P	K	Ca	Mg	Lignin	Cellulose	Ether extractive	
Colliguaya salicifolia (Euphorbiaceae)	1.34	0.16	0.64	1.83	0.40	4.2	14.4	7.96	222
Colliguaya integerrima (Euphorbiaceae)	1.67	0.22	0.74	0.75	0.26	6.9	10.2	2.56	164
Kageneckia angustifolia (Rosaceae)	1.09	0.15	1.08	1.30	0.29	4.9	20.3	5.50	370
Valenzuelia trinervis (Sapindaceae)	1.74	0.19	1.12	1.04	0.34	3.9	17.3	4.0	195
Mean	1.46	0.18	0.90	1.23	0.32	6.2	15.5	5.00	238

Table 18. Foliar analyses: Chile.
Vegetation structure: Hygrophilous forest.
Location: Cordillera de la Costa (collected by G. Montenegro and P.W. Rundel).
Reference: Rundel, P.W., 1981. The matorral zone of central Chile. pp. 175–201. In: di Castri, F., Goodall, D.W. and Specht, R.L. (eds). Mediterranean-type Shrublands. Elsevier, Amsterdam.

Species	% oven dry weight								Sclero-phyll index
	N	P	K	Ca	Mg	Lignin	Cellulose	Ether extractive	
Azara lanceolata (Flacourtiaceae)	1.47	0.13	0.68	0.76	0.36	10.0	11.6	3.30	235
Azara dentata (Flacourtiaceae)	1.80	0.19	1.12	1.00	0.42	8.4	17.8	0.86	233
Persea lingue (Lauraceae)	1.17	0.11	0.59	0.46	0.14	14.1	22.8	4.14	505
Flotovia diacanthoides (Asteraceae)	1.76	0.14	0.48	2.36	0.69	9.9	20.4	1.90	275
Aristotelia macqui (Elaeocarpaceae)	1.96	0.17	0.78	1.95	0.19	3.7	9.5	5.08	108
Peumus boldus (Monimiaceae)	1.92	0.22	1.35	0.51	0.22	7.6	16.1	7.62	198
Crinodendron patagua (Elaeocarpaceae)	1.69	0.15	0.85	1.91	0.30	7.9	13.5	1.16	203
Aextoxicon punctatum (Aextoxicaceae)	1.09	0.26	0.36	1.63	0.28	15.1	32.0	2.84	691
Drimys winteri (Winteraceae)	1.17	0.21	1.20	0.75	0.40	9.2	18.7	5.50	382
Rhaphithamnus spinosus (Verbenaceae)	1.65	0.15	0.68	1.34	0.23	14.6	18.1	1.16	317
Aristotelia chilensis (Elaeocarpaceae)	1.49	0.20	0.76	2.28	0.23	6.1	15.6	9.54	233
Mean	1.56	0.18	0.80	1.36	0.31	9.7	17.8	3.92	307

Table 19. Foliar analyses: France.
Vegetation structure: Maquis.
Location: Marseille (collected by C. Roux).

Species	% oven dry weight								Sclero-phyll index	Leaf specific weight (mg cm^{-2})
	N	P	K	Ca	Mg	Lignin	Cellulose	Ether extractives		
Arbutus unedo (Ericaceae)	1.23	0.14	0.62	1.73	0.19	6.7	10.9	6.6	229	19.0
Cistus albidus (Cistaceae)	1.30	0.13	0.73	0.86	0.18	10.0	18.4	5.5	350	25.7
Phillyrea angustifolia (Oleaceae)	1.50	0.11	0.64	0.92	0.10	8.3	14.4	7.6	242	19.8
Pistacia lentiscus (Anacardiaceae)	1.56	0.13	0.94	1.05	0.16	3.7	15.5	4.5	197	22.3
Rosmarinus officinalis (Lamiaceae)	1.23	0.11	1.28	1.48	0.25	6.0	19.2	15.1	328	ND
Juniperus phoenicea (Juniperaceae)	1.05	0.11	0.50	1.77	0.09	6.1	19.8	10.3	394	ND
Lonicera implexa (Caprifoliaceae)	2.03	0.19	1.17	2.06	0.26	1.5	9.3	3.9	85	12.5
Rhamnus alaternus (Rhamnaceae)	2.16	0.12	0.91	2.26	0.15	2.5	10.2	2.7	94	22.3
Quercus coccifera (Fagaceae)	1.14	0.19	1.12	1.04	0.14	10.3	18.1	4.0	261	19.2
Bupleurum fruticosum (Apiaceae)	1.50	0.14	1.36	2.33	0.26	3.7	11.9	4.9	166	15.8
Mean	1.53	0.15	0.93	1.55	0.36	5.9	14.8	6.5	203	19.6

Table 20. Foliar analyses: France.
Vegetation structure: Maquis arboré on ancient siliceous rocks.
Location: Cauro, Corsica (collected by R.L. Specht and L. Amandier, May 1983).
Reference: Brun, B. & L., Conrad, M. and Gamisans J., 1975. La Nature en France: Corse. Horizons de France, Strasbourg, France.

Species	N (%)	P (%)	K (%)	Ca (%)	Mg (%)	C (%)	Si (%)	S (%)	Fe (p.p.m.)	Mn (p.p.m.)	Leaf specific weight (mg cm^{-2})
Trees (3–5 m)											
Arbutus unedo (Ericaceae)	0.92	0.108	0.57	1.27	0.12	51.1	3.28	0.05	435	57	17.5
Quercus ilex (Fagaceae)	1.26	0.082	0.54	0.80	0.13	52.3	2.72	0.05	309	907	18.2
Shrubs (<1 m)											
Phillyrea angustifolia (Oleaceae)	1.00	0.078	0.71	0.92	0.11	54.9	2.74	0.09	247	53	20.4
Erica arborea (Ericaceae)	0.92	0.086	0.48	0.44	0.23	53.8	3.14	0.05	317	164	18.5
Cistus monspeliensis (Cistaceae)	1.30	–	–	–	–	47.5	3.46	–	–	–	25.6
Phillyrea latifolia (Oleaceae)	0.86	0.071	0.43	1.37	0.11	50.8	2.24	0.08	253	51	18.5
Lavandula stoechas (Lamiaceae)	1.35	–	–	–	–	49.6	–	–	–	–	26.5
Pistacia lentiscus (Anacardiaceae)	0.86	0.092	0.70	1.18	0.17	54.2	2.66	–	152	41	21.7
Cistus salviifolius (Cistaceae)	1.45	0.229	2.06	1.52	0.37	49.4	3.70	0.10	571	194	31.2

Table 21. Foliar analyses: France.
Vegetation structure: Mediterranean-type shrublands, maquis arboré, oak forest.
Location: Herault and Provence, France.
Reference: Specht, R.L. and Moll, E.J., 1983. Mediterranean-type heathlands and sclerophyllous shrublands of the world: an overview.
pp. 41–65. In: Mediterranean-Type Ecosystems: The Role of Nutrients. Springer-Verlag, Berlin.

Species	Mineral nutrient content											
	% oven dry weight								p.p.m. oven dry weight			
	N	P	S	K	Ca	Mg	Na	Cl	Fe	Cu	Mn	Zn
Mt. Auriol (Beziers) – Collected L. Traubaud and R.L. Specht 1979												
Calluna vulgaris	0.93	0.06	0.11	0.32	0.42	0.10	0.06	–	400	–	1000	–
Cistus crispus	1.36	0.10	0.11	0.53	0.90	0.18	0.03	–	300	–	200	–
Cistus monspeliensis	1.41	0.10	0.13	0.57	0.71	0.19	0.06	–	300	–	100	–
Erica arborea	1.05	0.05	0.11	0.42	0.33	0.15	0.04	–	200	–	100	–
Erica cinerea	0.89	0.05	0.16	0.40	0.33	0.19	0.11	–	200	–	200	–
Lavandula stoechas	1.04	0.07	0.12	1.11	0.61	0.20	0.08	–	–	–	–	–
St. Chinian – Collected L. Trabaud and R.L. Specht 1979												
Calicotome spinosa	1.65	0.10	0.07	0.53	0.25	0.17	0.01	–	200	–	200	–
Cistus ladaniferus	1.21	0.11	0.07	0.55	0.63	0.21	0.02	–	300	–	1000	–
Cistus monspeliensis	1.45	0.13	0.05	0.55	0.70	0.24	0.03	–	400	–	200	–
Erica arborea	1.02	0.06	0.11	0.53	0.33	0.14	0.05	–	300	–	200	–
Lavandula stoechas	1.39	0.10	0.10	1.19	0.67	0.26	0.07	–	300	–	200	–
Quercus coccifera	0.90	0.07	0.06	0.44	0.89	0.26	0.01	–	200	–	700	–
Ulex parviflorus	1.41	0.06	0.04	0.71	0.22	0.14	0.03	–	200	–	100	–
Vidauban – Collected R. Loisel & L. Trabaud 1980 Tree stratum												
Quercus suber	1.53	0.08	–	0.65	0.86	0.20	0.05	–	–	–	–	–
Arbutus unedo	1.25	0.09	–	0.77	1.15	0.20	0.01	–	–	–	–	–
Erica arborea	1.28	0.04	–	0.56	0.24	0.13	0.26	–	–	–	–	–
Shrub stratum –												
Calluna vulgaris	1.07	0.06	–	0.50	0.39	0.14	0.11	–	–	–	–	–
Cistus albidus	1.36	0.16	–	0.75	1.42	0.14	0.06	–	–	–	–	–
Cistus monspeliensis	1.49	0.21	–	0.55	1.02	0.19	0.17	–	–	–	–	–
Cistus salviifolius	1.46	0.19	–	0.74	1.59	0.21	0.51	–	–	–	–	–
Erica scoparia	1.58	0.07	–	0.36	0.41	0.11	0.33	–	–	–	–	–
St. Gély du Fesc – Collected M. Rapp & P. Loissaint 1968 *Quercetum cocciferae* (garrigue)												
Quercus coccifera	1.35	0.08	–	0.81	1.15	0.14	0.04	–	60	10	260	30
Quercetum ilicis (oak forest) Tree stratum												
Quercus ilex 1.	1.43	0.15	0.13	0.62	0.54	0.14	0.02	–	150	–	290	–
Quercus ilex 2.	1.17	0.16	0.13	0.61	1.30	0.12	0.02	–	210	–	440	–
Shrub stratum												
Acer monspessulanum	2.18	0.10	–	1.28	2.09	0.33	0.02	–	220	–	670	–
Arbutus unedo	1.58	0.08	–	1.26	2.39	0.21	0.02	–	100	–	40	–
Juniperus oxycedrus	1.53	0.07	–	0.59	2.60	0.13	0.02	–	–	–	70	–
Phillyrea angustifolia	1.59	0.06	–	0.85	1.85	0.14	0.02	–	170	–	60	–
Phillyrea media	1.57	0.06	–	1.03	2.22	0.19	0.02	–	150	–	10	–
Rhamnus alaternus	3.11	0.09	–	1.88	3.74	0.23	0.02	–	220	–	50	–
Viburnum tinus	1.20	0.09	–	1.31	2.43	0.37	0.02	–	200	–	130	–

Table 22. Foliar analyses: Greece.
Vegetation structure: maquis and phrygana.
Location: Greece (mainland and islands).
(1) Astakos   38° 32′ N, 21° 04′ E
(2) Attica   38° 00′ N, 23° 50′ E
(3) Kandhila   37° 46′ N, 22° 22′ E
(4) Karistos   38° 01′ N, 24° 25′ E
(5) Monemvasia   36° 41′ N, 23° 03′ E
(6) Paros   37° 04′ N, 25° 06′ E
(7) Volos   39° 22′ N, 22° 57′ E
Reference: Margaris, N.S., Adamandiadou, S., Siafaca, L. and Diamantopoulos, J., 1984. Nitrogen and phosphorus content in plant species of Mediterranean ecosystems in Greece. Vegetatio 55: 29–35.

Species	N (%)	P (%)	Sample location
Evergreen sclerophyll leaves (maquis)			
Arbutus unedo (Ericaceae)	1.02	0.06	(2)
Ceratonia siliqua (Fabaceae)	1.52	0.07	(2)
Cupressus sempervirens (Cupressaceae)	0.79	0.07	(2)
Juniperus oxycedrus (Cupressaceae)	0.81	0.07	(3)
Juniperus phoenicea (Cupressaceae)	0.66	0.05	(4)
Laurus nobilis (Lauraceae)	1.51	0.10	(7)
Myrtus communis (Myrtaceae)	0.95	0.05	(7)
Nerium oleander (Apocynaceae)	1.20	0.06	(4)
Olea europaea (Oleaceae)	1.31	0.07	(2)
Phillyrea latifolia (Oleaceae)	1.31	0.07	(2)
Pinus halepensis (Pinaceae)	0.94	0.06	(2)
Pistacia lentiscus (Anacardiaceae)	0.92	0.06	(2)
Quercus coccifera (Fagaceae)	1.15	0.06	(2)
Smilax aspera (Smilacaceae)	0.79	0.04	(4)
Mean (14 species)	1.06	0.06	
Deciduous leaves (maquis)			
Cercis siliquastrum (Fabaceae)	2.31	0.18	(2)
Crataegus laevigata (Rosaceae)	2.53	0.20	(3)
Ficus carica (Moraceae)	3.88	0.22	(2)
Platanus orientalis (Platanaceae)	2.01	0.12	(2)
Punica granatum (Punicaceae)	1.10	0.06	(7)
Pyrus amygdaliformis (Rosaceae)	2.00	0.12	(2)
Rosa canina (Rosaceae)	2.86	0.27	(3)
Rubus canescens (Rosaceae)	2.75	0.16	(2)
Mean (8 species)	2.43	0.17	
Seasonally dimorphic leaves (phrygana)			
Acinos alpinus (Lamiaceae)	1.61	0.08	(7)
Anthyllis hermanniae (Fabaceae)	2.05	0.13	(2)
Ballota acetabulosa (Lamiaceae)	2.37	0.22	(2)
Calicotome villosa (Fabaceae)	1.89	0.09	(2)
Cichorium spinosum (Asteraceae)	1.87	0.22	(7)
Cistus incanus ssp. *creticus* (Cistaceae)	1.88	0.12	(2)
Cistus monspeliensis (Cistaceae)	1.41	0.13	(2)
Cistus parviflorus (Cistaceae)	1.51	0.08	(2)
Cistus salviifolius (Cistaceae)	1.24	0.11	(2)
Echium vulgare (Boraginaceae)	2.36	0.17	(2)
Euphorbia acanthothamnos (Euphorbiaceae)	2.15	0.14	(2)
Euphorbia characias (Euphorbiaceae)	2.03	0.11	(2)
Genista acanthoclada (Fabaceae)	1.71	0.17	(2)
Helianthemum nummularium (Cistaceae)	2.05	0.23	(2)
Helichrysum stoechas (Asteraceae)	1.51	0.16	(2)
Lavandula stoechas (Lamiaceae)	1.49	0.09	(4)
Medicago arborea (Fabaceae)	3.42	0.13	(2)
Micromeria nervosa (Lamiaceae)	1.62	0.09	(2)
Onosma frutescens (Boraginaceae)	1.81	0.14	(2)
Phagnalon graecum (Asteraceae)	1.61	0.13	(2)
Phlomis fruticosa (Lamiaceae)	1.82	0.16	(2)
Prasium majus (Lamiaceae)	2.28	0.11	(2)
Psoralea bituminosa (Fabaceae)	3.69	0.18	(2)

Table 22. (Continued).

Species	N (%)	P (%)	Sample location
Rosmarinus officinalis (Lamiaceae)	1.64	0.09	(4)
Sarcopoterium spinosum (Rosaceae)	2.02	0.10	(2)
Teucrium polium (Lamiaceae)	1.51	0.12	(2)
Thymelaea hirsuta (Thymelaeaceae)	1.72	0.10	(2)
Thymelaea tartonraira (Thymelaeaceae)	1.74	0.09	(2)
Thymus capitatus (Lamiaceae)	1.62	0.07	(2)
Mean (29 species)	1.92	0.13	
Herbaceous phrygana species			
Alkanna tinctoria (Boraginaceae)	1.12	0.10	(2)
Arisarum vulgare (Araceae)	2.68	0.15	(2)
Asphodelus aestivus (Liliaceae)	2.03	0.14	(2)
Briza maxima (Poaceae)	1.02	0.09	(2)
Carduus pycnocephalus (Asteraceae)	0.82	0.14	(2)
Convolvulus althaeoides (Convolvulaceae)	1.53	0.08	(2)
Ecballium elaterium (Cucurbitaceae)	3.78	0.25	(2)
Euphorbia helioscopia (Euphorbiaceae)	1.66	0.14	(2)
Galium aparine (Rubiaceae)	1.72	0.14	(2)
Lagurus ovatus (Poaceae)	1.19	0.09	(2)
Lavatera cretica (Malvaceae)	2.67	0.13	(2)
Malva silvestris (Malvaceae)	2.89	0.16	(2)
Medicago orbicularis (Fabaceae)	2.61	0.12	(2)
Muscari commutatum (Liliaceae)	1.02	0.11	(2)
Muscari comosum (Liliaceae)	1.45	0.10	(2)
Onobrychis ebenoides (Fabaceae)	2.37	0.10	(2)
Senecio vulgaris (Asteraceae)	0.87	0.19	(2)
Thapsia garganica (Apiaceae)	1.94	0.12	(2)
Tordylium apulum (Apiaceae)	1.32	0.10	(2)
Tragopogon sp. (Asteraceae)	1.96	0.14	(2)
Trifolium stellatum (Fabaceae)	1.99	0.10	(2)
Umbilicus rupestris (Crassulaceae)	1.67	0.10	(3)
Urginea maritima (Liliaceae)	2.50	0.21	(2)
Verbascum undulatum (Scrophulariaceae)	1.63	0.16	(2)
Vicia villosa (Fabaceae)	2.96	0.10	(2)
Mean (25 species)	1.90	0.13	

Table 23. Foliar analyses: South Africa.
Vegetation structure: Mountain fynbos.
Location: Kogelberg, Cape Province (collected by P.W. Rundel, October 1980).
Reference: Boucher, C. 1978. Cape Hangklip area. II. The vegetation. Bothalia 12: 455–497.

Species	% oven dry weight								Sclero-phyll index	Leaf specific weight (mg cm^{-2})
	N	P	K	Ca	Mg	Lignin	Cellulose	Ether extractives		
Leptocarpus hyalinus (Restionaceae)										
stem and leaves	0.46	0.01	0.23	0.10	0.05	10.9	28.8	0.2	1885	–
Restio egregius (Restionaceae)										
stems	0.32	0.004	0.36	0.11	0.09	8.3	29.4	0.44	1885	–
Elegia parviflora (Restionaceae)										
stems and leaves	0.49	0.03	0.71	0.28	0.24	5.3	26.2	1.48	–	9.6
Thamnochortus dichotomus (Restionaceae)										
leaves	0.80	0.03	0.19	0.15	0.06	4.9	24.8	0.66	–	22.0
stems	0.49	0.02	0.27	0.18	0.09	5.3	28.6	0.64	1107	24.8
Hypodiscus cristatus (Restionaceae)										
stems and leaves	0.46	0.02	0.35	0.13	0.07	3.4	29.2	0.78	1134	20.0
Staberoha cernua (Restionaceae)										
stems and leaves	0.52	0.02	0.53	0.23	0.03	5.6	24.4	2.46	923	19.7
Berzelia lanuginosa (Bruniaceae)	1.16	0.03	0.37	1.33	0.17	6.2	23.0	6.92	403	17.4
Erica corifolia (Ericaceae)	0.90	0.03	0.32	0.69	0.23	2.1	19.9	5.06	391	25.5
Erica pulchella (Ericaceae)	0.76	0.03	0.34	0.43	0.09	3.2	18.9	6.24	465	36.0
Sympieza articulata (Ericaceae)										
leaf and fine branch	0.34	0.011	0.14	0.11	0.05	16.4	46.9	1.74	–	–
Aulax cneorifolia (Proteaceae)										
male	0.79	0.05	0.52	0.49	0.22	6.8	24.3	1.24	630	22.1
female	0.71	0.04	0.40	0.74	0.29	9.0	24.7	1.14	759	26.6
Leucadendron laureolum (Proteaceae)										
male	0.30	0.03	0.15	0.42	0.23	7.5	20.6	–	1499	30.7
female	0.33	0.03	0.19	0.73	0.34	10.9	22.3	–	1610	33.2

Table 24. Foliar analyses: South Africa.
Vegetation: Mountain fynbos.
Location: Jonkershoek State Forest, Cape Province (collected by P.W. Rundel, October 1980).
Reference: Kruger, F.J., 1979. South African heathlands. pp. 19-80. In: Specht, R.L. (ed). Heathlands and Related Shrublands. Descriptive Studies. Elsevier, Amsterdam.

Species	% oven dry weight								Sclero-phyll index	Leaf specific weight (mg cm^{-2})
	N	P	K	Ca	Mg	Lignin	Cellulose	Ether extractives		
Brabejum stellatifolium (Proteaceae)	0.92	0.05	0.25	0.54	0.24	6.0	29.1	2.12		24.8
Leucadendron salignum (Proteaceae)										
male (3 years since fire)	0.31	0.03	0.19	0.81	0.16	9.0	20.7	2.78	1532	21.9
female (3 years since fire)	0.36	0.03	0.18	0.78	0.18	13.6	24.8	3.02	1707	23.8
male (22 years since fire)	0.40	0.04	0.23	0.57	0.23	11.8	22.0	2.90	1352	25.2
female (22 years since fire)	0.39	0.04	0.19	0.43	0.16	13.3	23.3	1.68	1502	27.1
Protea nitida (Proteaceae)										
(38 years since fire)	0.68	0.04								
(3 years since fire)	0.75	0.06								
Protea neriifolia (Proteaceae)										
(38 years since fire)	0.63	0.03								
(22 years since fire)	0.79	0.05								
Leucospermum truncatulum (Proteaceae)	0.85	0.06								

Vegetation, nutrition and climate – data-tables

(3) Species richness

Co-ordinator: W.E. Westman

Contributors

M.D. Fox, R.L. Specht (Southeastern and South Australia)
D.T. Bell, A.J.M. Hopkins, L.E. Watson (Southwestern Australia)
Z. Naveh, A.R. Troeger, W.E. Westman, R.H. Whittaker (California and Mexico)
G. Montenegro, S. Teillier (Chile)
L. Amandier, C. Martinez, R.L. Specht (Corsica)
J. Lepart, Z. Naveh, L. Olsvig-Whittaker, L. Trabaud (France)
E. Ecomonidou (Greece)
Z. Naveh, L. Olsvig-Whittaker, R.H. Whittaker (Israel)
W. Bond, R.M. Cowling (South Africa)
T. Uslu (Turkey)

Contents

Australia – Southeastern and South (Tables 1–3)	82
Australia – Southwestern (Tables 4–5)	84
California and Mexico (Table 6)	86
Chile – Central (Table 7)	87
Mediterranean Basin	
– Corsica, France, Greece, Israel, Turkey (Table 8)	88
– France (Tables 9–10)	89
South Africa – Cape Province (Tables 11–12)	90

R.L. Specht (ed.) Mediterranean-type Ecosystems. ISBN 90-6193-652-7.
© Kluwer Academic Publishers.

Table 1. Species richness of plant communities in southwestern New South Wales, Australia (compiled by M.D. Fox).

Locality	Vegetation structure	Age since fire (yrs)*	Total vascular species						
			$100 m^2$	$200 m^2$	$400 m^2$	$600 m^2$	$800 m^2$	$1000 m^2$	Total**
Kinchega N.P.	1. Chenopod shrubland	?30 83		25	33	36	39	40	86(6)
(32° 30′ S 142° 10′ E)	2. *Casuarina* woodland	?30 83		20	37	41	42	42	82(6)
	3. *Eucalyptus* woodland	?30 83		21	29	34	36	39	91(5)
Yathong N.R.	1. *Triodia* mallee	8 85	11	15	20	25	25	26	
(32° 35′ S 145° 30′ E)	2. Mallee	8 85	7	9	10	15	17	17	
Matakana	1. *Triodia* mallee	9 86						14	
(33° 00′ S 145° 55′ E)									
'Nanya'	1. *Triodia* mallee	84	6	7	11	14	17	18	
(33° 15′ S 141° 18′ E)									
'Pan Ban Lake'	1. Low shrubland	83		20	21	21	23	23	
(33° 18′ S 143° 10′ E)	*Chenopodium nitrariaceum*								
'No Mans Land'	1. *Triodia* mallee	83		28	30	35	39	39	
(33° 40′ S 142° 35′ E)									
Mungo N.P.	1. Chenopod shrubland	?30		18	24	27	28	28	64(5)
(33° 45′ S 143° 00′ E)	*Maireana pyramidata*								
	2. Chenopod shrubland	?30 83		13	22	24	27	28	62(5)
	Maireana sedifolia								
	3. *Triodia* mallee	?10 83		22	32	34	35	40	81(5)
	4. *Casuarina* woodland	?10 84	10	12	15	17	18	21	27(2)
'Narweena'	1. *Triodia* mallee	84	10	13	23	23	25	25	
(33° 50′ S 141° 08′ E)									
'Tareena'	1. Mallee	84	11	12	16	19	22	24	
(33° 53′ S 141° 05′ E)									
Wentworth	1. *Atriplex vesicaria*								
(34° 05′ S 141° 50′ E)	*Pachycornia triandra*	?30 84	15	18	20	22	24	24	
Pulletop N.R.	1. Mallee	?30 83		41	47	52	55	57	83(4)
(34° 10′ S 146° 05′ E)									
Mallee Cliff N.P.	1. Mallee	83		16	19	24	27	30	
(34° 20′ S 142° 35′ E)	2. *Triodia* mallee	83		19	31	38	40	41	
	3. *Casuarina* woodland	83		16	18	20	21	22	
Hay Plain									
(34° 21′ S 144° 45′ E)	1. *Atriplex vesicaria*	84	12	21	24	28	30	30	
(34° 25′ S 144° 25′ E)	1. *Atriplex vesicaria*	84	11	13	15	19	21	22	
(34° 25′ S 144° 50′ E)	1. *Atriplex vesicaria*		10	13	17	21	23	23	
(34° 40′ S 144° 37′ E)	1. *Atriplex vesicaria*		12	15	17	19	21	22	
(34° 42′ S 144° 05′ E)	1. Low shrubland	84	18	19	19	22	25	28	
	Maireana aphylla								
Balranald	1. Chenopod shrubland	?30 83		27	34	40	44	49	84(6)
(34° 30′ S 143° 45′ E)	2. Mallee	?10 83		16	21	25	27	32	50(6)
	3. *Casuarina* woodland	?30 83		36	44	50	51	54	87(6)
	4. *Eucalyptus* woodland	?30 83		42	50	53	57	57	104(5)
W. of Balranald	1. Mallee	84	10	12	18	18	25	26	
(34° 36′ S 143° 21′ E)									
'Zara'	1. *Callitris* open-forest	?30 83		34	39	41	42	48	80(3)
(35° 10′ S 144° 40′ E)									
Deniliquin	1. *Eucalyptus* open-forest	? 5		18	23	27	28	28	59(3)
(35° 30′ S 145° 00′ E)	2. Chenopod shrubland	?30 83		21	23	24	25	26	56(3)
	3. *Callitris* open-forest	?30 83		19	22	24	27	29	66(4)
Jerilderie	1. *Eucalyptus* woodland	?30 83		33	41	43	46	47	81(4)
(35° 30′ S 145° 40′ E)									

* The year given is that in which the site was sampled.
** The 'Total' is the cumulative richness recorded at the site over a number of years, number in parentheses.

Table 2. Species richness of plant communities in southeastern Australia (Specht and Specht in press).

Locality	Vegetation structure	Age since fire (yrs)	Total vascular species			
			$1 m^2$	$10 m^2$	$100 m^2$	$1000 m^2$
Dark Island Soak, S.A. (36° 02' S, 140° 31' E)	1. Dry heathland	10	15	28	40	53
	2. Dry heathland	30	13	24	34	45
Wilson's Prom. Vic. (39° 08' S, 145° 25' E)	1. Dry heathland	10	18	29	39	50
	2. Dry heathland	32	11	23	35	48
	3. Wet/dry heathland	10	15	25	35	46
Mt Lofty, S.A. (34° 59' S, 138° 43' E)	Heathy open-forest	?	14	23	33	42
Para Wirra R.P., S.A. (34° 43' S, 138° 50' E)	Heathy open-forest	?	5	15	25	35
Brisbane Ranges, Vic. (37° 55' S, 144° 20' E)	Heathy open-forest	?	14	26	38	50
Mt Dandenong, Vic. (37° 50' S, 145° 21' E)	Heathy open-forest	?	13	21	30	39
Wilson's Prom. Vic. (39° 08' S, 145° 25' E)	Heathy open-forest	32	8	20	33	45
Keith, S.A. (36° 06' S, 140° 31' E)	Savanna woodland	?	7	23	39	55
Salter Springs, S.A. (34° 11' S, 138° 38' E)	Savanna woodland	?	11	21	31	42
Gellibrand Hill, Vic. (37° 40' S, 144° 48' E)	Savanna woodland	?	9	24	38	52
Dark Island Soak, S.A. (36° 02' S, 140° 31' E)	Mallee-broombush	30	8	19	29	39
Warrenben C.P., S.A. (35° 08' S, 137° 02' E)	Mallee-broombush	30?	9	17	25	33
Rankins Springs, N.S.W.* (33° 51' S, 146° 16' E)	Mallee-broombush	50?	5	16	26	37

* Whittaker, R.H., Niering, W.A. and Crisp, M.D. 1979. Structure, pattern and diversity of a mallee community in New South Wales. Vegetatio 39: 65–76.

Reference

Specht, R.L. and Specht, A., in press. Species richness of sclerophyll (heathy) plant communities in Australia – The influence of overstorey cover. Aust. J. Bot.

Table 3. Species richness of heathland communities in post-fire succession at Dark Island Soak (36° 02' S, 140° 31' E), near Keith, South Australia (Specht et al. 1958).

Age after fire (yrs)	1	2.5	9	15	25	50+
Number of spp. ($200 m^2$)	38	36	27	25	20	10
Foliage projective cover (%)						
Upper stratum	–	1.9	7.9	2.6	11.3	?
Mid stratum	–	32.5	54.8	49.1	53.8	?
Ground stratum	–	26.8	29.5	24.8	17.5	?
Bare ground	–	45.9	29.9	31.5	32.3	?

Reference

Specht, R.L., Rayson, P. and Jackman, M.E., 1958. Dark Island heath (Ninety-Mile Plain, South Australia). VI. Pyric succession: Changes in composition, coverage, dry weight, and mineral nutrient status. Aust. J. Bot. 6: 59–88.

Table 4. Species richness of plant communities in southwest Western Australia. (Compiled by D.T. Bell and L.E. Watson.)

Locality	Vegetation structure	Age since fire (yrs)	Total vascular species			
			$1\,m^2$	$10\,m^2$	$100\,m^2$	$1000\,m^2$
Jarrahdale	1. jarrah forest	?	8	19	34	47
(32°21′ S, 116°03′ E)	2. jarrah forest	?	13	26	39	55
Ravensthorpe (60 km NE)	heathland	?	7	24	42	48
(33°32′ S, 120°33′ E)						
Mt Ney	mallee-heath	?	9	17	25	50
(33°24′ S, 122°27′ E)						
Scadden-Grasspatch	1. Salt Complex					
(33°19′ S, 121°51′ E)	(*Stipa* dune)	?	12	17	25	29
	2. Salt Complex					
	(*Atriplex* dune)	?	6	13	16	21
	3. Salt Complex					
	(Samphire flat)	?	5	7	13	17
	4. Shrubland					
	(*Melaleuca* paperbark)	?	6	16	19	24
	5. Shrubland					
	(Sandy *Melaleuca* thicket)	?	6	13	22	33
	6. Tree mallee					
	(*Eucalyptus occidentalis* swamp)	?	2	6	9	17
	7. Shrub mallee	?	7	15	33	45
	8. Tree mallee	?	6	14	22	33
	9. Woodland	?	6	9	14	24
	10. Mixed woodland/Tree mallee	?	5	13	20	29
Star Swamp, Watermans	1. *Melaleuca* (paperbark)	?	4	9	15	?
(33°51′ S, 115°45′ E)	shrubland					
	2. Tuart woodland	?	9	25	45	?
	3. *Banksia* low woodland	before fire	9	23	49	?
		2	7	20	35	?
Tutanning N.R.	1. Kwongan (on laterite)	29	12	34	56	78
(32°31′ S, 117°23′ E)		41	?	21	38	61
	2. Kwongan (on sand)	21	?	?	40	86
		41	9	32	55	78
	3. Kwongan (on duplex soil)	15	?	31	58	95
		16	27	51	77	103
		41	10	38	66	94
		49	22	45	69	93

Table 5. Species richness of heathlands in southwest Western Australia (George *et al.*, 1979).

Locality	Vegetation structure	Age since fire (yrs)	Total vascular species			
			$1\,m^2$	$10\,m^2$	$100\,m^2$	$1000\,m^2$
Corackerup N.R. (34° 16′ S, 118° 42′ E)	Heathland (inland)	?	9	20	32	43
East Yuna N.R. (28° 20′ S, 115° 00′ E)	Heathland (inland)	?	1	3	23	43
Geraldton (east) (28° 46′ S, 114° 37′ E)	Heathland (inland)	?	8	28	48	?
Kalbarri N.P. (27° 52′ S, 114° 30′ E)	1. Heathland (inland)	6	9	23	37	51
	2. Heathland (inland)	7	10	24	38	52
Lime Lake (33° 25′ S, 117° 23′ E)	Heathland (inland)	20	11	22	34	45
Tutanning N.R. (32° 31′ S, 117° 23′ E)	Heathland (inland)	?	14	25	36	47
Two People Bay N.R. (34° 58′ S, 118° 11′ E)	1. Heathland (inland)	15	2	17	33	49
	2. Heathland (inland)	10	9	23	38	52
	3. Heathland (inland)	6–10	7	20	33	47
Augusta District (34° 19′ S, 115° 09′ E)	Heathland (coastal)	5	9	24	40	55
Nornalup N.P. (35° 00′ S, 116° 44′ E)	Heathland (coastal)	6	5	22	39	56
Torndirrup N.P. (35° 07′ S, 117° 53′ E)	Heathland (coastal)	10	18	32	46	60
Badgingarra N.P. (30° 10′ S, 115° 25′ E)	Heathland (on laterite)	6–8	7	33	59	85
Cape Riche (34° 36′ S, 118° 46′ E)	Heathland (on laterite)	?	2	28	54	80
Cheyne Bay (34° 34′ S, 118° 46′ E)	Heathland (on laterite)	?	18	40	62	84
Corackerup N.R. (34° 16′ S, 118° 42′ E)	Heathland (on laterite)	?	29	41	53	65
Eneabba N.R. (29° 05′ S, 115° 18′ E)	1. Heathland (on laterite)	?	24	48	77	99
	2. Heathland (on laterite)	8	8	35	61	87
	3. Heathland (on laterite)	?	35	56	77	98
Geraldton (east) (28° 46′ S, 114° 37′ E)	Heathland (on laterite)	10	4	22	48	74
Kalbarri N.P. (27° 52′ S, 114° 30′ E)	Heathland (on laterite)	7	8	31	54	77
Mt Lesueur (30° 11′ S, 115° 12′ E)	Heathland (on laterite)	?	48	55	62	69
Stirling Range N.P. (34° 20′ S, 118° 00′ E)	1. Heathland (on laterite)	20	14	35	56	77
	2. Heathland (on laterite)	6	1	32	63	94

Reference

George, A.S., Hopkins, A.J.M. and Marchant, N.G., 1979. The heathlands of Western Australia. In: Ecosystems of the World. 9A. Heathlands and Related Shrublands (R.L. Specht ed.) pp. 211–230. Elsevier, Amsterdam.

Table 6. Species richness of plant communities in California and northwest Mexico (Naveh and Whittaker 1979, Troeger 1983, Westman, 1981, 1983).

Locality	Vegetation structure	Age since fire (yrs)	Total vascular species			
			$1\,m^2$	$10\,m^2$	$100\,m^2$	$1000\,m^2$
Santa Barbara (34°25′N, 119°41′W)	Oak woodland (6 comm.)		11	23	35	47 ± 6
Carmel (36°34′N, 121°56′W)	1. Oak woodland (protected) (6 comm.)		14	28	43	58 ± 10
	2. Oak woodland (grazed) (6 comm.)		16	32	48	64 ± 15
San Jacinto Mts (33°45′N, 116°45′W)	Chaparral (*Adenostoma*) (5 comm.)		7	13	19	24 ± 8
Santa Ynez Mts (34°30′N, 120°00′W)	Chaparral (*Ceanothus*) (5 comm.)		4	13	22	32 ± 3
Santa Rosa Mts (33°30′N, 116°15′W)	Coastal sage scrub (Riversidian)		1	5	12	24 ?
Mt Diablo-Santa Barbara (36–38°N, 121–122°W)	Coastal sage scrub (Diablan) - 13 comm.	7+	?	?	?	30 ± 3.4* (5+11+13+1)**
Santa Barbara-Los Angeles (33–35°N, 118–121°W)	Coastal sage scrub (Venturan) - 34 comm.	7+	?	?	?	19 ± 1.9* (5+7+6+1)**
	1 comm.	9	17	31	40	47
	1 comm.	23	6	11	17	27
Los Angeles-San Bernardino-San Diego (32–34°30′N, 116–119°W)	Coastal sage scrub (Riversidian) - 19 comm.	7+	?	?	?	29 ± 2.0* (7+6+14+2)**
Orange Co.-San Diego-Baja Cal. (31–34°N, 116°30′–118°30′W)	Coastal sage scrub (Diegan) - 12 comm.	7+	?	?	?	26 ± 3.0* (8+7+8+3)**
Ensenada-San Quintin (Baja Cal.) (30°–31°30′N, 115°30′–117°W)	Coastal succulent scrub (Martirian) - 8 comm.	7+	?	?	?	41 ± 4.3* (8+6+22+5)**
El Rosario (Baja Cal.) (30°02′N, 115°46′W)	Coastal succulent scrub (Vizcainan) - 7 comm.	7+	?	?	?	33 ± 2.3* (6+5+15+7)**

* Vascular spp. recorded in 625 m² quadrats ± s.e.
** (woody spp. + herbaceous perennials + annuals + vines and succulents).

References

Naveh, Z. and Whittaker, R.H., 1979. Structural and floristic diversity of shrublands and woodlands in northern Israel and other Mediterranean areas. Vegetatio 41: 171–190.

Troeger, A.R., 1983. Methods of analysis of microscale pattern and diversity in Californian shrublands. M.A. Thesis, Dept. Geography, University of California, Los Angeles.

Westman, W.E., 1981. Diversity relations and succession in Californian coastal sage scrub. Ecology 62: 170–184.

Westman, W.E., 1983. Xeric Mediterranean-type shrubland associations of Alta and Baja California and the community/continuum debate. Vegetatio 52: 3–19.

Table 7. Species richness of plant communities in the mediterranean-climate region of Chile. (Compiled by G. Montenegro and S. Teillier.)

Locality	Vegetation structure	Age since fire (yrs)	Total vascular species			
			1 m²	10 m²	100 m²	1000 m²
Papudo (32° 35' S, 71° 28' W)	Coastal matorral	20	9 (1+6+2)*	25 (8+11+6)*	43 (14+18+11)*	85 (32+33+20)*
Fundo Santa Laura (33° 04' S, 71° 00' W)	Typical matorral	25	8 (1+3+4)*	44 (6+17+21)*	94 (22+31+41)*	114 (29+39+46)*
Paso Marchant (33° 52' S, 70° 10' W)	Montane matorral	25	6 (2+3+1)*	17 (3+6+8)*	46 (9+21+16)*	102 (26+44+32)*
Polpaico (33° 14' S, 70° 40' W)	Espinal (*Acacia caven*) steppe	25	7 (0+3+4)*	12 (1+3+8)*	20 (2+3+15)*	25 (5+5+15)*

* (woody spp. + herbaceous perennials + annuals).

Table 8. Species richness of plant communities in the Mediterranean Basin.

Locality	Vegetation structure	Age since fire (yrs)	Total vascular species				Authors*
			$1\,m^2$	$10\,m^2$	$100\,m^2$	$1000\,m^2$	
Israel							
Mt Gilboa (32° 30′ N, 35° 25′ E)	shrub grassland	lightly grazed	29	42	105	179 (32+46+101)**	ZN,RHW
Mt Carmel (Forty Oaks) (32° 45′ N, 35° 00′ E)	open shrubland	disturbed	20	34	75	119 (24+39+56)**	ZN,RHW
Mt Carmel (Muhraqa) (32° 45′ N, 35° 00′ E)	maquis (closed shrubland)	undisturbed (30+ yrs)	?	?	?	51 (13+38+0)**	ZN,RHW
Mt Meron (33° 00′ N, 35° 25′ E)	maquis (closed shrubland)	protected	5	8	26	67 (32+19+16)**	ZN,RHW
Allonim (32° 43′ N, 35° 09′ E)	oak woodland	grazed	21	39	88	135 (15+23+97)**	ZN,RHW
Allone Abba (32° 43′ N, 35° 13′ E)	oak woodland	grazed	23	34	76	137 (25+30+82)**	ZN,RHW
Neve Ya'ar (32° 43′ N, 35° 12′ E)	oak woodland	ungrazed (1 year)	10	29	36	47	ZN,LOW
		grazed	9	23	43	50	ZN,LOW
		grazed/burnt	10	28	36	?	ZN,LOW
Mt Carmel (32° 45′ N, 35° 00′ E)	batha (*Sarcopoterium*)	?	14	31	48	65	ZN,LOW
Turkey	phrygana (19 communities)	?	1	9	18	27	TU
Greece							
Skiathos Is. (39° 10′ N, 23° 30′ E)	phrygana (4 communities)	?	?	?	18	?	EE
France							
St Clément (Montpellier) (43° 43′ N, 3° 51′ E)	*Pinus halepensis* open-forest	1	7	18	29	30	ZN,LOW
		15+	6	17	27	30	ZN,LOW
Puechabon (Montpellier) (43° 44′ N, 3° 35′ E)	*Quercus ilex* dense coppice	1	9	27	45	64	ZN,LOW
		10+	6	18	33	51	ZN,LOW
Mt Maures (43° 20′ N, 6° 29′ E)	*Quercus ilex–Q. suber* dense coppice	1	6	19	23	28	ZN,LOW
		unburnt	6	10	13	20	ZN,LOW
	Pinus pinaster woodland	15	5	11	17	19	ZN,LOW
Mt de la Gardiole (Montpellier) (43° 31′ N, 4° 46′ E)	*Quercus coccifera* garrigue	1	8	18	25	29	ZN,LOW
		5	8	18	23	28	ZN,LOW
Corsica							
Cauro (41° 55′ N, 8° 55′ E)	maquis arboré – dense (FPC 80%)	15	12	29	45	?	RLS,LA,CM
	– gap (FPC 0+50%)	15	21	42	62	?	RLS,LA,CM
Désert des Agriates (42° 40′ N, 9° 10′ E)	maquis (FPC 27+36%)	12	31	44	58	?	RLS,LA,CM

* EE – E. Ecomonidou, CM – C. Martinez, LA – L. Amandier, LOW – L. Olsvig-Whittaker, RHW – R.H. Whittaker, RLS – R.L. Specht, TU – T. Uslu, ZN – Z. Naveh.
** (woody spp. + herbaceous perennials + annuals).

References

Ecomonidou, E., 1969. Geobotanical Research on the Island of Skiathos. Dissertation, University of Athens, Athens, Greece.

Naveh, Z. and Whittaker, R.H., 1979. Structural and floristic diversity of shrublands and woodlands in northern Israel and other mediterranean areas. Vegetatio 41: 171–190.

Table 9. Species richness of plant communities in the Mediterranean Basin (France-Trabaud and Lepart 1980).

Locality	Vegetation structure	Age since fire (yrs)	Total vascular species (100 m²)
France			
Bas Languedoc (43° 30′ N, 3° 15′ E)	1. *Quercus ilex* forest (open) (6 communities)	11	30 (18+12+0)*
	2. *Quercus ilex* forest (dense) (7 communities)	9	29 (16+13+0)*
	3. *Pinus halepensis* woodland (7 communities)	11	39 (14+25+0)*
	4. *Rosmarinus officinalis* garrigue (6 communities)	6	30 (13+16+1)*
	5. *Brachypodium ramosum* grassland (3 communities)	7	42 (9+25+8)*
	6. *Brachypodium phoenicoides* grassland (3 communities)	6	51 (14+33+4)*

* (woody spp. + herbaceous perennials + annuals).

Reference

Trabaud, L. and Lepart, J., 1980. Diversity and stability in garrigue ecosystems after fire. Vegetatio 43: 49–57.

Table 10. Species richness of plant communities in post-fire succession near Montpellier, France. (Compiled by L. Trabaud.)

1. *Aniane*, near *Puechabon* (43° 44′ N, 3° 35′ E)	*Quercus ilex* coppice forest (40 years after cutting)			
Age (yrs) since fire	1	2	5	10
Overstorey F.P.C. (%)**	23	39	88	100
No. vascular spp. (100 m²)	34 (14+13+7)*	42 (16+16+10)*	27 (17+10+0)*	30 (18+12+0)*
2. *St Gély du Fesc* (43° 41′ N, 3° 48′ E)	*Quercus coccifera* garrigue			
Age (yrs) since fire	2	6	18	30
Overstorey F.P.C. (%)**	90	100	99	96
No. vascular spp. (100 m²)	34 (15+16+3)*	31 (15+16+0)*	30 (16+14+0)*	31 (17+14+0)*

* (woody spp. + herbaceous perennials + annuals).
** F.P.C. = Foliage Projective Cover.

Table 11. Species richness of plant communities in the southern Cape Mountains, South Africa (Bond 1983).

Locality	Vegetation structure	Total vascular species			
		1 m^2	100 m^2	500 m^2	1000 m^2
Swartberg Mts (33°22′ S, 21°55′ E)	1. Tall open proteoid shrubland - heath understorey	13	43	58	65
	2. Tall open proteoid shrubland - heath understorey	19	58	68	76
	3. Tall open proteoid shrubland - heath understorey	17	37	46	54
	4. Low closed *Erica* spp. heath	17	60	76	90
	5. Tall open proteoid shrubland - heath understorey	16	54	65	74
	6. Tall open proteoid shrubland - restioid heath	10	26	34	41
	7. Tall open proteoid shrubland - restioid heath	12	37	45	51
	8. Low open heathland - arid fynbos	11	42	56	59
	9. Arid fynbos-renosterveld transition	10	35	43	49
	10. Renosterveld	6	21	25	28
	11. Succulent karoo	4	24	30	36
Baviaanskloof Mts (33°30′ S, 24°15′ E)	1. Tall open proteoid shrubland - heath understorey	14	40	55	64
	2. Tall closed proteoid shrubland - heath understorey	27	63	74	77
	3. Tall proteoid shrubland - grassy heath understorey	13	51	66	75
	4. Low open grassy heathland	16	47	55	65
	5. Karoid shrubland	8	33	39	49
	6. Valley bushveld	14	55	71	79
Outeniqua Mts (33°45′ S, 21°50′ E and 33°50′ S, 23°00′ E)	1. Closed proteoid shrubland - heath understorey	13	43	52	55
	2. Tall open proteoid shrubland - heath understorey	17	55	61	65
	3. Tall open proteoid shrubland - heath understorey	20	50	59	66
	4. Low open heathland - arid fynbos	10	43	54	60
	5. Tall open shrubland - heath understorey (waboomveld)	18	68	91	104
Tsitsikama Mts (33°50′ S, 23°50′ E)	1. Tall open shrubland - heath understorey	16	35	40	45
	2. Tall open shrubland - grassy heath understorey	20	54	61	69
Knysna 'island' (34°03′ S, 23°03′ E)	Open scrub	5	15	19	21
George (33°57′ S, 22°28′ E)	Moist forest	6	38	49	52

Reference

Bond, W. 1983. On alpha diversity and the richness of the Cape flora: a study in southern Cape fynbos. In: F.J. Kruger, D.T. Mitchell and J.U.M. Jarvis, (eds.) Mediterranean-Type Ecosystems. The Role of Nutrients. pp. 337–356. Springer-Verlag, Berlin.

Table 12. Species richness of plant communities in Cape Province, South Africa. (Compiled by R.M. Cowling.)

Locality	Vegetation structure	Age since fire (yrs)	Total vascular species			
			$1\,m^2$	$100\,m^2$	$500\,m^2$	$1000\,m^2$
Bontevok N.P., Swellendam (34° 10′ S, 20° 30′ E)	South Coast renosterveld (*Elytropappus-Aspalathus*)	> 10	13	49	57	60
Burtkraal, near Grahamstown (33° 20′ S, 26° 30′ E)	South Coast renosterveld (*Elytropappus-Aspalathus*)	> 10	19	71	88	95
Cape Town (Signal Hill) (33° 57′ S, 18° 30′ E)	West Coast strandveld (*Hyparrhenia-Leysera*)	?	21	70	90	99
Elandsberg Mts (33° 45′ S, 25° E)	1. South Eastern mountain fynbos (proteoid-restioid)	15	13	26	36	41
	2. South Eastern mountain fynbos ((*Leucospermum-Tetraria*)	15	–	27	–	–
	3. Knysna Afromontane forest (*Rapanea-Canthium*)	> 300	11	33	46	53
	4. Knysna Afromontane forest (*Rapanea-Ocotea*)	> 300	–	21	–	–
Gamtoos (33° 50′ S, 25° E)	1. Grassy fynbos (*Themeda-Passerina*)	10	–	36	–	–
	2. Kaffrarian succulent thicket (*Sideroxylon-Euphorbia*)	> 100	13	36	53	61
	3. Kaffrarian succulent thicket (*Euclea-Brachylaena*)	> 100	5	45	80	98
Humansdorp (34° 07′ S, 24° 47′ E)	1. Grassy fynbos (*Thamnochortus-Erica*)	3–12	16	34	42	45
	2. Grassy fynbos (*Protea-Clutia*)	2–10	25	44	72	87
	3. Grassy fynbos (*Erica-Trachypogon*)	5–15	13–17	40	(60)	60–83
	4. Grassy fynbos (*Thamnochortus-Tristachya*)	1–2	22	47	65	74
	5. South Coast dune fynbos (*Restio-Agathosma*)	1–10	16	32	50	60
	6. South Coast dune fynbos (*Restio-Maytenus*)	2–15	7	34	48	55
	7. South Coast fynbos (*Themeda-Stenotaphrum*)	1–3	16	40	74	93
	8. South Coast renosterveld (*Themeda-Cliffortia*)	1–3	5–23	34	(54)	57–74
	9. South Coast renosterveld (*Elytropappus-Eustachys*)	1–10	19–21	52	78	87–95
	10. South Coast renosterveld (*Elytropappus-Metalasia*)	1–12	–	41	–	–
	11. South Coast renosterveld (*Elytropappus-Relhamia*)	3–6	9	43	71	85
	12. Kaffrarian thicket (*Pterocelastrus-Gonioma*)	> 100	15	37	59	71
	13. Kaffrarian thicket (*Pterocelastrus-Euclea*)	> 100	10	37	47	52
	14. Kaffrarian thicket (*Cassine-Cussonia*)	> 100	9	25	33	37
	15. Kaffrarian thicket (*Cassine-Schottia*)	> 100	–	30	–	–
Koeberg (33° 46′ S, 18° 28′ E)	West Coast strandveld (*Putterlickia-Rhus*)	> 50	7	25	32	35
Tygerberg (33° 50′ S, 18° 30′ E)	West Coast strandveld (*Elytropappus-Aristida*)	15	21	67	91	103

Vegetation, nutrition and climate – data-tables

(4) Climate

Co-ordinator: R.L. Specht

Contributors

R.L. Specht (Australia)
R.L. Specht, P.H. Zedler (California and Arizona)
R. Gajardo, E.J. Hajek, L.B. Hutley, J.D. Molina,
R.L. Specht (Chile)
M. Arianoutsou, L.B. Hutley, R.L. Specht (Greece)
R. Berliner, L.B. Hutley, R.L. Specht (Israel)
L. Ahdali, Ph. Daget (Mediterranean Basin)
F.M. Catarino, A.I.D. Correia (Portugal)
E.J. Moll, R.L. Specht (South Africa)
C.A. Gracia (Spain)

Contents

Australia	
– New South Wales – southwest (Tables 1–3)	94
– South Australia (Tables 4–6)	96
– Victoria – western (Tables 7–9)	100
– Western Australia – southwest (Tables 10–12)	104
United States of America	
– California (Tables 13–15)	106
– Arizona (Tables 16–18)	110
Chile (Tables 19–21)	113
Mediterranean Basin (Tables 22–23)	117
– Greece (Tables 24–26)	125
– Israel (Tables 27–29)	126
– Portugal (Tables 30–31)	129
– Spain (Tables 32–34)	131
South Africa – Cape Province (Tables 35–37)	134

R.L. Specht (ed.) Mediterranean-type Ecosystems. ISBN 90-6193-652-7.
© Kluwer Academic Publishers.

Table 1. Climatic stations in southwest New South Wales; R.L. Specht.

Climate station	Latitude (South)	Longitude (East)	Altitude (m)	Vegetation Type
Western (Lower Darling)				
Broken Hill	31° 58'	141° 27'	303.9	*Acacia aneura* low woodland
Lake Victoria	34° 03'	141° 16'	42.7	*Casuarina cristata* low woodland
Wentworth	34° 06'	141° 55'	36.6	*E. socialis – E. dumosa* open scrub
Western (Southwest Plains)				
Balranald	34° 38'	143° 34'	61.0	*E. socialis – E. dumosa* open scrub
Euston	34° 35'	142° 44'	61.0	*E. socialis – E. dumosa* open scrub
Ivanhoe	32° 54'	144° 18'	88.4	*Casuarina cristata* low woodland
Mt. Hope	32° 51'	145° 53'	182.9	*E. socialis – E. dumosa* open scrub
Riverina (West)				
Griffith	34° 17'	146° 02'	125.6	*E. microcarpa* woodland
Hay	34° 30'	144° 51'	93.6	*Atriplex* low shrubland
Yenda	34° 15'	146° 11'	128.9	*E. microcarpa* woodland
Riverina (East)				
Deniliquin	35° 31'	144° 56'	91.4	*E. camaldulensis* woodland
Narrandera	34° 45'	146° 33'	151.5	*E. microcarpa* woodland
Tocumwal	35° 49'	145° 34'	111.3	*E. microcarpa* woodland
Wagga	35° 10'	147° 28'	214.3	*E. microcarpa* woodland
Central Western Plains				
Condobolin	33° 05'	147° 09'	189.0	*E. microcarpa* woodland
Central Western Slopes				
Dubbo	32° 15'	148° 36'	262.4	*E. microcarpa* woodland
Forbes	33° 23'	148° 01'	244.8	*E. microcarpa* woodland
Parkes	33° 08'	148° 11'	338.6	*E. microcarpa* woodland
Wellington	32° 34'	148° 57'	303.3	*E. microcarpa* woodland
Southwest Slopes (North)				
Cootamundra	34° 39'	148° 02'	329.8	*E. microcarpa* woodland
Grenfell	33° 54'	148° 10'	384.0	*E. microcarpa* woodland
Temora	34° 25'	147° 32'	292.0	*E. microcarpa* woodland
Wyalong	33° 56'	147° 15'	253.3	*E. behriana* open scrub *E. microcarpa* woodland
Young	34° 19'	148° 18'	457.2	*E. microcarpa* woodland
Southwest Slopes (South)				
Adelong	35° 19'	148° 05'	332.8	*E. macrorhyncha – E. rossii* open forest
Albury	36° 04'	146° 56'	182.9	*E. camadulensis* woodland
Hume Reservoir	36° 06'	147° 02'	183.5	*E. dives – E. macrorhyncha – E. radiata* open forest
Aust. Capital Territory				
Canberra	35° 19'	149° 12'	570.6	*E. macrorhyncha – E. rossii* open forest

References

Aust. Govt., Bureau of Meteorology, 1975. Climatic Averages New South Wales. Aust. Govt. Publ. Service, Canberra.

Carnahan, J.A., 1976. Natural vegetation. In: Atlas of Australian Resources. Second Series. Div. National Mapping, Dept. National Resources, Canberra.

Hayden, E.J., 1971. The Natural Plant Communities of New South Wales. M.Sc. Thesis, Australian National University, Canberra.

Specht. R.L. (ed.), in press. Major Plant Communities in Australia: An Objective Assessment. C.S.I.R.O. Aust., Melbourne.

Table 2. Climatic data for southwest New South Wales – Classification; R.L. Specht.

Climate station	Mean annual precipitation (mm)	Mean temperature (°)		Emberger pluviothermic quotient (Q)	Climatic type	
		hottest month max.	coldest month min.,		Emberger	Köppen
Western (Lower Darling)						
Broken Hill	241	32.1	5.4	31	temperate, arid	BWk
Lake Victoria	268	32.3	5.0	34	temperate, arid	BWk/BSk
Wentworth	288	33.3	4.5	34	temperate, arid	BSk
Western (Southwest Plains)						
Balranald	317	32.7	3.2	37	temperate/cool, arid/semi-arid	BSk
Euston	311	33.1	2.2	35	cool, arid/semi-arid	BSk
Ivanhoe	290	34.3	3.6	32	temperate, arid	BSh
Mt. Hope	382	33.3	4.1	45	temperate, semi-arid	BSh
Riverina (west)						
Griffith	441	31.1	2.4	53	cool, semi-arid	BSk
Hay	360	33.1	3.7	42	temperate, semi-arid/arid	BSk
Yenda	410	32.9	2.7	47	cool, semi-arid	BSk
Riverina (East)						
Deniliquin	427	31.9	3.1	51	temperate, semi-arid	BSk
Narrandera	411	32.9	3.9	49	temperate, semi-arid	BSk
Tocumwal	451	31.9	3.1	54	temperate, semi-arid	BSk
Wagga	571	31.1	2.6	69	cool, sub-humid/semi-arid	Cfa
Central Western Plains						
Condobolin	439	32.8	3.5	52	temperate, semi-arid	BSh
Central Western Slopes						
Dubbo	584	31.6	2.9	70	cool/temperate, sub-humid/semi-arid	Cfa
Forbes	526	31.9	2.3	61	cool, semi-arid/sub-humid	Cfa
Parkes	569	31.4	4.2	72	temperate, sub-humid/semi-arid	Cfa
Wellington	614	31.7	1.6	70	cool, sub-humid	Cfa
Southwest Slopes (North)						
Cootamundra	616	31.0	2.0	73	cool, sub-humid	Cfa
Grenfell	635	31.2	2.5	76	cool, sub-humid	Cfa
Temora	523	31.4	1.4	60	cool, sub-humid/semi-arid	BSk/Cfa
Wyalong	480	32.2	2.0	55	cool, semi-arid	BSk
Young	657	30.7	0.6	76	cool, sub-humid	Cfa
Southwest Slopes (South)						
Adelong	778	30.8	0.0	88	cool/cold, sub-humid	Cfb
Albury	680	31.1	2.2	81	cool, sub-humid	Cfa
Hume Reservoir	691	30.6	3.4	88	temperate, sub-humid	Cfa
Aust. Capital Territory						
Canberra (Airport)	633	27.5	−0.5	79	cool/cold, sub-humid	Cfb

Table 3. Climatic Data for southwest New South Wales – Growth Indices; R.L. Specht.

Climatic station	Annual precip. (mm)	Annual pan evap. (mm)	Evap. coeff. (k)	Smax (cm)	Moisture Index		Summer drought (mo)*	T.I. × M.I. (Ann. Total)	
					Annual	Summer		Temp.	Subtrop.
Western (Lower Darling)									
Broken Hill	234	1,811	0.035	5	0.15	0.07	4 (8)	0.62	0.34
Lake Victoria	245	1,515	0.041	6	0.19	0.08	4 (6)	0.61	0.46
Wentworth	274	1,715	0.037	6	0.18	0.09	3 (6)	0.65	0.44
Western (Southwest Plains)									
Ivanhoe	281	2,021	0.030	7	0.16	0.07	4 (8)	0.49	0.34
Riverina (West)									
Griffith	385	1,498	0.042	10	0.31	0.12	0 (4)	0.85	0.75
Hay	347	1,612	0.040	9	0.26	0.10	3 (5)	0.86	0.58
Yenda	374	1,567	0.038	9	0.28	0.13	0 (4)	0.80	0.60
Riverina (East)									
Deniliquin	393	1,555	0.039	11	0.31	0.11	1 (4)	0.98	0.75
Wagga	544	1,495	0.041	14	0.44	0.18	0 (3)	1.30	1.01
Central Western Slopes									
Dubbo	531	1,675	0.038	12	0.36	0.20	0 (1)	0.79	0.88
Forbes	489	1,648	0.038	12	0.34	0.16	0 (3)	1.07	0.75
Wellington	567	1,670	0.038	12	0.38	0.18	0 (2)	1.00	0.89
Southwest Slopes (North)									
Cootamundra	583	1,420	0.044	14	0.48	0.20	0 (2)	1.24	0.99
Temora	473	1,387	0.045	11	0.39	0.22	0 (1)	1.03	0.94
Southwest Slopes (South)									
Albury	703	1,561	0.040	19	0.55	0.19	0 (3)	1.63	1.21
Hume Reservoir	651	1,335	0.044	19	0.59	0.22	0 (0)	1.57	1.26
Aust. Capital Territory									
Canberra (Acton)	592	1,214	0.053	12	0.55	0.28	0 (0)	1.36	0.87

* Months with M.I. <0.10 and, in parentheses, M.I. <0.20.

Table 4. Climate Stations in South Australia; R.L. Specht.

Climate station	Latitude (South)	Longitude (East)	Altitude (m)	Vegetation type
Northwest District				
Tarcoola	30° 43'	134° 34'	118.9	*Acacia aneura* tall shrubland
Yudnapinna	32° 07'	137° 09'	121.9	*Acacia papyrocarpa* low woodland
Far North District				
Farina	30° 05'	138° 08'	92.7	*Acacia aneura – A. ramulosa* tall shrubland
Marree	29° 39'	138° 04'	49.7	*Atriplex rhagodioides* low shrubland
Oodnadatta	27° 33'	135° 28'	112.5	*Atriplex rhagodioides* low shrubland
William Creek	28° 55'	136° 21'	76.3	*Eremophila – Acacia* tall open-shrubland
Western Agric. District				
Ceduna	32° 08'	133° 42'	21.9	*E. socialis – E. gracilis* open scrub
Cook	30° 37'	130° 25'	121.3	*Atriplex vesicaria – Maireana sedifolia* low shrubland
Fowlers Bay	31° 59'	132° 27'	4.9	*E. socialis – E. gracilis* open scrub
Kimba	33° 09'	136° 25'	263.3	*E. incrassata – Melaleuca uncinata* open scrub
Kyancutta	33° 08'	135° 34'	58.2	*E. incrassata – Melaleuca uncinata* open scrub
Port Lincoln	34° 44'	135° 51'	4.0	*E. diversifolia* open scrub
Streaky Bay	32° 48'	134° 13'	13.1	*E. diversifolia* open scrub
Whyalla	33° 02'	137° 35'	14.3	*Acacia papyrocarpa* low woodland
Upper North District				
Angorichina	31° 07'	138° 35'	616.1	*Cassia-Eremophila-Acacia* tall open shrubland
Hawker	31° 53'	138° 25'	314.9	Arid grassland
Port Augusta	32° 33'	137° 47'	8.5	*Atriplex vesicaria – Maireana sedifolia* low shrubland
Yongala	33° 02'	138° 45'	514.2	*Lomandra* spp. tussock grassland
North east District				
Yunta	32° 35'	139° 34'	303.3	*Atriplex vesicaria – Maireana sedifolia* low shrubland

Table 4. (Continued).

Climate station	Latitude (South)	Longitude (East)	Altitude (m)	Vegetation type
Lower North District				
Bundaleer F.R.	33°17'	138°35'	458.1	*E. leucoxylon* woodland
Clare	33°50'	138°37'	397.8	*E. leucoxylon* woodland
Georgetown	33°22'	138°24'	273.1	*E. leucoxylon* woodland
Port Pirie	33°11'	138°01'	4.0	*E. socialis – E. gracilis* open scrub
Snowtown	33°47'	138°13'	102.7	*E. socialis – E. gracilis* open scrub
Yorke Peninsula				
Kadina	33°58'	137°43'	44.2	*E. socialis – E. gracilis* open scrub
Maitland	34°23'	137°41'	185.9	*E. porosa* woodland
Warooka	34°59'	137°25'	53.0	*Melaleuca lanceolata* woodland
Kangaroo Island				
Cape Borda	35°45'	136°36'	145.4	*E.diversifolia* open scrub
Cape de Couedic	36°04'	136°43'	88.4	*E. diversifolia* open scrub
Kingscote	35°40'	137°38'	30.5	*E. cneorifolia – Melaleuca uncinata* open scrub
Adelaide Plains				
Adelaide	34°56'	138°35'	42.7	*E. porosa* woodland
Roseworthy	34°31'	138°41'	114.3	*E. socialis – E. gracilis* open scrub
Waite Institute	34°58'	138°38'	121.9	*E. microcarpa* woodland
County Light District				
Kapunda	34°21'	138°55'	244.8	*E. porosa* woodland
Nuriootpa	34°29'	139°01'	274.3	*E. leucoxylon* woodland
Mt. Lofty Ranges				
Belair	35°00'	138°37'	304.8	*E. microcarpa* woodland
Mt. Barker	35°04'	138°52'	330.1	*E. obliqua* open forest; *E. leucoxylon* woodland
Mt. Crawford Forest	34°43'	138°57'	403.9	*E. obliqua – E. goniocalyx* open forest
Myponga	35°24'	138°28'	225.6	*E. leucoxylon* woodland
Strathalbyn	35°15'	138°54'	71.0	*E. leucoxylon* woodland
Stirling West	35°00'	138°43'	496.2	*E. obliqua – E. baxteri* open forest
Murray Valley-Upper				
Berri	34°17'	140°36'	30.5	*E. socialis – E. gracilis* open forest
Renmark	34°10'	140°45'	20.1	*E. socialis – E. gracilis* open forest
Murray Valley-Lower				
Eudunda	34°11'	139°05'	414.8	*E. porosa* woodland
Meningie	35°41'	139°20'	2.7	*E. diversifolia* open scrub
Murray Bridge	35°07'	139°17'	25.9	*E. socialis – E. gracilis* open scrub
Upper South East				
Bordertown	36°18'	140°46'	81.7	*E. leucoxylon* woodland; *Allocasuarina luehmannii* woodland
Keith	36°06'	140°21'	30.8	*E. leucoxylon – E. fasciculosa* woodland
Lameroo	35°20'	140°31'	97.8	*E. socialis – E. gracilis* open scrub
Lower South East				
Cape Northumberland	38°04'	140°40'	36.3	*E. obliqua – E. ovata* woodland
Kybybolite	36°53'	140°55'	91.1	*E. camaldulensis* woodland
Lucindale	36°59'	140°22'	31.1	*E. baxteri – E. fasciculosa* woodland/open forest
Mt. Burr Forest	37°33'	140°26'	64.0	*E. obliqua* open forest
Mt. Gambier	37°49'	140°46'	64.9	*Acacia mearnsii* woodland (cleared)
Naracoorte	36°58'	140°44'	57.6	*E. baxteri* open forest
Robe	37°10'	139°45'	3.7	*E. diversifolia* open scrub

References

Aust. Govt., Bureau of Meteorology 1975. Climatic Averages South Australia and Northern Territory. Aust. Govt. Publ. Service, Canberra.

Specht, R.L. (1972) The Vegetation of South Australia. Second Edition. Govt. Printer, Adelaide.

Specht, R.L. (ed.) in press. Major Plant Communities in Australia: An Objective Assessment. C.S.I.R.O. Aust., Melbourne.

Table 5. Climatic data for South Australia – Classification; R.L. Specht.

Climate station	Mean annual precipitation (mm)	Mean temperature (°)		Emberger pluviothermic quotient (Q)	Climatic type	
		hottest month max.	coldest month min.		Emberger	Köppen
Northwest District						
Tarcoola	170	34.9	4.7	19	temperate, arid	BWh
Yudnapinna	206	33.8	4.6	24	temperate, arid	BWh
Far North District						
Farina	146	35.4	4.1	16	temperate, arid/very arid	BWh
Marree	155	38.0	4.6	16	temperate, arid/very arid	BWh
Oodnadatta	155	38.2	5.8	16	temperate, arid/very arid	BWh
William Creek	127	35.9	4.7	14	temperate, arid	BWh
Western Agric. District						
Ceduna	321	28.5	5.7	49	temperate, semi-arid	BSk
Cook	165	33.6	4.6	20	temperate, arid	BWk/BWh
Fowlers Bay	299	25.7	6.6	54	temperate/warm, semi-arid	BSks
Kimba	346	31.7	4.7	44	temperate, semi-arid	BSk
Kyancutta	330	33.0	4.5	40	temperate, semi-arid/arid	BSk
Port Lincoln	486	25.5	8.3	98	warm, sub-humid	Csb
Streaky Bay	378	29.7	7.9	59	warm, semi-arid	Csa
Whyalla	273	28.8	7.2	43	warm/temperate, semi-arid	BSk
Upper North District						
Angorichina	302	32.8	4.4	37	temperate, arid	BSk
Hawker	301	33.2	4.0	35	temperate, arid	BSk
Port Augusta	256	32.3	7.1	35	temperate, arid	BWh
Yongala	369	30.8	2.0	44	cool, semi-arid	BSk
Northeast District						
Yunta	227	32.5	3.1	27	temperate/cool, arid	BWk
Lower North District						
Bundaleer F.R.	554	30.5	3.2	70	temperate/cool, sub-humid, semi-arid	Csa
Clare	632	30.1	3.4	82	temperate, sub-humid	Csa
Georgetown	468	31.3	4.4	60	temperate, semi-arid	BSk
Port Pirie	343	31.4	7.5	49	warm, semi-arid	BSh
Snowtown	407	31.4	4.7	52	temperate, semi-arid	BSk
Yorke Peninsula						
Kadina	396	30.4	5.7	55	temperate, semi-arid	BSk
Maitland	509	28.7	6.6	79	temperate/warm, sub-humid	Csb
Warooka	450	27.5	7.2	76	warm/temperate, sub-humid/semi-arid	Csb
Kangaroo Island						
Cape Borda	630	23.0	8.4	150	warm, humid/sub-humid	Csb
Cape de Couedic	639	22.6	9.3	166	warm, humid	Csb
Kingscote	491	24.8	8.0	101	warm, sub-humid	Csb
Adelaide Plains						
Adelaide	533	28.5	7.9	89	warm, sub-humid	Csa/Csb
Roseworthy	439	30.1	6.0	63	temperate, semi-arid	Csa
Waite Institute	627	28.5	8.1	106	warm, sub-humid	Csb
County Light District						
Kapunda	496	30.7	5.1	67	temperate, semi-arid	Csa
Nuriootpa	506	29.0	4.2	71	temperate, sub-humid/semi-arid	Csa
Mt. Lofty Ranges						
Belair	755	27.1	7.1	131	warm/temperate, sub-humid	Csb
Mt. Barker	781	27.0	4.3	119	temperate, sub-humid	Csb
Mt. Crawford Forest	784	27.7	2.9	110	cool/temperate, sub-humid	Csb
Myponga	763	27.0	4.2	116	temperate, sub-humid	Csb
Strathalbyn	495	27.9	5.6	77	temperate, sub-humid	Csb
Stirling West	1,121	25.7	4.9	187	temperate, humid	Csb
Murray Valley-Upper						
Berri	269	32.1	5.5	35	temperate, semi-arid/arid	BSk
Renmark	263	32.8	5.1	33	temperate, arid	BSk
Murray Valley-Lower						
Eudunda	441	28.6	4.9	64	temperate, semi-arid	BSk
Meningie	470	26.6	6.1	79	temperate, sub-humid	Csb
Murray Bridge	341	28.9	5.5	50	temperate, semi-arid	BSk
Upper South East						
Bordertown	541	29.6	4.5	74	temperate, sub-humid	Csb
Keith	471	29.9	5.5	66	temperate, semi-arid	Csb
Lameroo	393	31.1	4.1	50	temperate, semi-arid	BSk
Lower South East						
Cape Northumberland	709	21.6	6.7	166	temperate/warm, humid	Csb
Kybybolite	552	28.6	3.9	77	temperate, sub-humid	Csb
Lucindale	624	28.7	5.0	91	temperate, sub-humid	Csb
Mt. Burr Forest	788	25.4	5.0	134	temperate, sub-humid	Csb
Mt. Gambier	712	25.6	4.7	118	temperate, sub-humid	Csb
Naracoorte	586	28.4	4.6	85	temperate, sub-humid	Csb
Robe	630	22.7	8.3	152	warm, humid	Csb

Table 6. Climatic Data for South Australia – Growth Indices; R.L. Specht.

Climatic station	Annual precip. (mm)	Annual pan evap. (mm)	Evap. coeff. (k)	Smax (cm)	Moisture Index		Summer drought (mo)*	T.I. × M.I. (Ann. Total)	
					Annual	Summer		Temp.	Subtrop.
Northwest District									
Tarcoola	156	1,991	0.034	3	0.09	0.05	8 (12)	0.30	0.21
Yudnapinna	190	1,752	0.036	5	0.13	0.06	6 (10)	0.42	0.29
Far North District									
Farina	146	2,121	0.031	3	0.08	0.05	10 (12)	0.24	0.15
Marree	143	2,282	0.029	3	0.07	0.05	12 (12)	0.20	0.14
William Creek	127	2,240	0.031	2	0.06	0.04	12 (12)	0.16	0.13
Western Agric. District									
Cook	153	1,693	0.041	3	0.10	0.07	6 (12)	0.37	0.31
Fowlers Bay	296	1,210	0.070	5	0.29	0.10	2 (6)	1.50	0.49
Kyancutta	329	1,743	0.038	9	0.23	0.07	4 (6)	0.84	0.57
Port Lincoln	463	1,136	0.062	13	0.51	0.11	3 (3)	2.53	0.47
Streaky Bay	371	1,443	0.045	13	0.35	0.07	3 (5)	1.74	0.70
Upper North District									
Angorichina	302	1,810	0.035	7	0.19	0.10	3 (6)	0.63	0.44
Port Augusta	236	1,773	0.039	5	0.16	0.07	5 (8)	0.65	0.41
Yongala	370	1,529	0.041	10	0.30	0.09	3 (4)	0.61	0.52
Lower North District									
Clare	612	1,365	0.047	20	0.57	0.13	0 (3)	1.33	0.88
Port Pirie	330	1,603	0.042	8	0.25	0.09	3 (5)	1.19	0.65
Snowtown	390	1,481	0.044	12	0.34	0.08	3 (5)	1.22	0.61
Kangaroo Island									
Cape Borda	624	1,041	0.063	27	0.70	0.15	0 (3)	3.56	0.38
Cape de Couedic	629	964	0.075	24	0.72	0.19	0 (3)	4.15	0.30
Kingscote	498	923	0.076	16	0.64	0.15	0 (3)	3.02	0.45
Adelaide Plains									
Adelaide	536	1,606	0.040	15	0.44	0.15	1 (4)	1.86	0.80
Roseworthy	420	1,344	0.046	13	0.41	0.10	3 (4)	1.53	0.83
Waite Institute	643	1,389	0.046	19	0.58	0.16	0 (3)	2.28	1.00
County Light District									
Kapunda	491	1,345	0.048	15	0.47	0.11	0 (4)	1.46	0.74
Mt. Lofty Ranges									
Belair	815	1,239	0.049	30+	0.72	0.19	0 (2)	2.88	1.14
Mt. Barker	784	1,143	0.057	30+	0.74	0.27	0 (2)	2.46	0.76
Strathalbyn	491	1,164	0.056	13	0.53	0.13	0 (3)	2.23	0.57
Stirling West	1,183	1,021	0.069	30+	0.87	0.61	0 (0)	2.83	0.66
Murray Valley-Upper									
Berri	250	1,559	0.040	7	0.19	0.07	4 (6)	0.66	0.48
Upper South East									
Keith	457	1,395	0.042	14	0.41	0.11	0–2 (4)	1.66	0.55
Lower South East									
Cape Northumberland	695	885	0.081	25	0.82	0.45	0 (0)	3.97	0.31
Kybybolite	513	1,121	0.055	15	0.57	0.13	0 (3)	1.63	0.33
Mt. Burr Forest	785	966	0.078	30	0.85	0.56	0 (0)	3.49	0.38
Mt. Gambier	682	1,010	0.066	24	0.75	0.26	0 (0)	2.92	0.42
Robe	622	994	0.067	24	0.71	0.18	0 (2)	3.38	0.30

* Months with M.I. <0.10 and, in parentheses, M.I.<0.20.

Table 7. Climatic stations in Western Victoria; R.L. Specht.

Climate station	Latitude (South)	Longitude (East)	Altitude (m)	Vegetation Type
North Mallee District				
Merbein	34° 10'	142° 04'	56.4	*E. socialis – E. gracilis* open scrub
Mildura	34° 14'	142° 05'	50.3	*E. socialis – E. gracilis* open scrub
Ouyen	35° 04'	142° 19'	50.3	*E. socialis – E. gracilis* open scrub
Walpeup	35° 07'	142° 00'	106.7	*E. socialis – E. gracilis* open scrub
South Mallee District				
Beulah	35° 57'	142° 25'	88.4	*E. behriana* open scrub
Swan Hill	35° 21'	143° 33'	70.1	*E. socialis – E. gracilis* open scrub
North Wimmera District				
Donald	36° 24'	143° 00'	117.7	*Stipa* grassland (+ *E. microcarpa*)
Nhill	36° 21'	141° 39'	129.2	*Allocasuarina luehmannii* woodland
Serviceton	36° 22'	140° 59'	118.9	*Allocasuarina luehmannii* woodland
Warracknabeal	36° 16'	142° 24'	113.4	*Stipa* grassland (+ *E. microcarpa*)
Wycheproof	36° 06'	143° 12'	110.0	*E. microcarpa* woodland
South Wimmera District				
Horsham	36° 43'	142° 12'	138.4	*E. microcarpa* woodland
				Allocasuarina luehmanni woodland
St Arnaud	36° 37'	143° 16'	239.0	*E. macrorhyncha – E. polyanthemos* woodland
Stawell	37° 06'	142° 48'	231.3	*E. macrorhyncha – E. polyanthemos* woodland
Upper North District				
Boort	36° 06'	143° 42'	93.3	*E. socialis – E. gracilis* open scrub/*Stipa* grassland
Charlton	36° 17'	143° 21'	117.7	*E. microcarpa* woodland
Echuca	36° 08'	144° 45'	96.0	*E. camaldulensis* wetland
Kerang	35° 44'	143° 55'	77.7	Tussock grassland
Numurkah	36° 06'	145° 24'	107.9	*E. microcarpa* woodland
Lower North District				
Avoca	37° 05'	143° 29'	242.0	*E. camaldulensis* woodland
Bendigo	36° 45'	144° 17'	225.2	*E. macrorhyncha – E. polyanthemos* woodland
				E. incrassata – E. viridis open scrub
Dookie	36° 23'	145° 42'	171.6	*E. microcarpa* woodland
Shepparton	36° 23'	145° 24'	113.4	*E. microcarpa* woodland
Yarrawonga	36° 01'	146° 01'	128.0	*E. microcarpa* woodland
Upper Northeast District				
Beechworth	36° 22'	146° 41'	548.6	*E. dives – E. macrorhyncha – E. radiata* open forest
Benalla	36° 33'	145° 59'	169.5	*E. microcarpa* woodland
Myrtleford	36° 34'	146° 44'	222.5	*E. dives – E. macrorhyncha – E. radiata* open forest
Rutherglen	36° 05'	146° 29'	167.0	*E. microcarpa* woodland
Wangaratta	36° 22'	146° 18'	150.0	*E. microcarpa* woodland
West Coast District				
Camperdown	38° 13'	143° 09'	164.9	Tussock grassland
				E. obliqua open forest
Cape Otway	38° 15'	143° 30'	82.9	*E. obliqua – E. viminalis* open forest
Colac	38° 20'	143° 36'	134.1	Tussock grassland
				E. obliqua open forest
Hamilton	37° 45'	142° 02'	186.8	*E. camaldulensis* woodland
Portland (Cape Nelson)	38° 26'	141° 33'	45.4	*E. baxteri* open forest
				E. diversifolia open scrub
Warrnambool	38° 23'	142° 29'	21.3	*E. camaldulensis* woodland
Western Plains District				
Ararat	37° 17'	142° 57'	332.2	*E. macrorhyncha – E. polyanthemos* woodland
Ballarat	37° 32'	143° 49'	459.6	*E. macrorhyncha – E. polyanthemos* open forest
West Central District				
Ballan	37° 36'	144° 12'	442.0	*E. macrorhyncha – E. polyanthemos* open forest
Durdiwarrah	37° 49'	144° 13'	365.8	*E. macrorhyncha – E. polyanthemos* open forest
				E. behriana open scrub
Geelong	38° 07'	144° 22'	17.1	*Themeda* grassland
Laverton	37° 53'	144° 45'	14.3	*Themeda* grassland
Queenscliffe	38° 16'	144° 40'	15.2	*E. baxteri* open forest
Werribee	37° 54'	144° 41'	45.7	*Themeda* grassland

Table 7. (Continued).

Climate station	Latitude (South)	Longitude (East)	Altitude (m)	Vegetation Type
East Central District				
Cape Schanck	38°30'	144°53'	79.2	*E. baxteri* open forest
Frankston	38°07'	145°11'	27.4	*Eucalyptus – Banksia* low woodland
Kew	37°49'	145°02'	60.7	(urban)
Melbourne	37°49'	144°58'	34.7	(urban)
Mornington	38°14'	145°02'	45.7	*E. baxteri* open forest
Mt Dandenong	37°50'	145°21'	631.5	*E. obliqua – E. viminalis* open forest
Mt Eliza	38°12'	145°06'	106.7	*Eucalyptus – Banksia* low woodland
North Central District				
Creswick	37°25'	143°54'	452.3	*E. macrorhyncha – E. polyanthemos* open forest
Castlemaine	37°04'	144°13'	280.7	*E. macrorhyncha – E. polyanthemos* woodland
Maryborough	37°03'	143°44'	253.6	*E. macrorhyncha – E. polyanthemos* woodland
Seymour	37°01'	145°09'	141.7	*E. microcarpa* woodland

References

Aust. Govt., Bureau of Meteorology, 1975. Climatic Averages Victoria. Aust. Govt. Publ. Service, Canberra.

Carnahan, J.A., 1976. Natural vegetation. In: Atlas of Australian Resources. Second Series. Div. National Mapping, Dept. National Resources, Canberra.

Specht R.L. (ed.) in press. Major Plant Communities in Australia: An Objective Assessment. C.S.I.R.O. Aust., Melbourne.

Table 8. Climatic data for Western Victoria – Classification; R.L. Specht.

Climate station	Mean annual precipitation (mm)	Mean temperature (°)		Emberger pluviothermic quotient (Q)	Climatic type	
		hottest month max.	coldest month min.		Emberger	Köppen
North Mallee District						
Merbein	275	31.8	4.4	35	temperate, arid/semi-arid	BSk
Mildura	294	32.1	4.3	36	temperate, arid/semi-arid	BSk
Ouyen	337	32.5	4.2	41	temperate, semi-arid	BSk
Walpeup	348	32.0	4.7	44	temperate, semi-arid	BSk
South Mallee District						
Beulah	379	31.4	3.8	47	temperate, semi-arid	BSk
Swan Hill	345	31.5	3.7	43	temperate, semi-arid	BSk
North Wimmera District						
Donald	421	30.3	3.7	55	temperate, semi-arid	BSk
Nhill	423	30.4	3.4	54	temperate, semi-arid	BSk
Serviceton	497	29.8	3.5	65	temperate, semi-arid	BSk/Csb
Warracknabeal	490	30.5	3.4	62	temperate, semi-arid	BSk
Wycheproof	374	30.3	2.9	47	cool/temperate, semi-arid	BSk
South Wimmera District						
Horsham	449	30.8	3.9	58	temperate, semi-arid	BSk/Cfb
St. Arnaud	504	29.9	3.0	65	temperate/cool, semi-arid	BSk/Cfb
Stawell	533	29.9	4.2	72	temperate, sub-humid/semi-arid	Cfb
Upper North District						
Boort	420	31.2	3.6	52	temperate, semi-arid	BSk
Charlton	427	31.1	3.3	53	temperate, semi-arid	BSk
Echuca	436	31.2	3.8	55	temperate, semi-arid	BSk
Kerang	367	31.3	4.0	46	temperate, semi-arid	BSk
Numurkah	453	31.7	3.1	55	temperate, semi-arid	BSk
Lower North District						
Avoca	542	29.1	2.4	70	cool, sub-humid	Cfb
Bendigo	546	29.2	3.5	73	temperate, sub-humid	Cfb
Dookie	547	30.0	3.8	72	temperate, sub-humid	Cfa
Shepparton	509	31.0	3.0	63	temperate/cool, semi-arid	Cfa
Yarrawonga	512	31.3	3.6	64	temperate, semi-arid	BSk
Upper Northeast District						
Beechworth	929	27.4	2.5	130	cool, humid	Cfb
Benalla	670	31.2	2.9	82	cool/temperate, sub-humid	Cfa
Myrtleford	903	31.6	1.4	103	cool, humid/sub-humid	Cfa
Rutherglen	583	31.4	1.5	67	cool, sub-humid	Cfa
Wangaratta	640	31.3	2.8	77	cool/temperate, sub-humid	Cfa
West Coast District						
Camperdown	777	26.8	4.3	120	temperate, sub-humid	Cfb
Cape Otway	887	21.2	7.3	222	warm, per-humid/humid	Cfb
Colac	721	26.2	3.2	109	temperate, sub-humid	Cfb
Hamilton	692	26.4	4.7	111	temperate, sub-humid	Cfb
Portland (Cape Nelson)	782	20.8	7.8	209	warm, humid	Csa/Csb
Warrnambool	726	23.2	5.8	145	temperate, humid	Cfb
Western Plains District						
Ararat	616	27.7	4.0	90	temperate, sub-humid	Cfb
Ballarat	719	25.5	3.2	112	temperate, sub-humid	Cfb
West Central District						
Ballan	573	26.0	2.7	86	cool, sub-humid	Cfb
Durdiwarrah	684	24.9	3.6	112	temperate, sub-humid	Cfb
Geelong	542	24.0	5.1	100	temperate, sub-humid	Cfb
Laverton	571	26.1	4.6	92	temperate, sub-humid	Cfb
Queenscliffe	610	23.3	6.2	124	temperate, sub-humid	Cfb
Werribee	543	26.0	4.3	87	temperate, sub-humid	Cfb
East Central District						
Cape Schanck	752	22.0	7.5	180	warm, humid	Cfb
Frankston	769	26.8	4.9	122	temperate, sub-humid	Cfb
Kew	734	26.3	5.7	123	temperate, sub-humid	Cfb
Melbourne	658	26.5	6.2	112	temperate, sub-humid	Cfb
Mornington	736	25.4	6.9	138	temperate/warm, humid/sub-humid	Cfb
Mt. Dandenong	1,192	23.4	3.7	211	temperate, per-humid	Cfb
Mt. Eliza	696	25.4	7.2	132	warm/temperate, sub-humid/humid	Cfb
North Central District						
Creswick	787	27.3	2.2	109	cool, sub-humid	Cfb
Castlemaine For.	624	28.9	2.0	80	cool, sub-humid	Cfb
Maryborough	526	29.3	2.9	69	cool/temperate, sub-humid/semi-arid	Cfb
Seymour	596	29.9	2.7	76	cool, sub-humid	Cfb

Table 9. Climatic data for western Victoria – Growth Indices; R.L. Specht.

Climatic station	Annual precip. (mm)	Annual pan evap. (mm)	Evap. coeff. (k)	Smax (cm)	Moisture Index		Summer drought (mo)*	T.I. × M.I. (Ann. Total)	
					Annual	Summer		Temp.	Subtrop.
North Mallee District									
Merbein	247	1,541	0.041	6	0.19	0.08	4 (6)	0.61	0.50
Mildura	264	1,620	0.039	6	0.19	0.08	4 (6)	0.54	0.51
Walpeup	317	1,491	0.041	8	0.25	0.09	3 (6)	0.85	0.59
South Mallee District									
Swan Hill	333	1,523	0.041	9	0.26	0.10	2 (5)	0.85	0.66
North Wimmera District									
Nhill	393	1,375	0.045	11	0.35	0.10	2 (4)	1.24	0.55
Serviceton	478	1,218	0.050	14	0.49	0.13	1 (4)	1.45	0.60
Wycheproof	374	1,305	0.046	10	0.35	0.12	1 (4)	0.82	0.69
South Wimmera District									
Horsham	446	1,354	0.045	13	0.41	0.12	1 (4)	1.18	0.66
St. Arnaud	479	1,310	0.046	15	0.47	0.14	0 (3)	1.19	0.72
Stawell	523	1,256	0.049	15	0.52	0.14	0 (3)	1.53	0.65
Upper North District									
Boort	393	1,494	0.042	10	0.32	0.11	1 (4)	1.05	0.66
Charlton	427	1,371	0.045	12	0.39	0.13	1 (4)	0.93	0.71
Echuca	425	1,462	0.042	12	0.36	0.12	1 (4)	1.21	0.77
Kerang	363	1,475	0.043	9	0.30	0.11	1 (4)	0.96	0.64
Numurkah	438	1,382	0.044	12	0.39	0.13	0 (3)	1.08	0.77
Lower North District									
Avoca	540	1,252	0.049	16	0.54	0.17	0 (3)	1.54	0.67
Bendigo	515	1,328	0.047	14	0.48	0.17	0 (3)	1.32	0.77
Dookie	542	1,418	0.042	16	0.49	0.16	0 (3)	1.35	0.92
Shepparton	507	1,356	0.044	14	0.46	0.16	0 (3)	1.41	0.88
Yarrawonga	509	1,511	0.043	13	0.41	0.16	0 (3)	1.35	0.87
Upper Northeast District									
Beechworth	997	1,241	0.051	30+	0.81	0.47	0 (0)	2.02	1.73
Myrtleford	874	1,352	0.043	29	0.72	0.28	0 (0)	1.98	1.25
Rutherglen	608	1,318	0.046	16	0.56	0.21	0 (1)	1.43	1.02
Wangaratta	650	1,475	0.041	19	0.55	0.19	0 (3)	1.63	0.99
West Coast District									
Camperdown	721	1,051	0.057	24	0.75	0.34	0 (0)	2.72	0.73
Cape Otway	854	896	0.100	27	0.95	0.88	0 (0)	5.02	0.38
Colac	693	957	0.067	23	0.78	0.40	0 (0)	3.09	0.54
Hamilton	676	1,048	0.059	21	0.73	0.26	0 (0)	2.41	0.38
Portland (Cape Nelson)	850	920	0.100	29	0.93	0.82	0 (0)	5.01	0.59
Warrnambool	655	890	0.081	19	0.79	0.38	0 (0)	3.16	0.34
Western Plains District									
Ararat	607	1,159	0.054	18	0.63	0.19	0 (1)	2.13	0.34
Ballarat	695	1,013	0.062	20	0.76	0.35	0 (0)	2.58	0.31
West Central District									
Ballan	518	899	0.069	11	0.65	0.310	0 (0)	1.78	0.16
Durdiwarrah	658	1,001	0.062	15	0.73	0.34	0 (0)	2.44	0.29
Geelong	542	1,047	0.062	11	0.59	0.26	0 (1)	2.13	0.65
Laverton	549	1,135	0.057	11	0.55	0.25	0 (1)	2.05	0.60
Queenscliffe	631	949	0.074	13	0.72	0.35	0 (0)	3.27	0.50
Werribee	496	1,034	0.062	9	0.54	0.26	0 (0)	2.05	0.53
East Central District									
Cape Schanck	741	925	0.075	21	0.84	0.52	0 (0)	4.13	0.52
Frankston	652	1,232	0.049	16	0.61	0.27	0 (0)	2.37	0.61
Kew	710	1,202	0.051	14	0.67	0.31	0 (0)	2.49	0.90
Melbourne	658	1,205	0.057	11	0.61	0.32	0 (0)	2.76	0.98
Mornington	720	992	0.061	21	0.79	0.41	0 (0)	3.23	0.95
Mt. Eliza	670	1,007	0.061	16	0.74	0.34	0 (0)	3.39	0.65
North Central District									
Castlemaine	522	1,288	0.048	14	0.50	0.15	0 (3)	1.25	0.58
Maryborough	510	1,257	0.048	15	0.50	0.16	0 (3)	1.38	0.66
Seymour	563	1,304	0.049	15	0.54	0.18	0 (3)	1.52	0.85

* Months with M.I. <0.10 and, in parentheses, M.I. <0.20.

Table 10. Climate stations in southwest Western Australia; R.L. Specht.

Climate station	Latitude (South)	Longitude (East)	Altitude (m)	Vegetation type
North Coastal District				
Carnamah	29° 41'	115° 53'	267.9	York and salmon gum woodland/*Casuarina* thicket
Eneabba	29° 48'	115° 21'	228.6	Scrub heath (Kwongan)
Geraldton	28° 48'	114° 42'	33.2	*Acacia – Banksia* shrubland
Wongan Hills P.O.	30° 53'	116° 43'	285.6	Mallee and *Casuarina* thicket
Central Coastal District				
Badgingarra	30° 20'	115° 33'	257.6	Scrub heath mosaic
Fremantle	32° 03'	115° 45'	14.3	Tuart woodland
Jurien Bay	30° 18'	115° 02'	1.5	Heath on limestone
Kalamunda	32° 00'	116° 04'	317.0	Jarrah forest
Perth	31° 57'	115° 51'	18.6	Tuart – jarrah woodland
Rottnest Island	32° 00'	115° 31'	46.3	Cypress pine low forest
W.A.I.T. (South Bentley)	32° 00'	115° 52'	18.3	*Banksia* low woodland
South Coastal District				
Albany	34° 57'	117° 48'	68.9	Jarrah-marri forest
Bridgetown	33° 58'	116° 08'	154.8	Jarrah-marri forest
Bunbury	33° 19'	115° 38'	3.7	*Agonis* scrub
Cape Naturaliste	33° 32'	115° 01'	111.9	*Acacia* thicket
Collie	33° 26'	116° 09'	190.2	Jarrah-marri forest
Denmark Res. Stn	34° 56'	117° 20'	18.3	Jarrah/karri forest
Donnybrook	33° 34'	115° 49'	63.4	Jarrah-marri forest
Dwellingup	32° 47'	116° 02'	263.3	Jarrah-marri forest
Eclipse Island	35° 11'	117° 53'	102.7	Scrub/heath
Esperance	33° 51'	121° 53'	4.3	Coastal dune scrub
Karridale	34° 12'	115° 08'	47.9	Karri tall forest
Manjimup	34° 14'	116° 09'	279.5	Jarrah-marri forest
Mount Barker	34° 38'	117° 40'	253.6	Jarrah-marri forest
Pemberton Forest	34° 27'	116° 01'	171.0	Karri-marri tall forest
North Central District				
Bencubbin	30° 49'	117° 52'	350.5	York gum – salmon gum woodland/*Acacia* thicket
Kellerberrin	31° 38'	117° 43'	247.2	York gum – salmon gum woodland
Merredin	31° 28'	118° 19'	314.6	Mallee + woodland patches
Muresk	31° 45'	116° 41'	152.4	York gum woodland
Northam	31° 39'	116° 40'	149.4	York gum woodland
York	31° 53'	116° 46'	173.4	York gum woodland
South Central District				
Katanning	33° 41'	117° 33'	311.5	York gum – wandoo woodland
Lake Grace	33° 06'	118° 28'	288.0	Salmon gum woodland/Mallee
Narrogin	32° 56'	117° 10'	339.5	York gum – wandoo woodland
Pingelly	32° 32'	117° 05'	296.6	York gum – wandoo woodland
Wandering	32° 40'	116° 41'	335.3	Wandoo woodland
Eucla District				
Balladonia	32° 27'	123° 52'	153.0	*E. flocktoniae – E. oleosa* woodland
Eucla	31° 41'	128° 53'	86.9	Myall with saltbush or bluebush
Forrest	30° 50'	128° 06'	157.0	Open bluebush plains
Mandura Motel	31° 54'	127° 00'	21.3	Myall with saltbush or bluebush
Rawlinna	31° 01'	125° 20'	184.1	Open bluebush plains
South East District				
Booylgoo	27° 45'	119° 54'	609.6	Mulga low woodland
Kalgoorlie	30° 47'	121° 27'	359.7	Goldfield blackbutt woodland
Laverton	28° 37'	122° 24'	457.2	Mulga low woodland
Menzies	29° 41'	121° 02'	427.6	Mulga low woodland
Salmon Gums	32° 59'	121° 39'	248.7	Mallee with patches of salmon gum and gimlet
Southern Cross	31° 13'	119° 19'	356.6	Salmon gum – morrell woodland

References

Aust. Govt., Bureau of Meteorology, 1975. Climatic Averages Western Australia. Aust. Govt. Publ. Service, Canberra.

Beard, J.S., 1975, 1981. Vegetation Survey of Western Australia. Sheet 4. Nullarbor. Sheet 7. Swan (1:1,000,000 Vegetation Series). University of Western Australia Press, Perth.

Table 11. Climatic data for southwest Western Australia – Classification; R.L. Specht.

Climate station	Mean annual precipitation (mm)	Mean temperature (°)		Emberger pluviothermic quotient (Q)	Climatic type	
		hottest month max.	coldest month min.		Emberger	Köppen
North Coastal District						
Carnamah	397	35.8	7.2	47	warm/temperate, semi-arid	Csa
Eneabba	590	35.7	6.8	69	temperate/warm, semi-arid	Csa
Geraldton	477	32.2	8.7	69	warm, semi-arid	Csa
Wongan Hills P.O.	389	34.0	5.8	47	temperate, semi-arid	Csa
Central Coastal District						
Badgingarra	633	34.8	6.3	76	temperate, sub-humid	Csa
Fremantle	775	28.0	9.6	144	warm, sub-humid	Csa
Jurien Bay	519	30.2	9.2	84	warm, semi-arid	Csa
Kalamunda	1,069	30.0	7.4	162	warm, humid	Csa
Perth	837	31.4	7.7	121	warm, sub-humid	Csa
Rottnest Island	736	26.7	11.4	165	hot, sub-humid	Csa
W.A.I.T. (South Bentley)	768	31.9	8.3	111	warm, sub-humid	Csa
South Coastal District						
Albany	815	25.8	6.8	148	temperate/warm, humid	Csb
Bridgetown	856	29.9	4.4	116	temperate, sub-humid	Csb
Bunbury	881	27.9	8.3	154	warm, humid	Csa
Cape Naturaliste	838	25.8	9.8	180	warm/hot, humid	Csb
Collie	988	31.1	4.7	129	temperate, sub-humid	Csa
Denmark Res. Stn	1,012	26.1	6.6	179	temperate, humid	Csb
Donnybrook	1,019	30.8	5.9	141	temperate, humid	Csb
Dwellingup	1,306	29.6	5.0	183	temperate, humid	Csa
Eclipse Island	951	21.8	9.2	262	warm, per-humid	Csb
Esperance	675	25.9	7.2	125	warm/temperate, sub-humid	Csb
Manjimup	1,055	27.6	6.0	169	temperate, humid	Csb
Mount Barker	756	27.4	6.0	122	temperate, sub-humid	Csb
Pemberton Forest	1,255	26.3	7.0	225	warm/temperate, per-humid/humid	Csb
Karridale	1,210	24.7	8.1	252	warm, per-humid	Csb
North Central District						
Bencubbin	323	34.4	5.3	38	temperate, arid/semi-arid	BShs
Kellerberrin	339	33.9	5.6	41	temperate, semi-arid/arid	BShs
Merredin	327	33.2	4.6	39	temperate, arid/semi-arid	BSks
Muresk	463	33.9	4.7	54	temperate, semi-arid	Csa
Northam	435	34.1	5.3	52	temperate, semi-arid	Csa
York	454	34.0	4.8	53	temperate, semi-arid	Csa
South Central District						
Katanning	490	30.7	5.4	67	temperate, semi-arid	Csa
Lake Grace	358	32.2	4.8	45	temperate, semi-arid	BSk
Narrogin	507	30.9	5.0	67	temperate, semi-arid	Csa
Pingelley	460	31.5	5.2	60	temperate, semi-arid	Csa
Wandering	633	32.0	3.8	77	temperate, semi-humid	Csa
Eucla District						
Balladonia	245	31.9	4.9	31	temperate, arid	BSk
Eucla	254	25.9	7.0	46	temperate/warm, semi-arid	BSk
Forrest	180	32.7	4.2	22	temperate, arid	BWh
Mandura Motel	247	33.3	6.1	31	temperate, arid	BSk
Rawlinna	183	33.6	4.8	22	temperate, arid	BWh
South East District						
Booylgoo	206	36.7	4.8	22	temperate, arid	BWh
Kalgoorlie	252	33.6	5.1	30	temperate, arid	BWh
Laverton	221	36.3	5.4	24	temperate, arid	BWh
Menzies	239	35.2	5.3	27	temperate, arid	BWh
Salmon Gums	341	31.1	4.2	44	temperate, semi-arid	BSk
Southern Cross	279	35.0	4.4	31	temperate, arid	BSh

Table 12. Climatic data for southwest Western Australia – Growth Indices; R.L. Specht.

Climatic station	Annual precip. (mm)	Annual pan evap. (mm)	Evap. coeff. (k)	Smax (cm)	Moisture Index Annual	Moisture Index Summer	Summer drought (mo)*	T.I. × M.I. (Ann. Total) Temp.	T.I. × M.I. (Ann. Total) Subtrop.
North Coastal District									
Carnamah	401	1,900	0.031	17	0.30	0.06	4 (5)	1.59	0.61
Geraldton	472	1,506	0.050	14	0.42	0.07	3 (5)	3.62	1.04
Central Coastal District									
Fremantle	791	1,293	0.054	30+	0.67	0.10	2 (3)	4.85	1.72
Kalamunda	1,087	1,486	0.041	30+	0.66	0.10	1 (3)	2.94	1.44
Perth	914	1,480	0.045	30+	0.65	0.09	2 (3)	3.99	1.44
Rottnest Island	750	1,237	0.057	30+	0.66	0.09	3 (3)	5.30	1.89
South Coastal District									
Albany	1,008	1,017	0.085	30+	0.89	0.64	0 (0)	4.94	1.38
Bridgetown	879	1,303	0.048	30+	0.69	0.14	1 (3)	2.37	0.97
Bunbury	844	1,158	0.056	30+	0.70	0.13	1 (3)	3.95	1.64
Cape Naturaliste	821	1,136	0.060	30+	0.72	0.16	1 (2)	4.64	1.19
Collie	1,006	1,299	0.047	30+	0.70	0.13	1 (3)	1.74	1.47
Donnybrook	1,059	1,479	0.043	30+	0.67	0.12	1 (3)	2.25	1.33
Dwellingup	1,292	1,132	0.054	30+	0.75	0.20	1 (2)	2.40	1.23
Eclipse Island	913	1,006	0.080	30+	0.86	0.56	0 (0)	5.52	0.89
Esperance	679	1,098	0.070	22	0.70	0.18	0 (2)	4.25	1.00
Karridale	1,210	997	0.071	30+	0.83	0.47	0 (1)	4.33	1.11
Manjimup	1,081	1,072	0.059	30+	0.79	0.35	0 (1)	3.06	0.92
Mount Barker	768	1,039	0.062	30+	0.78	0.37	0 (1)	3.04	0.95
North Central District									
Kellerberrin	353	1,669	0.037	13	0.29	0.07	2 (5)	1.03	0.52
Merredin	305	1,705	0.036	11	0.25	0.06	3 (6)	0.75	0.46
Muresk	471	1,608	0.036	22	0.44	0.06	3 (5)	1.39	0.70
York	457	1,555	0.039	19	0.43	0.07	3 (5)	1.46	0.69
South Central District									
Katanning	494	1,256	0.047	20	0.54	0.10	2 (3)	1.38	0.69
Eucla District									
Balladonia	227	1,487	0.045	5	0.18	0.08	2 (7)	0.69	0.47
Eucla	253	1,297	0.063	4	0.22	0.12	0 (5)	1.08	0.68
Rawlinna	168	1,785	0.039	3	0.11	0.07	6 (12)	0.41	0.32
South East District									
Booylgoo	202	2,187	0.030	4	0.11	0.09	5 (12)	0.22	0.27
Kalgoorlie	240	1,818	0.035	6	0.16	0.09	3 (8)	0.60	0.39
Laverton	220	2,209	0.030	5	0.11	0.08	5 (12)	0.34	0.26
Menzies	237	2,035	0.032	6	0.14	0.08	3 (10)	0.49	0.28
Southern Cross	283	1,784	0.035	9	0.21	0.08	2 (7)	0.62	0.36

* Months with M.I.<0.10 and, in parentheses, M.I.<0.20.

Table 13. Climate stations in California, U.S.A. P.H. Zedler.

Climate station	Latitude (North)	Longitude (West)	Altitude (m)	Vegetation type*
North Coast Drainage				
Butler Valley Ranch	40° 46'	123° 54'	107	redwood forest
Duttons Landing	38° 12'	122° 18'	5	coastal prairie
Ferndale	40° 36'	124° 17'	3	coastal prairie
Trinity Dam Hatchery	40° 43'	122° 48'	635	Oregon oak forest; mixed evergreen forest with chinquapin
Tulelake	41° 58'	121° 28'	1,025	sagebrush steppe
Warm Springs Dam	38° 43'	123° 00'	57	blue oak-digger pine forest
Willow Creek	40° 57'	123° 38'	117	mixed evergreen forest with chinquapin
Sacramento Drainage				
Auburn Dam	38° 53'	121° 04'	323	blue oak-digger pine forest
Berryessa Lake	38° 33'	122° 14'	117	blue oak-digger pine forest; chaparral
Brannan Island	38° 07'	121° 42'	8	Tule marsh
Chico Univ. Farm	39° 42'	121° 49'	47	blue oak-digger pine forest; riparian forest
Davis Expt. Farm	38° 32'	121° 46'	15	California prairie; riparian forest

Table 13. (Continued).

Climate station	Latitude (North)	Longitude (West)	Altitude (m)	Vegetation type*
Fall River Mills	41° 01'	121° 28'	848	Sierran montane forest
Folsom Dam	38° 42'	121° 10'	89	blue oak-digger pine forest; riparian forest
Grizzly Island	38° 09'	121° 58'	<1	Tule marsh
Lakeshore	40° 53'	122° 23'	273	northern yellow pine forest
Lake Solano	38° 30'	122° 00'	36	blue oak-digger pine forest; California prairie; riparian forest
Lake Spaulding	39° 19'	120° 38'	1,310	Sierra montane forest
Markley Cove	38° 30'	122° 07'	122	blue oak-digger pine forest; chaparral
Monticello Dam	38° 30'	122° 07'	128	blue oak-digger pine forest; chaparral
Placerville IFG	38° 44'	120° 44'	700	blue oak-digger pine forest; Sierra yellow pine forest
Shasta Dam	40° 43'	122° 25'	273	northern yellow pine forest
Turntable Creek	40° 46'	122° 18'	271	northern yellow pine forest
Walnut Grove	38° 14'	121° 31'	6	Tule marsh
Whiskey Town	40° 37'	122° 32'	329	northern yellow pine forest
Willows	39° 32'	122° 12'	34	California prairie
North East Interior Basins				
Boca	39° 23'	120° 06'	1,416	Sierra montane forest
Fleming	40° 22'	120° 19'	1,016	sagebrush steppe
Tahoe City	39° 10'	120° 08'	1,582	Sierra montane forest
Central Coast Drainage				
Burlingame	37° 35'	122° 21'	3	mixed hardwood forest; coastal prairie-scrub mosaic
Nacimiento Dam	35° 46'	120° 53'	196	blue oak-digger pine forest
Newark	37° 31'	122° 02'	3	coastal prairie-scrub mosaic; coastal salt marsh
San Joaquin Drainage				
Antioch	37° 59'	121° 44'	15	California prairie
Avenal	35° 54'	120° 03'	135	California prairie
Camp Pardee	38° 15'	120° 51'	167	blue oak-digger pine forest
Friant Govt. Camp	36° 59'	119° 43'	104	blue oak-digger pine forest; California prairie
Hetch Hetchy	37° 57'	119° 47'	983	Sierra yellow pine
Kettleman City	36° 00'	119° 58'	64	California prairie
Knights Ferry	37° 48'	120° 39'	80	blue oak-digger pine forest; California prairie
Little Panoche Dam	36° 47'	120° 48'	172	California prairie
Lodi	38° 07'	121° 17'	10	California prairie
Los Banos	37° 01'	120° 56'	103	California prairie
Mandeville Island	38° 02'	121° 34'	3	Tule marsh
Manteca	37° 48'	121° 12'	10	California prairie
New Melones Dam	37° 57'	120° 32'	197	blue oak-digger pine forest
Oakdale Woodward Dam	37° 52'	120° 52'	55	
Salt Springs	38° 30'	120° 13'	940	Sierran montane forest
San Luis Dam	37° 03'	121° 04'	70	California prairie
Stockton Mowry Bridge	37° 51'	121° 23'	4	California prairie
Tracey Pumping Plant	37° 48'	121° 35'	15	California prairie
South Coast Drainage				
Beaumont Pump Plant	33° 59'	116° 58'	773	coastal sagebrush (inland)
Cachuma Lake	34° 35'	119° 59'	198	southern oak forest; chaparral
Camp Pendleton	33° 20'	117° 30'	28	coastal sagebrush
Chula Vista	32° 36'	117° 06'	2	coastal sagebrush
Cuyamaca Lake	32° 59'	116° 35'	1,179	southern Jeffrey pine forest
Echo Valley	32° 55'	116° 40'	254	chaparral
Mount Laguna	32° 50'	116° 26'	516	southern Jeffrey pine forest
Riverside Citrus Expt. Stn	33° 58'	117° 21'	250	coastal sagebrush (inland)
San Dimas (Tanbark Flat)	34° 12'	117° 46'	711	chaparral
Twitchell Dam	34° 59'	120° 19'	148	southern oak forest
Warner Springs	33° 17'	116° 38'	970	chaparral
South East Desert Basins				
Backus Ranch	34° 57'	118° 11'	672	Mojave creosote bush
Death Valley	36° 28'	116° 52'	−49	desert saltbush
Indio Date Garden	33° 44'	116° 15'	3	Sonoran creosote bush
Mojave	35° 03'	118° 10'	695	Mojave creosote bush

* Kuchler, A.W. (1977) *Natural Vegetation of California.* Map In: *Terrestrial Vegetation of California.* M.G. Barbour and J. Major (eds). John Wiley, New York.

Table 14. Climatic data for California, U.S.A. – Classification; R.L. Specht.

Climate station	Mean annual precipitation (mm)	Mean annual snow (m)	Mean temperature (°) hottest month max.	Mean temperature (°) coldest month min.	Emberger pluviothermic quotient (Q)	Climatic type Emberger	Köppen
North Coast Drainage							
Butler Valley Ranch	1,640	trace	16.3	5.0	512	temperate, per-humid	Csb
Duttons Landing	501	–	31.4	2.7	60	cool, semi-arid/sub-humid	Csb
Ferndale	1,080	trace	18.3	4.8	281	temperate, per-humid	Csb
Trinity Dam	821	(0.27)*	35.0	– 1.1	78	cold, sub-humid	Csa/b
Tulelake	277	0.56	29.6	– 7.1	27	very cold, semi-arid	BSks
Warm Springs Dam	1,092	trace	32.2	1.7	123	cool, humid	Csb
Willow Creek	1,393	(0.25)	35.0	1.1	141	cool, humid	Csa/b
Sacramento Drainage							
Auburn Dam	865	trace	34.4	2.2	92	cool, sub-humid	Csa
Berryessa Lake	783	trace	31.7	2.8	93	cool/temperate, sub-humid	Csa
Brannan Island	398	–	31.7	2.8	47	cool/temperate, semi-arid	Csa
Chico Univ. Farm	681	–	36.3	2.2	68	cool, sub-humid/semi-arid	Csa
Davis Expt. Farm	439	–	35.2	1.9	45	cool, semi-arid	Csa
Fall River Mills	457	0.80	35.0	– 5.0	40	very cold, sub-humid/semi-arid	Csb
Folsom Dam	614	trace	33.9	1.7	66	cool, sub-humid	Csa
Grizzly Island	363	–	32.2	3.3	43	temperate/cool, semi-arid	Csb
Lakeshore	1,801	(0.30)	35.4	0.2	176	cool/cold, humid	Csa
Lake Solano	570	–	35.0	2.2	60	cool, semi-arid	Csa
Lake Spaulding	1,767	(4.00)	27.2	– 4.6	195	very cold, per-humid	Csb
Markley Cove	645	–	33.3	2.8	73	cool/temperate, sub-humid	Csa
Monticello Dam	870	–	33.3	3.3	100	temperate/cool, sub-humid	Csa
Placerville IFG	995	(2.00)	29.4	– 2.2	110	cold, humid	Csa
Shasta Dam	1,605	0.24	35.5	3.3	170	temperate/cool, humid	Csa
Turntable Creek	1,635	(0.25)	34.3	2.6	177	cool, humid	Csa
Walnut Grove	411	–	33.9	2.2	45	cool, semi-arid	Csa
Whiskeytown	1,598	(0.25)	35.6	3.3	169	temperate/cool, humid	Csa
Willows	456	–	35.7	2.1	46	cool, semi-arid	Csa
North East Interior Basins							
Boca	580	3.50	28.2	– 13.4	50	ext. cold, sub-humid	Csb
Fleming	231	(1.50)	31.1	– 6.1	22	very cold, semi-arid	BSks
Tahoe City	842	5.42	25.8	– 8.4	87	ext. cold, sub-humid/humid	Csb
Central Coast Drainage							
Burlingame	530	–	23.3	4.8	100	temperate, sub-humid	Csb
Nacimiento Dam	367	–	33.9	0.6	38	cool, semi-arid	Csa
Newark	380	–	31.1	3.3	47	temperate/cool, semi-arid	Csb
San Joaquin Drainage							
Antioch	311	–	32.9	2.9	36	cool/temperate, arid	Csa
Avenal	181	–	37.8	3.9	18	temperate, arid	BSks
Camp Pardee	530	–	36.0	3.2	55	temperate/cool, semi-arid	Csa
Friant Govt. Camp	346	–	38.6	2.2	32	cool, arid	Csa
Hetch Hetchy	898	1.93	30.9	– 2.8	93	cold/very cold, humid	Csb
Kettleman City	181	–	37.6	3.7	18	temperate, arid	BSks
Knights Ferry	439	(0.10)	34.4	– 0.6	43	cold, semi-arid	Csa
Little Panoche Dam	185	–	35.6	2.2	19	cool, arid	BSks
Lodi	415	–	33.2	2.6	47	cool, semi-arid	Csa
Los Banos	205	–	38.5	2.4	19	cool, arid	BSks
Mandeville Island	311	–	33.3	2.8	35	cool, arid	Csa
Manteca	301	–	33.9	2.2	33	cool, arid/semi-arid	BSks
New Melones Dam	767	(0.19)	35.0	1.1	78	cool, sub-humid	Csa
Oakdale Woodward Dam	344	–	35.6	1.8	35	cool, semi-arid	Csa
Salt Springs	1,153	–	30.6	0.3	132	cool/cold, humid	Csa
San Luis Dam	259	–	38.3	2.2	24	cool, arid	BSks
Stockton Mowry Bridge	323	–	33.9	2.1	35	cool, arid/semi-arid	Csa
Tracy Pumping Plant	291	–	32.8	2.2	33	cool, arid/semi-arid	BSks
South Coast Drainage							
Beaumont Pump Plant	486	–	34.8	2.5	52	cool, semi-arid	Csa
Cachuma Lake	543	–	23.9	4.4	97	temperate, sub-humid	Csb
Camp Pendleton	204	–	22.8	7.2	45	hot/temperate, semi-arid	BSks
Chula Vista	226	–	22.5	5.6	47	temperate, semi-arid	BSks
Cuyamaca Lake	1,081	0.97	29.6	– 2.2	119	cool, humid	Csb
Echo Valley	474	–	33.9	5.0	56	temperate, semi-arid	Csb
Mount Laguna	449	(1.00)	28.3	– 2.8	51	cold, sub-humid	Csa
Riverside Citrus Expt. Stn	252	–	34.5	3.2	28	temperate/cool, arid	BSks
San Dimas (Tanbark Flat)	693	trace	33.9	2.2	75	cool, sub-humid	Csa/b
Twitchell Dam	465	–	33.9	0.6	48	cool, semi-arid	Csb
Warner Springs	412	(0.50)	33.9	– 1.1	41	cold, semi-arid	Csb
South East Desert Basins							
Backus Ranch	163	(0.60)	37.0	– 1.2	15	cold, arid/very arid	BWks
Death Valley	46	–	46.7	5.0	4	temperate, very arid	BWhs
Indio Date Garden	76	–	41.6	3.4	7	temperate, very arid	BWhs
Mojave	147	(0.60)	37.2	– 1.1	13	cold, arid/very arid	BWks

* Estimated annual snow fall, shown in parentheses.

Table 15. Climatic data for California, U.S.A. – Growth Indices; R.L. Specht.

Climatic station	Annual precip. (mm)	Annual pan evap. (mm)	Evap. coeff. (k)	Smax (cm)	Moisture Index Annual	Moisture Index Summer	Summer drought (mo)**	T.I. × M.I. (Ann. Total) Temp.	T.I. × M.I. (Ann. Total) Subtrop.
North Coast Drainage									
Butler Valley Ranch	1,640	898	0.056	>30	0.78	0.27	1 (1)	1.94	0.33
Dutton's Landing	501	1,646	0.041	>30	0.47	0.02	4 (5)	0.82	0.09
Ferndale	1,080	795	0.084	>30	0.80	0.38	1 (1)	1.18	0.09
Trinity Dam	821	1,433	0.039	>30	0.61	0.07	2 (3)	1.19	0.29
Tulelake	277 (incl. 56 cm snow)	1,186 (5 mo)+	0.042	15–20	0.28*	0.04	3 (4)	0.35	0.09
Warm Springs Dam	1,092	1,671	0.036	>30	0.60	0.04	3 (4)	1.32	0.61
Willow Creek	1,393	1.075	0.046	>30	0.74	0.14	1 (3)	1.76	0.87
Sacramento Drainage									
Auburn Dam	865	1,745	0.033	>30	0.57	0.06	3 (4)	1.12	0.58
Berryessa Lake#	783	1,976	0.029	>30	0.45	0.01	3 (6)	0.73	0.41
Brannan Is.	398	2,089	0.029	>30	0.31	0.01	4 (6)	0.64	0.18
Chico Univ. Farm	681	1,964	0.029	>30	0.49	0.02	3 (5)	1.27	0.45
Davis Expt. Farm	439	1,774	0.036	>30	0.39	0.01	4 (5)	1.11	0.29
Fall River Mill	457 (incl. 80 cm snow)	1,593	0.033	>30	0.42	0.05	3 (4)	0.52	0.09
Folsom Dam	614	1,627	0.036	>30	0.53	0.03	3 (4)	1.28	0.48
Grizzly Is.	363	1,685	0.037	10–15	0.33	0.03	4 (5)	0.72	0.24
Lakeshore	1,801	1,392	0.040	>30	0.75	0.18	1 (2)	2.13	1.41
Lake Solano	570	2,037	0.031	>30	0.42	0.01	5 (5)	1.25	0.25
Lake Spaulding	1,767	816 (6 mo)	0.042	>30	0.54*	0.25	2 (2)	1.03	0.26
Markley Cove	645	1,741	0.032	>30	0.51	0.02	3 (4)	1.12	0.36
Monticello Dam#	870	1,592	0.036	>30	0.50	0.02	3 (5)	0.86	0.46
Placerville IFG	995	1,359	0.043	>30	0.66	0.10	1 (3)	1.43	0.94
Shasta Dam	1,605 (incl. 24 cm snow)	1,863	0.031	>30	0.66	0.12	1 (3)	1.13	1.00
Turntable Creek	1,635	1,862	0.036	>30	0.68	0.12	1 (3)	1.69	1.03
Walnut Grove	411	1,624	0.038	>30	0.38	0.01	4 (6)	1.13	0.30
Whiskeytown	1,598	1,479	0.036	>30	0.69	0.12	1 (3)	1.36	0.93
Willows	456	2,007	0.033	15–20	0.34	0.02	5 (5)	0.77	0.27
Northeast Interior Basins									
Boca	580 (incl. 350 cm snow)	1,200 (6 mo)	0.038	>30	0.35*	0.09	2 (4)	0.37	0.07
Fleming	231	1,248 (5 mo)	0.040	15–20	0.20*	0.05	4 (6)	0.41	0.13
Tahoe City	842 (incl. 542 cm snow)	604 (6 mo)	0.073	>30	0.61*	0.42	1 (2)	1.31	0.17
Central Coast Drainage									
Burlingame	530	1,508	0.045	>30	0.50	0.02	4 (4)	1.32	0.11
Nacimiento Dam	367	1,812	0.034	15–20	0.31	0.02	5 (6)	0.63	0.19
Newark	380	1,425	0.047	10–15	0.40	0.01	5 (5)	1.08	0.13
San Joaquin Drainage									
Antioch	311	1,895	0.032	10–15	0.25	0.01	5 (6)	0.56	0.20
Avenal	181	2,891	0.021	5–10	0.09	0.004	7 (10)	0.29	0.14
Camp Pardee	530	1,559	0.033	>30	0.51	0.03	3 (4)	1.12	0.65
Friant Govt. Camp	346	2,285	0.024	15–20	0.23	0.01	4 (6)	0.65	0.29
Hetch Hetchy	898 (incl. 193 cm snow)	962 (6 mo)	0.043	>30	0.46*	0.18	2 (2)	1.34	0.66
Kettleman City	181	2,352	0.027	5–10	0.12	0.003	6 (8)	0.30	0.15
Knights Ferry	439	1,726	0.034	>30	0.41	0.02	3 (5)	0.56	0.41
Little Panoche Dam	185	2,825	0.021	5–10	0.11	0.01	6 (10)	0.29	0.14
Lodi	415	1,715	0.034	>30	0.38	0.01	4 (5)	0.81	0.30
Los Banos	205	2,797	0.021	10–15	0.12	0.01	6 (8)	0.33	0.13
Mandeville Is.	311	1,920	0.034	10–15	0.24	0.01	6 (6)	0.67	0.19
Manteca	301	1,803	0.034	10–15	0.26	0.02	5 (5)	0.48	0.22
New Melones Dam	767	1,942	0.030	>30	0.50	0.01	3 (4)	1.04	0.35
Oakdale Woodward Dam	344	1,936	0.027	15–20	0.27	0.01	4 (6)	0.65	0.30
Salt Springs	1,153	1,744	0.036	>30	0.64	0.08	2 (3)	1.31	0.52
San Luis Dam	259	2,761	0.021	10–15	0.15	0.01	6 (8)	0.43	0.14
Stockton Mowry Bridge	323	1,746	0.034	15–20	0.28	0.02	5 (6)	0.81	0.22
Tracey Pump Pl.	291	2,539	0.023	15–20	0.18	0.01	5 (7)	0.56	0.19
South Coast Drainage									
Beaumont Pump Plant	486	2,105	0.032	10–15	0.31	0.04	4 (5)	0.60	0.24
Cachuma Lake	543	1,744	0.042	15–20	0.42	0.02	5 (6)	1.38	0.18
Camp Pendleton	204	1,631	0.049	2–5	0.16	0.01	6 (7)	0.78	0.07
Chula Vista	226	1,609	0.051	2–5	0.18	0.01	6 (7)	0.67	0.05
Cuyamaca Lake	1,081 (incl. 965 cm snow)	1,492	0.038	>30	0.61	0.11	1 (3)	0.81	0.32
Echo Valley	474	2,941	0.023	10–15	0.21	0.03	3 (6)	0.44	0.13
Mount Laguna	449	1,737	0.035	10–15	0.33	0.12	1 (5)	0.21	0.36
Riverside Citrus Expt. Stn.	252	1,646	0.043	5–10	0.21	0.02	5 (7)	0.86	0.17
San Dimas (Tanbark Flat)	693	1,613	0.041	>30	0.57	0.06	2 (4)	1,27	0.30
Twitchel Dam	465	1,752	0.044	10–15	0.35	0.03	5 (5)	1.33	0.14

Table 15. (Continued).

Climatic station	Annual precip. (mm)	Annual pan evap. (mm)	Evap. coeff. (k)	Smax (cm)	Moisture Index		Summer drought (mo)**	T.I. × M.I. (Ann. Total)	
					Annual	Summer		Temp.	Subtrop.
South East Desert Basins									
Backus Ranch	163	2,950	0.021	5–10	0.08	0.006	8 (11)	0.20	0.07
Death Valley	46	3,879	0.017	1–2	0.02	0.005	12 (12)	0.09	0.00
Indio Date Garden	76	2,808	0.025	2–5	0.04	0.017	12 (12)	0.17	0.06
Mojave	147	2,956	0.022	5–10	0.07	0.006	8 (12)	0.07	0.07

* Mean Moisture Index for non-frozen months.
# Data for only 1–2 years.
+ Number months evap. pan frozen.
** Months with M.I.<0.10 and, in parentheses, M.I.<0.20.

Table 16. Climate stations in Arizona, U.S.A.; R.L. Specht.

Climate	County	Latitude (North)	Longitude (West)	Altitude (m)	Vegetation type
North East					
Fort Valley	Coconino	35° 16′	111° 44′	1,866	yellow pine – douglas fir forest
Grand Canyon NP	Coconino	36° 03′	112° 09′	1,723	pinyon pine – ponderosa pine – juniper forest
Hawley Lake	Apache	33° 59′	109° 45′	2,078	ponderosa pine forest
Many Farms	Apache	36° 22′	109° 37′	1,347	sagebrush
McNary	Apache	34° 04′	109° 51′	1,859	ponderosa pine forest
Page	Coconino	36° 56′	111° 27′	1,085	juniper – pinyon pine forest/sagebrush
Snowflake 15W	Navajo	34° 30′	110° 20′	1,544	sagebrush/plains grass
Wahweap	Coconino	36° 59′	111° 29′	947	juniper – pinyon pine forest/sagebrush
Whiteriver	Navajo	33° 50′	109° 58′	1,341	ponderosa pine forest/chaparral/oak woodland
East Central					
Roosevelt 1WNW	Gila	33° 40′	111° 09′	560	chaparral/oak woodland
San Carlos Res.	Gila	33° 10′	110° 31′	643	creosote bush/salt bush
Sierra Ancha	Gila	33° 48′	110° 58′	1,295	chaparral/oak woodland
South West					
Yuma Citrus Station	Yuma	32° 37′	114° 39′	49	creosote bush/mesquite/palo verde
South Central					
Bartlett Dam	Maricopa	33° 49′	111° 38′	419	chaparral/oak woodland/creosote bush
Mesa Exp. Stn	Maricopa	33° 25′	111° 52′	312	chaparral/oak woodland/creosote bush/palo verde
Stewart Mtn	Maricopa	33° 34′	111° 32′	361	chaparral/oak woodland/creosote bush/salt bush
Tempe Citrus Exp. Stn	Maricopa	33° 23′	111° 58′	300	chaparral/oak woodland/creosote bush
Winkelman 9S	Pinal	32° 52′	110° 43′	538	chaparral/oak woodland/creosote bush/mesquite
South East					
Black River Pumps	Graham	33° 29′	109° 46′	1,543	chaparral/oak woodland
Douglas	Cochise	31° 21′	109° 32′	1,026	chaparral/oak woodland/creosote bush/mesquite
Nogales 6N	Santa Cruz	31° 25′	110° 57′	904	chaparral/oak woodland/mesquite
Tucson Univ. of Arizona	Pima	32° 15′	110° 57′	621	chaparral/oak woodland/creosote bush
North West					
Davis Dam No. 2	Mohave	35° 12′	114° 34′	167	creosote bush/salt bush

References

Huphrey, R.R., 1963. Arizona Natural Vegetation (Map): University of Arizona, Tucson, Agric. Exp. Stn. Bull. A-45.

Sellers, W.D. and Hill, R.H. (eds), 1974. Arizona Climate 1931–1972. The University of Tuscon Press, Tucson, Arizona, 616 pp.

Table 17. Climatic data for Arizona, U.S.A. – Classification; R.L. Specht.

Climate station	Mean annual precip. (mm)	Mean annual snow (mm)	Mean temperature (°C)		Emberger pluviothermic quotient (Q)	Emberger climatic type
			hottest month max.	coldest month min.		
North East						
Fort Valley	559	2,466	27.3	−13.1	49.4	non-medit.
Grand Canyon NP	367	1,648	29.3	−6.9	35.7	non-medit.
Hawley Lake	778	5,006	24.0	−14.3	73.1	non-medit.
Many Farms	193	165	33.6	−9.8	15.6	non-medit.
McNary	632	2,593	26.9	−8.9	62.6	non-medit.
Page	146	124	36.2	−5.8	12.1	non-medit.
Snowflake 15W	282	406	29.9	−7.4	26.6	non-medit.
Wahweap	115	48	37.3	−4.2	9.6	very cold, arid
Whiteriver	441	533	33.1	−5.6	39.7	very cold, sub-humid
East Central						
Roosevelt 1WNW	359	13	38.9	2.8	33.8	cool, semi-arid
San Carlos Res.	359	3	37.7	0.3	32.9	cool, semi-arid
Sierra Ancha	629	290	33.5	−0.5	63.9	cold, sub-humid
South West						
Yuma Citrus Stn	70	trace	41.3	3.4	6.3	temperate, very arid
South Central						
Bartlett Dam	301	trace	40.6	4.5	28.2	temperate, arid
Mesa Exp. Stn	191	trace	40.2	2.0	17.0	cool, arid
Stewart Mtn	291	trace	40.3	2.2	26.0	cool, arid
Tempe Exp. Stn	196	trace	40.6	1.4	17.0	cool, arid
Winkelman 9S	322	trace	39.5	−1.7	26.8	cold, semi-arid
South East						
Black River Pumps	441	612	30.9	−6.8	41.0	non-medit.
Douglas	331	trace	34.4	−1.9	31.5	cold, semi-arid
Nogales 2N	408	5	34.4	−3.0	37.8	cold, semi-arid
Tucson (Univ.)	273	15	38.4	2.6	26.0	cool, arid
North West						
Davis Dam No. 2	106	3	42.3	5.4	9.7	temperate, very arid

Table 18. Climatic data for Arizona, U.S.A. – Growth Indices; R.L. Specht.

Climate station	Annual precip. (mm)	Annual snow (mm)	Annual pan evap. (mm)	Evap. pan frozen (mo)	Evap. coeff. (k)	Smax (cm)	Moisture Index		T.I. × M.I. (Ann. Total)	
							Annual	Summer	Temp.	Subtrop.
North East										
Fort Valley	559	2,466	713	7	0.054	30+	0.66*	–	1.35	0.17
Grand Canyon NP	367	1,648	1,209	6	0.038	30+	0.31*	–	0.73	0.43
Hawley Lake	778	5,006	863	7	0.044	30+	0.71	–	1.05	0.11
Many Farms	193	165	2,208	–	0.030	2–5	0.10	0.10	0.09	0.17
McNary	632	2,593	1.047	5	0.047	30+	0.54*	–	1.54	0.26
Page	146	124	2,083	2	0.029	2–5	0.09*	–	0.14	0.09
Snowflake 15W	282	406	1,329	6	0.028	10–15	0.26*	–	0.33	0.48
Wahweap	115	48	2,624	2	0.023	2–5	0.05*	–	0.06	0.06
Whiteriver	441	533	1,863	2	0.033	10–15	0.27*	–	0.32	0.52
East Central										
Roosevelt 1WNW	359	13	2,238	–	0.028	5–10	0.21	0.12	0.38	0.33
San Carlos Res.	359	3	2,482	–	0.027	5–10	0.18	0.14	0.30	0.25
Sierra Ancha	629	290	1,922	–	0.036	10–15	0.40	0.29	0.64	0.43
South West										
Yuma Citrus Stn	70	trace	2,441	–	0.030	1–2	0.04	0.02	0.13	0.04
South Central										
Bartlett Dam	301	trace	3,133	–	0.023	5–10	0.11	0.09	0.38	0.13
Mesa Exp. Stn	191	trace	2,160	–	0.034	2–5	0.11	0.08	0.31	0.15
Stewart Mtn	291	trace	2,684	–	0.027	5–10	0.13	0.10	0.35	0.15
Tempe Exp. Stn	196	trace	1,822	–	0.036	2–5	0.13	0.10	0.29	0.21
Winkelman 9S	322	trace	2,614	–	0.026	2–5	0.13	0.16	0.21	0.16
South East										
Black River Pumps	441	612	1,338	5	0.039	30+	0.34*	–	0.50	0.72
Douglas	331	trace	1,887	5	0.029	5–10	0.18*	–	0.36	0.27
Nogales 2N	408	5	2,389	–	0.027	5–10	0.19	0.33	0.25	0.45
Tucson (Univ.)	273	15	2,238	–	0.030	2–5	0.14	0.15	0.39	0.22
North West										
Davis Dam No. 2	106	3	3,912	–	0.020	1–2	0.03	0.02	0.15	0.08

* Mean Moisture Index for non-frozen months.

Table 19. Climate stations in Chile; E.J. Hajek (climate), R. Gajardo (native vegetation) and J.D. Molina (man-made landscape, M-ML)

Climate station	Latitude (South)	Longitude (West)	Altitude (m)	Vegetation type
Perhumid				
Concepción	36° 50'	73° 02'	15	4-B – *Nothofagus obliqua* + *Persea lingue* M-ML – Afforestations
Contulmo	38° 02'	73° 13'	30	?
Coyhaique	45° 29'	71° 33'	140	6-B – *Nothofagus pumilio*
Cullinco	38° 22'	72° 15'	377	4-B – *Nothofagus obliqua* + *Laurelia sempervirens*
Lonquimay	38° 26'	71° 15'	900	6-A – *Nothofagus obliqua*
Osorno	40° 35'	73° 09'	24	4-B – *Nothofagus obliqua* + *Laurelia sempervirens* M-L – Urban, farms
Punta Carranza	35° 36'	72° 38'	30	4-A – *Nothofagus glauca*
Punta Tumbes	36° 37'	73° 06'	120	4-B – *Nothofagus obliqua*
Rio Bueno	40° 19'	72° 55'	58	4-B – *Nothofagus obliqua* + *Laurelia sempervirens* M-ML – Agropastoral
Temuco	38° 45'	72° 35'	114	4-B – *Nothofagus obliqua* + *Laurelia sempervirens* M-ML – Urban, agropastoral
Humid				
Angol	37° 49'	72° 39'	77	4-B – *Nothofagus obliqua* + *Cryptocaria alba* M-ML – Urban, agropastoral
Balmaceda	45° 54'	71° 43'	520	8 – *Festuca pallescens* + *Mulinum spinosum*
Chillán	36° 36'	72° 02'	118	3-C – *Cryptocaria alba* M-ML – Urban, agropastoral
Constitución	35° 20'	72° 56'	7	4-A – *Nothofagus glauca* M-ML – Afforestation, plantations, farms
Linares	35° 51'	71° 36'	157	3-C – *Cryptocaria alba* M-ML – Urban, agropastoral
Los Angeles	37° 28'	72° 21'	130	4-B – *Nothofagus obliqua* + *Cryptocaria alba* M-ML – Urban, agropastoral
Molina	35° 05'	71° 16'	235	3-C – *Cryptocaria alba* + *Quillaja saponaria* M-ML – Urban, agropastoral
Panimávida	35° 46'	71° 24'	197	3-C – *Cryptocaria alba* + *Quillaja saponaria*
Punta Lavapié	37° 08'	73° 35'	46	4-B – *Nothofagus obliqua* M-ML – Afforestation, plantations, farms
Rio Cisnes	44° 45'	72° 00'	700	6-B – *Nothofagus pumilio*
Sewell (El Teniente)	34° 06'	70° 22'	2,134	2-B – *Shuquiraga oppositifolia*
Traiguén	38° 15'	72° 40'	170	4-B – *Nothofagus obliqua* + *Cryptocaria alba* M-ML – Agropastoral
Victoria	38° 13'	72° 21'	360	4-B – *Nothofagus obliqua* M-ML – Agropastoral
Sub-humid				
Cauquenes	35° 59'	72° 22'	177	3-B – *Acacia caven* + *Maytenus boaria*
Curicó	34° 58'	72° 13'	225	3-C – *Cryptocaria alba* + *Peumus boldus* M-ML – Urban, agropastoral
Punta Angeles (Valparaiso)	33° 01'	71° 38'	41	3-C – *Cryptocaria alba* + *Peumus boldus* M-ML – Urban, agropastoral
Rancagua	34° 10'	70° 45'	500	3-C – *Quillaja saponaria* M-ML – Urban, agropastoral
Rengo	34° 24'	70° 52'	139	3-C – *Cryptocaria alba* + *Peumus boldus* M-ML – Urban, agropastoral
San Antonio	33° 34'	71° 37'	5	3-C – *Cryptocaria alba* + *Peumus boldus*
San Fernando	34° 35'	71° 00'	342	3-C – *Cryptocaria alba* + *Peumus boldus* M-ML – Urban, agropastoral
San José de Maipo	33° 39'	70° 22'	1,060	3-B – *Lithraea caustica* + *Quillaja saponaria*
Talca	35° 26'	71° 40'	97	3-C – *Quillaja saponaria* M-ML – Urban, agropastoral
Zapallar	32° 33'	71° 30'	30	3-C – *Cryptocaria alba* + *Peumus boldus* M-ML – Afforestation, plantations, farms
Semi arid				
Baños de Jahuel	32° 41'	70° 39'	1,180	3-B – *Lithraea caustica* + *Porlieria chilensis*
Chile Chico	46° 36'	71° 43'	382	8 – *Festuca pallescens* + *Acaena splendens*
Colina	33° 12'	70° 40'	542	3-B – *Prosopis chilensis* + *Acacia caven*
El Belloto	33° 03'	71° 24'	121	3-C – *Cryptocaria alba* + *Peumus boldus* M-ML – Urban, agropastoral

Table 19. (Continued).

Climate station	Latitude (South)	Longitude (West)	Altitude (m)	Vegetation type
El Bosque	33° 34'	70° 41'	580	3-C – *Quillaja saponaria* + *Lithraea caustica* M-ML – Urban
Juncal	32° 52'	70° 10'	2,250	2-B – *Chuquiraga oppositifolia*
Llay-Llay	32° 50'	70° 59'	385	3-C – *Cryptocaria alba* + *Quillaja saponaria* M-ML – Urban, agropastoral
Los Andes	32° 50'	70° 37'	616	?
Los Cerrillos	33° 30'	70° 42'	506	3-C – *Quillaja saponaria* + *Lithraea caustica* M-ML – Urban
Peñablanca	33° 04'	71° 23'	154	3-C – *Cryptocaria alba* + *Lithraea caustica*
Quillota	32° 53'	71° 16'	128	3-C – *Cryptocaria alba* + *Peumus boldus* M-ML – Urban, agropastoral
Quintero	32° 47'	71° 32'	2	3-C – *Cryptocaria alba* + *Peumus boldus*
Santiago	33° 27'	70° 42'	520	3-C – *Quillaja saponaria* + *Lithraea caustica* M-ML – Urban
Arid				
La Serena	29° 54'	71° 15'	32	3-A – *Flourensia thurifera* + *Heliotropium stenophyllum*
Ovalle	30° 36'	71° 12'	220	3-A – *Acacia caven*
Punta Tortúga	29° 55'	71° 22'	25	3-A – *Flourensia thurifera* + *Heliotropium stenophyllum*
Vicuña	30° 02'	70° 44'	620	3-A – *Flourensia thurifera* + *Heliotropium stenophyllum*
Per-arid				
Antofagasta	23° 42'	70° 24'	94	?
Caldera	27° 03'	70° 58'	28	1-C – *Skytanthus acutus*
Cerro Moreno	23° 29'	70° 26'	119	1-C – *Cassia brogniartii*
Chañaral	26° 20'	70° 37'	9	1-C – *Skytanthus acutus*
Copiapó	27° 21'	70° 24'	370	1-C – *Skytanthus acutus*
Potrerillos	26° 30'	69° 27'	2,850	1-B – *Atriplex atacamensis*
Refresco	25° 19'	69° 52'	1,850	1-A – Absolute desert
Taltal	25° 25'	70° 34'	39	1-C – *Skytanthus acutus*
Vallenar	28° 35'	70° 46'	470	1-D – *Balsamocarpon brevifolium*

References

Gajardo, R., 1983. Sistema básico de clasificación de la vegetación nativa chilena. Corporación Nacional Forestal. Santiago. 315 pp. + Annex 21 pp. + 15 maps.

Quintanilla, V.G., 1981. Carta de la formaciones vegetales de Chile. Contribuciones Cientificas y Tecnológicas, Area Geociencias, Universidad de Santiago 11 (47): 5–32 + 1 map.

Table 20. Climatic data for Chile – Classification; E.R. Hajek.

Climate station	Mean annual precipitation (mm)	Mean temperature (°) hottest month max.	Mean temperature (°) coldest month min.	Emberger pluviothermic quotient (Q)	Climatic type Emberger	Climatic type Köppen
Per-humid					*Per-humid*	
Concepción	1,308	24.4	4.4	228	temperate	Csb_2
Coyhaique	1,164	18.8	−0.2	217	cold	BSk
Cullinco	1,558	24.8	2.3	241	cool	Csb_3
Lonquimay	1,851	25.4	−2.6	244	cold	Csb_3
Osorno	1,217	22.4	2.5	214	cool	Csb_3
Punta Carranza	825	18.7	7.1	253	hot	Csb_2
Punta Tumbes	907	18.7	6.7	246	temperate	Csb_2
Rio Bueno	1,235	23.7	3.4	212	temperate	Csb_3
Temuco	1,190	25.3	4.1	217	temperate	Csb_3
Humid					*Humid*	
Angol (El Verg)	953	27.2	4.1	143	temperate	Csb_2
Balmaceda	572	19.1	−3.8	89	very cold	BSk
Chillan	1,034	28.8	3.5	141	temperate	Csb_2
Constitución	943	24.1	6.0	178	temperate	Csb_2
Linares	1,007	29.5	4.3	138	temperate	Csb_2
Los Angeles	1,311	29.0	4.8	187	temperate	Csb_3
Molina	921	29.2	2.5	119	cool	Csb_1
Panimávida	1,107	29.1	3.5	149	temperate	Csb_2
Punta Lavapié	804	20.4	7.2	212	hot	Csb_2
Rio Cisnes	702	19.1	−4.7	105	very cold	BSk
Sewell (El Teniente)	1,073	20.2	0.3	187	cool	Csb_1
Traiguén	1,208	26.8	3.8	187	temperate	Csb_3
Victoria	1,319	26.1	2.1	194	cool	Csb_3
Sub-humid					*Sub-humid*	
Cauquenes	717	31.3	4.6	92	temperate	Csb_2
Curicó	734	30.7	3.6	93	temperate	Csb_1
Punta Angeles (Valparaiso)	462	22.4	8.4	111	hot	Csb_1
Rancagua	563	27.5	1.7	76	cool	Csb_1
Rengo	601	29.8	2.5	76	cool	Csb_1
San Antonio	441	19.6	6.2	115	temperate	Csb_1
San Fernando	791	27.7	3.7	112	temperate	Csb_1
San José de Maipo	623	28.0	2.5	86	cool	Csb_1
Talca	735	30.8	3.8	91	temperate	Csb_1
Zapallar	384	22.6	8.4	94	hot	Csb_1
Semi-arid					*Semi-arid*	
Baños de Jahuel	305	30.8	5.7	42	temperate	BSt
Chile Chico	191	21.7	−0.4	30	cold	BSk
Colina	355	31.0	1.3	41	cool	BSt
El Belloto	334	26.4	4.4	48	temperate	Csb_1
El Bosque	330	29.2	3.0	44	cool	Csb_1
Juncal	287	22.1	−0.4	45	cold	EFH
Llay-Llay	387	29.0	2.5	51	cool	Csb_1
Los Andes	305	31.9	2.8	37	cool	BSt
Los Cerrillos	310	28.8	3.0	38	cool	Csb_1
Peñablanca	382	27.2	5.2	60	temperate	Csb_1
Quillota	436	26.9	5.5	69	temperate	Csb_1
Quintero	282	20.7	5.5	64	temperate	Csb_1
Santiago	367	29.5	3.5	47	temperate	Csb_1
Arid					*Arid*	
La Serena	118	22.8	8.2	30	hot	BSn
Ovalle	134	28.4	6.1	21	temperate	BSn
Punta Tortúga	106	20.8	9.2	31	hot	BSn
Vicuña	157	28.6	5.4	23	temperate	BSt
Per-arid					*Per-arid*	
Caldera	30	23.5	9.8	7	hot	BWn
Cerro Moreno	2	24.5	9.6	0.5	hot	BWn
Chañaral	2	23.2	8.8	0.4	hot	BWn
Copiapó	29	30.6	4.8	4	temperate	BWn
Potrerillos	62	18.7	4.0	12	temperate	BWH
Refresco	12	27.5	4.1	1.4	temperate	BWt
Taltal	25	27.3	11.2	5	hot	BWn
Vallenar	21	27.2	5.0	3	temperate	BWn

References

Hajek E.R. and di Castri F., 1975. Bioclimatografia de Chile. Dirección de Investigacion, Universidad Católica de Chile, Santiago, Chile, 225 pp.

Di Castri F. and Hajek E.R., 1976. Bioclimatología de Chile. Vicerrectoria Academica, Universidad Católica de Chile, Santiago, Chile, 129 pp.

Table 21. Climatic data for Chile – Growth Indices; E.R. Hajek, R.L. Specht and L.B. Hutley.

Climatic station	Annual precip. (mm)	Annual pan evap.* (mm)	Evap. coeff. (k)	Smax (cm)	Moisture Index Annual	Moisture Index Summer	Summer drought (mo)†	T.I. × M.I. (Ann. Total) Temp.	T.I. × M.I. (Ann. Total) Subtrop.
Per-humid									
Concepción	1,308	827	0.086	30.0	0.89	0.85	0 (0)	3.59	0.32
Contulmo	1,961	886	0.100	30.0	1.00	1.00	0 (0)	4.66	0.41
Coyhaique	1,164	857	0.100	30.0	1.00	1.00	0 (0)	1.97	0.12
Cullinco	1,558	1,005	0.068	30.0	0.97	0.95	0 (0)	3.01	0.19
Lonquimay	1,851	954	0.100	30.0	1.00	1.00	0 (0)	2.27	0.19
Osorno	1,217	894	0.075	30.0	0.92	0.89	0 (0)	3.58	0.56
Punta Carranza	825	727	0.100	30.0	0.87	0.86	0 (0)	3.01	0.11
Punta Tumbes	907	718	0.100	30.0	0.94	0.99	0 (0)	3.19	0.11
Rio Bueno	1,235	852	0.100	30.0	1.00	1.00	0 (0)	3.73	0.30
Temuco	1,190	796	0.100	30.0	1.00	1.00	0 (0)	4.43	0.41
Humid									
Angol (El Verg)	953	1,216	0.052	30.0	0.71	0.35	1 (3)	2.27	0.29
Balmaceda	572	833	0.084	20.1	0.77	0.47	0 (0)	0.58	0.10
Chillan	1,034	1,222	0.050	30.0	0.72	0.35	1 (3)	1.93	0.84
Constitución	943	935	0.073	30.0	0.77	0.54	1 (2)	3.19	0.44
El Teniente (Sewell)	1,073	1,127	0.067	30.0	0.75	0.53	0 (2)	1.28	0.12
Linares	1,007	1,118	0.051	30.0	0.73	0.40	1 (3)	2.20	0.86
Los Angeles	1,311	1,127	0.049	30.0	0.78	0.50	0 (1)	2.21	1.03
Molina	921	1,212	0.051	30.0	0.68	0.29	2 (3)	1.78	0.62
Panimávida	1,107	1,146	0.051	30.0	0.74	0.41	1 (3)	2.02	0.84
Punta Lavapié	804	816	0.097	30.0	0.88	0.79	0 (0)	3.98	0.26
Rio Cisnes	702	829	0.092	30.0	0.86	0.79	0 (0)	1.19	0.11
Traiguén	1,208	870	0.078	30.0	0.90	0.85	0 (0)	2.94	0.60
Victoria	1,319	996	0.062	30.0	0.87	0.80	0 (0)	3.01	0.84
Sub-humid									
Cauquenes	717	1,400	0.042	30.0	0.61	0.16	3 (4)	1.81	0.61
Curicó	734	1,274	0.047	30.0	0.63	0.18	3 (3)	1.76	0.66
Rancagua	563	1,142	0.057	28.7	1.45	0.83	3 (3)	1.45	0.83
San Fernando	791	844	0.071	30.0	0.74	0.50	1 (2)	1.85	1.08
Talca	735	1,138	0.052	30.0	0.66	0.27	3 (3)	1.81	1.16
Valparaiso (Punta Angeles)	462	821	0.092	22.4	0.65	0.20	3 (4)	3.07	0.23
Semi-arid									
Los Andes	305	1,474	0.048	10.0	0.30	0.04	5 (6)	0.72	0.19
Quillota	436	944	0.075	21.8	0.58	0.09	3 (5)	1.99	0.22
Santiago	367	1,129	0.056	15.5	0.48	0.06	4 (5)	1.06	0.28
Arid									
La Serena	118	818	0.010	3.0	0.19	0.01	6 (7)	0.58	0.04
Punta Tortúga	106	776	0.010	2.7	0.18	0.01	6 (8)	0.60	0.02
Per-arid									
Antofagasta	10	1,026	0.008	0.2	0.01	0.00	12 (12)	0.07	0.00
Caldera	30	908	0.009	0.8	0.04	0.00	10 (12)	0.23	0.01
Potrerillos	62	1,392	0.007	0.5	0.05	0.01	9 (12)	0.04	0.00
Refresco	12	2,033	0.005	0.2	0.01	0.00	12 (12)	0.02	0.00

* Estimated by the technique of Fitzpatrick, E.A. (1963) *J. Appl. Met.* 2: 780–792.
† Months with M.I.<0.10, in parentheses, M.I.<0.20.

Table 22. Climate stations in the mediterranean basin; Ph. Daget and L. Ahdali. Vegetation types listed in column 6 of the Table are classified according to the following legend: Perennial formation (of shrubs, bushes, succulents, gramineae, lichens,[1] etc.) with or without ephemerophytes in subdesert and attenuated-desert climates.

[13] Mediterranean-biased formations
[14] Formations with a tropical bias
[15] Transitional formations
[16] Perennial formations with or without ephemerophytes in accentuated desert climates
[17] Ephemerophyte-dominated formations
[18] Sparse ephemerophyte formations or no vegetation
[19] Arbuscular shrub pseudosteppe in warm temperate climates
[20] Arbuscular shrub pseudosteppe in temperate climates
[21] Upland and lowland steppes, with or without shrubs, in cool temperate climates
[22] Shrub or tree pseudosteppes and open forests in very dry climates
[23] Shrub or tree pseudosteppes and open forests in less dry climates
[24] Temperate and cold temperate climates with shrub pseudosteppe or pistachio-almond tree pseudosteppe
[25] Mainly high steppe with or without trees or shrubs in cold and cool temperate climates
[26] Formations of the western Mediterranean evergreen oak stage
[27] Formations of the eastern Mediterranean evergreen oak stage
[28] Formations of the sub-humid Mediterranean stage
[29] Mediterraneo-Alpine steppe and grassland
[30] Mediterranean high mountain steppe and grassland
[31] Plateau and submountain steppe
[32] Lowland steppes
[33] Steppes or tree-steppes with pistachio, almond and juniper
[34] Steppe or tree steppes with juniper
[35] Oak and juniper forest stage formations
[36] Formations of the deciduous and semi-deciduous oak-forest stage
[37] Formations of the western sub-Mediterranean oak and pine stage
[38] Formations of the eastern sub-Mediterranean oak and pine stage
[39] Formations of the fir and cedar stage
[43] Formations of the dry subalpine stage
[45] Formations of the dry western mountain stage
[47] Formations of the western humid submountain stage
[50] Formations of the sub-Mediterranean submountain stage
[51] Steppe and desert formations
[67, 71, 73] Semi-permanent streams, with or without gallery forests, becoming wadis in the subdesert regions
[74] Mediterranean dunes
[101] Oases in general
[102] Irrigated regions

More geographic or botanic details can be found in: Gaussen, H., Emberger, L., Kassas, Ph., Lalande, L. 1969. Carte de la Végétation de la Région Méditerranéenne (Notice Explicative). Recherches sur la Zone Aride xxx, UNESCO-FAO, Paris, 90 p. (+ maps).

Table 22a.

Station	Country	Latitude	Longitude	Altitude	Vegetation type
per-humid					
La Coruna	Spain	43° 22'	08° 24' W	53	37
humid					
Horta	Portugal: Azores	68° 32'	28° 38' W	60	–
Ile d'Yeu	France	46° 42'	02° 20' W	20	47
Penmarch	France	47° 48'	04° 21' W	19	47
Saint Pons	France	43° 28'	02° 47' E	301	47
Medrignac	France	48° 11'	02° 22' W	160	47
sub-humid					
Poitiers	France	46° 35'	00° 21' E	117	47
Millau	France	44° 07'	03° 01' E	715	37
Briançon	France	44° 53'	06° 39' E	1406	43

Table 22b.

Station	Country	Latitude	Longitude	Altitude	Vegetation type
per-humid					
Qadmous	Syria	35° 06'	36° 09' E	750	28
Slenfer	Syria	35° 36'	36° 11' E	496	15
Albertacce	France: Corsica	42° 20'	08° 56' E	1074	43
humid					
Ponta Delgada	Portugal: Azores	37° 45'	25° 40' E	35	–
Larache	Morocco	35° 09'	06° 12' W	29	28
Coimbra	Portugal	40° 12'	08° 25' W	139	37
Trabzon	Turkey	41° 00'	39° 43' E	107	50
Pisa	Italy	43° 43'	10° 24' E	9	37
Isernia	Italy	41° 35'	14° 14' E	402	37
Braganca	Portugal	41° 49'	06° 47' W	730	47
Ifrane	Morocco	33° 51'	05° 07' W	1636	–
Nord-Salang	Afghanistan	35° 12'	69° 01' E	3336	30
sub-humid					
Orotava	Spain: Canaries	28° 25'	16° 32' W	100	–
Palermo	Italy	38° 07'	13° 19' E	107	26
Tizi-Ouzou	Algeria	36° 45'	04° 01' E	100	28
Rome	Italy	41° 48'	12° 36' E	114	37
Sartene	France: Corsica	41° 36'	08° 58' E	50	26
Vlore	Albania	40° 28'	19° 29' E	3	26
Keles	Turkey	39° 54'	29° 13' E	1000	37
semi-arid – upper					
Sta Cruz de Palma	Spain: Canaries	22° 44'	17° 46' W	10	23
Mostaganem	Algeria	35° 56'	00° 04' E	73	22
Mascara	Algeria	35° 24'	00° 08' E	590	23
Badajoz	Spain	38° 54'	06° 58' E	186	26
Aknoul	Morocco	34° 40'	04° 52' W	1210	26
Valladolid	Spain	38° 40'	04° 44' W	690	26
Mesudiye	Turkey	40° 28'	37° 46' E	1050	45
semi-arid – lower					
Gaza	Israel	31° 30'	34° 28' W	45	16
Berkane	Morocco	34° 56'	01° 56' W	145	28
Averroes	Morocco	33° 18'	07° 25' W	240	25
Sidi-bel-Abbes	Algeria	35° 11'	00° 38' E	476	19
Bab-bou-Idir	Morocco	34° 04'	04° 07' W	1570	39
arid – upper					
Sta. Cruz de Tenerife	Spain: Canaries	28° 22'	16° 17' E	10	15
Almeria	Spain	36° 50'	02° 26' W	67	22
Beer Sheva	Israel	31° 14'	34° 47' E	270	16
Murcia	Spain	37° 59'	01° 08' W	60	26
Qumren	Jordan	30° 06'	35° 28' E	1510	20
Eregli	Turkey	37° 50'	34° 04' E	1044	31
arid – lower					
Tefia	Spain: Canaries	28° 31'	14° 00' W	600	14
Sidi Barani	Egypt	31° 37'	25° 55' E	23	13
El Adem	Egypt	31° 51'	23° 55' E	160	15/16
Mafraq	Jordan	32° 22'	36° 15' E	686	20
En Nebek	Syria	34° 02'	36° 43' E	1330	25/20
per-arid – upper					
Jask	Iran	25° 45'	57° 45' E	3	15
Sallum	Egypt	31° 33'	25° 11' E	154	13
per-arid – mean					
Jeddah	Saudi Arabia	21° 30'	39° 12' E	11	13
per-arid – lower					
Masirah Island	Oman	20° 41'	58° 54' E	16	14
Kwam Umbu	Egypt	24° 29'	32° 56' E	102	16/17
Gat	Libya	24° 58'	10° 11' E	570	15

Table 22c.

Station	Country	Latitude	Longitude	Altitude	Vegetation type
per-humid					
Ain-el-Ksar	Algeria	36° 52'	06° 26' E	200	28
humid					
Anamur	Turkey	36° 06'	32° 50' E	3	19
Fethiye	Turkey	36° 37'	29° 06' E	3	26
Guercif	Morocco	34° 14'	03° 21' W	360	19
Desulo	Italy: Sardinia	40° 00'	09° 14' E	68	26
Titograd	Yugoslavia	42° 26'	19° 16' E	40	26
Gjirokaster	Albania	40° 05'	20° 09' E	193	26
sub-humid					
Karatas	Turkey	36° 34'	35° 20' E	5	26
Izmir	Turkey	38° 27'	27° 15' E	28	26
Bayramic	Turkey	39° 47'	26° 33' E	70	37
Bucak	Turkey	37° 28'	30° 36' E	850	26
Hadim	Turkey	36° 29'	32° 59' E	1500	37
Pulumur	Turkey	39° 28'	39° 52' E	1500	35
Kurkes	Albania	42° 02'	20° 24' E	354	37
semi-arid – upper					
Ramat David	Israel	32° 40'	35° 11' E	50	23
Fez	Morocco	34° 02'	05° 00' W	415	25
Boghari	Algeria	35° 55'	02° 43' E	910	23
Yunak	Turkey	38° 48'	31° 43' E	1000	31
Faizabad	Afghanistan	37° 07'	70° 31' E	1200	38/34
Tomarsa	Turkey	38° 26'	35° 48' E	1400	36
Baskale	Turkey	38° 03'	44° 01' E	2400	30
semi-arid – lower					
Hefzi Bah Gilib	Israel	32° 31'	35° 26' E	80	23
Nicosia	Cyprus	35° 09'	33° 17' E	218	26
Kasbah Tadla	Morocco	32° 26'	06° 16' W	505	26
Aleppo	Syria	36° 14'	37° 08' E	390	25
Aflou	Algeria	34° 07'	02° 06' E	1425	23
Konya	Turkey	37° 51'	32° 30' E	1025	31
Van	Turkey	38° 28'	43° 21' E	1733	31
arid – upper					
Bushire	Iran	28° 59'	50° 49' E	4	19/67
Bet Sehan	Israel	32° 30'	35° 30' E	110	13
Matmata	Tunisia	33° 32'	10° 00' E	441	13
Amman	Jordan	31° 51'	35° 57' E	25	20
Sednaya	Syria	33° 41'	36° 22' E	1250	25
Shiraz	Iran	29° 32'	52° 35' E	1491	20
Cihanbeyli	Turkey	38° 40'	32° 57' E	969	31
Zanjan	Iran	36° 41'	48° 29' E	1663	33
arid – lower					
Gabes	Tunisia	33° 53'	10° 07' E	2	13
Gafsa	Tunisia	34° 25'	08° 49' E	314	13
Laghouat	Algeria	33° 45'	02° 53' E	767	13
Inherm	Morocco	30° 06'	08° 28' W	1750	22
Mashad	Iran	36° 16'	59° 38' E	985	25
per-arid – upper					
Bahrain	Bahrain	26° 12'	50° 30' E	5	16
Dehibat	Tunisia	32° 00'	10° 42' E	395	13
Sakakah	Saudi Arabia	29° 58'	40° 12' E	574	18
H.4	Jordan	32° 30'	38° 12' E	686	13
Sefid Dast	Iran	29° 38'	55° 43' E	1743	31
Gonabad	Iran	34° 21'	58° 42' E	1150	33
per-arid – mean					
Medina	Saudi Arabia	24° 33'	39° 43' E	634	16
Cairo	Egypt	29° 52'	31° 20' E	116	18/102
El Sharqiyah	Libya	30° 47'	13° 36' E	487	13
Maian	Jordan	30° 10'	35° 47' E	1069	13
Jafr	Jordan	30° 17'	36° 02' E	865	13
per-arid – lower					
Menia	Egypt	28° 05'	30° 44' E	40	13
Saint Antoine	Egypt	28° 55'	32° 20' E	442	13
Miniyah	Egypt	28° 06'	30° 46' E	43	17
Tazerbo	Libya	25° 48'	21° 08' E	259	18

Table 22d.

Station	Country	Latitude	Longitude	Altitude	Vegetation type
sub-humid					
Shalahaddin	Iraq	36° 26'	44° 17' E	1088	36
Bingol	Turkey	38° 52'	40° 30' E	1177	35
Mus	Turkey	38° 44'	41° 31' E	1283	35
semi-arid – upper					
Malakiyen	Syria	37° 11'	42° 08' E	475	25
Siverek	Turkey	37° 45'	39° 19' E	801	36
Hinis	Turkey	39° 22'	41° 44' E	1720	35
semi-arid – lower					
Sinjar	Iraq	36° 19'	41° 50' E	476	25
Urfa	Turkey	37° 07'	38° 46' E	540	25
Diyarbakir	Turkey	37° 55'	40° 12' E	660	25
Elazig	Turkey	38° 40'	39° 13' E	1105	36
Erzigan	Turkey	39° 44'	39° 30' E	1215	35
Malazgirt	Turkey	39° 08'	42° 31' E	1565	31
arid – upper					
Mashed Soleyman	Iran	31° 59'	49° 16' E	363	20
Khanaqin	Iraq	33° 45'	45° 33' E	137	101/15
Birecik	Turkey	37° 02'	37° 58' E	347	25
Kunduz	Afghanistan	36° 40'	68° 55' E	433	25
Arak	Iran	34° 06'	49° 42' E	1767	33
Erivan	U.S.S.R.	40° 10'	44° 30' E	990	31
arid – lower					
Basra	Iraq	30° 40'	47° 47' E	2	67
Ahwaz	Iran	31° 20'	48° 40' E	18	67
Baghdad	Iraq	33° 20'	44° 24' E	33	101/15
El Kom	Syria	35° 12'	38° 51' E	460	20
Mazari Sharif	Afghanistan	36° 42'	67° 12' E	378	21
Sabzevar	Iran	36° 13'	57° 40' E	940	25
per-arid – upper					
Doha	Qatar	25° 17'	51° 34' E	8	16/73
Ahmadi	Koweit	29° 06'	48° 04' E	122	15
Hai	Iraq	32° 10'	46° 03' E	15	71
El Meghaim	Algeria	33° 39'	05° 59' E	23	13
T.3	Syria	34° 32'	38° 45' E	410	13
Semman	Iran	35° 33'	53° 24' E	1138	33
Isfahan	Iran	32° 34'	51° 44' E	1774	20
per-arid – mean					
Ras Tammurah	Saudi Arabia	26° 42'	50° 05' E	5	73
Al Hufuf	Saudi Arabia	25° 30'	49° 34' E	160	16
Rapha	Saudi Arabia	29° 38'	43° 29' E	441	16
Tagounite	Morocco	29° 58'	05° 37' W	600	15
Zabol	Iran	31° 02'	61° 29' E	487	16/102
per-arid – lower					
Reggane	Algeria	26° 43'	00° 09' E	267	17
Aoulef El Ared	Algeria	27° 04'	00° 44' E	275	17
Murzuq	Libya	25° 55'	13° 55' E	395	17
Awbari	Libya	26° 35'	12° 46' E	425	17

Table 22e.

Station	Country	Latitude	Longitude	Altitude	Vegetation type
arid – lower					
Aschad	U.S.S.R.	37° 57'	58° 30' E	226	51
per-arid – upper					
Najaf	Iraq	31° 59'	44° 19' E	31	16
Kazalinsk	U.S.S.R.	45° 46'	64° 06' E	63	51
per-arid – lower					
Adrar	Algeria	27° 52'	00° 17' E	286	74/16

Table 23. Climatic data for the Mediterranean Basin – Classification; Ph. Daget and L. Ahdali.
Rain: Mean annual precipitation in mm per yr; Snow: melting water in mm per yr: T(max): Mean temperature of the hottest month in
°C; T(min): Mean temperature of the coldest month in °C; Q2: Emberger's pluviothermic quotient; Variant: Winter thermic variant in
the Emberger's climatic type.

1 = very hot	4 = temperate	7 = very cold
2 = hot	5 = cool	8 = extremely cold
3 = warm	6 = cold	9 = icy

Köppen: Symbolism in a Köppen classification.

Table 23a.

	Rain	Snow	T(max)	T(min)	Q2	Variant	Köppen
per-humid							
La Coruna	962		18.6	9.8	215.8	3	Cfb
humid							
Horta	1023		22.7	14.4	251.8	1	Cfa
Ile d'Yeu	740		19.1	7.3	141.5	3	Cfb
Penmarch	767		17.2	6.6	167	4	Csb
Saint Pons	952		21.4	6	122.8	5	Csb
Medrignac	714		17.2	3.6	108.1	6	Csb
sub-humid							
Poitiers	638		18.9	3.9	91.4	5	Cfb
Millau	730	30	18.6	2	94.9	6	Cwb
Pec	913	104	22.3	−0.1	97.1	7	Cfa
Briançon	694	160	16.4	−3.1	80.8	8	Dfb

Table 23b.

	Rain	Snow	T(max)	T(min)	Q2	Variant	Köppen
per-humid							
Qadmous	1365		21.8	5.8	201	4	Csb
Slenfer	1362		20.9	4.1	200.7	5	Csb
Albertacce	1807	388	17.8	1.5	237.5	6	Csbn
humid							
Ponta Delgada	827		21.9	14.2	203.9	1	Csb
Larache	948		22.8	14.2	167	2	Csa
Coimbra	952		22.2	9.8	137.6	3	Csa
Trabzon	760		23.1	6.9	118.8	4	Cfa
Pisa	1048		22.8	6.9	134.9	4	Cfa
Braganca	1366	44	19.7	3.6	171.5	6	Csb
Ifrane	1113	297	21.1	1.9	109.3	7	Csb
Nord Salang	1121	857	9.6	−10	150.2	9	ETc
sub-humid							
Orotava	405		21.6	15.2	113.9	1	Csb
Palermo	708		26.4	11.3	108.5	2	Csa
Tizi Ouzou	855		27.6	10.6	98.3	3	Csa
Rome	653		24.4	8	82.6	4	Csa
Sartene	683		21.5	7.4	84.5	5	Csb
Vlore	945	14	25.8	8.5	87.2	6	Csa
Keles	834	236	19.5	0.4	97.3	7	Csa
semi-arid – upper							
Sta. Cruz de Palma	439		23.4	17.5	127.9	1	Csa
Mostaganem	435		25.7	12	67.3	2	Csa
Mascara	527		26.6	9.6	64.5	3	Csa
Badajoz	447		25.8	8.3	50.4	4	Csa
Aknoul	506		24.7	6.6	54	5	Csa
Valladolid	404	2	21.3	3.3	46.8	6	Csb
Mesudiye	518	83	17.6	−0.6	59.8	7	Csb
semi-arid – lower							
Gaza	371		26.1	13.5	55.8	2	BSsa
Berkane	350		26	11.6	43.9	3	BSsa
Averroes	382		24.7	10.8	43.6	4	Csa
Sidi-bel-Abbes	395		24.4	8	43.4	5	Csa
Bab-bou-Idir	325	15	21.7	2.6	35.4	6	Csb
arid – upper							
Sta. Cruz de Tenerife	264		25.2	17.5	59.7	1	BSsa
Almeria	222		25.8	11.9	35	2	BSsa
Beer Sheva	204		26.2	11.6	25.3	3	BSsa
Murcia	286		26.1	10	34.5	4	BSsa
Qumren	206		22.7	4.8	25.2	5	BSsa
Eregli	299	27	20.6	1.4	30.7	7	Csb
arid – lower							
Tefia	116		24.1	15.8	21.6	1	BWsa
Sidi Barani	169		25.6	13.1	28.1	2	BWsa
El Adem	101		25.6	11.6	13.9	3	BWsa
Mafraq	152		24.2	7.9	16.9	5	BWsa
En Nebek	136	6	21.7	2.9	15.1	6	BWsb
per-arid – upper							
Jask	116		32.7	19.4	19	1	BWsa
Sallum	91		25	12.2	13.8	2	BWsa
per-arid – mean							
Jeddah	56		32.7	24	10.5	1	BWsc
per-arid – lower							
Masirah Island	15		30.6	21.9	2.9	1	BWsc
Kwam Umbu	1		32.1	15.9	0.1	2	BWsc
Gat	11		33.9	13.3	1.1	3	BWfc

For legend, see Tab. 23a.

Table 23c.

	Rain	Snow	T(max)	T(min)	Q2	Variant	Köppen
per-humid							
Ain El Ksar	1595		23.8	6.3	213.1	5	Csa
humid							
Anamur	1032		33.8	12	137.6	2	Csa
Fethiye	993		27.8	11	118.2	3	Csa
Guercif	1462		28.6	10.2	148.6	4	Csa
Desulo	1181		22.2	4.4	151.8	5	Csa
Titograd	1733	5	26.9	4.8	173.7	6	Cfa
Gjirokaster	2062	210	24.6	5.3	164.4	7	Csa
sub-humid							
Karatas	787		27.6	9.8	103.9	3	Csa
Izmir	652		26.9	8.3	76	4	Csa
Bayramic	655		23.8	5.5	76.7	5	Csa
Bucak	744	11	24.7	3.6	78	6	Csa
Hadim	653	133	20.1	−0.6	74.5	7	Csb
Pulumur	792	292	20.6	−3.4	74.2	8	Dsb
Kurkes	1022	172	23.3	1.3	74	9	Cfa
semi-arid – upper							
Ramat David	483		27.6	11.4	58	3	Csa
Fez	536		27.2	9.8	56.8	4	Csa
Boghari	403		26.4	5.8	45.3	5	Csa
Yunak	452	33	21.9	1.3	53.9	6	Csb
Faizabad	521	45	25.7	1	44.9	7	Csa
Tomarsa	446	110	19	−2.7	43.6	8	Csb
Baskale	567	194	19.4	−6.9	54.6	9	Dfb
semi-arid – lower							
Hefzi Bah Gilib	405		29.6	13.4	49.5	2	BSsa
Nicosia	369		28.4	10	41.1	3	BSsa
Kasbah Tadla	409		29.9	10.6	38.6	4	BSsa
Aleppo	394		28.4	5.6	38.6	5	Csa
Aflou	342	4	24	3.5	34.5	6	Csa
Konya	330	55	22.5	−0.6	33	7	Csa
Van	395	131	21.1	−3.3	38.1	8	Dsb
arid – upper							
Bushire	275		32.5	14.2	36.4	1	BSsa
Bet Sehan	307		29.3	13.3	36.4	2	BSsa
Matmata	207		28.3	9.3	25.2	3	BSsa
Amman	277		25.2	8	33.6	4	BSsa
Sednaya	249		25.2	4.6	30.5	5	BSsa
Shiraz	305	5	28.2	5.4	27.5	6	BSsa
Cihanbeyli	293	29	21.5	0.6	31.3	7	Csb
Zanjan	318	51	23.4	−2.4	28.2	8	Csa
arid – lower							
Gabes	170		27.5	11.1	21.8	3	BWsa
Gafsa	152		29.7	9.1	15	4	BWsa
Laghouat	171		28.5	7.4	16.9	5	BWsa
Inherm	183		25.3	4.8	18.4	6	BSsa
Mashad	235	22	25.8	1.1	21.2	7	BSsa
per-arid – upper							
Bahrain	86		33.6	16.9	12	1	BWsa
Dehibat	15		30.8	10.9	9.5	3	BWsa
Sakakah	85		31.8	10.4	7.9	4	BWsa
H.4	76		28.3	8	7.3	5	BWsc
Sefid-Dast	126	7	28.1	4.8	11.1	6	BWsa
Gonabad	130	6	27.5	4.3	11.1	7	BWsa
per-arid – mean							
Medina	41		35.9	18.3	4.5	1	BWsc
Cairo	29		28.4	13.3	3.6	2	BWsc
El Sharqiyah	35		28.4	10.3	3.7	4	BWsc
Maian	43		25.6	7.7	4.5	5	BWsc
Jafr	36		26.1	7.2	3.4	6	BWsc
per-arid – lower							
Menia	4		28.5	12.3	0.4	4	BWsc
Saint Antoine	10		29.6	13.2	1.3	2	BWfc
Miniyah	2		29	13.1	0.4	3	BWsc
Tazerbo	3		29.7	12.1	0.3	4	BWfc

For legend, see Tab. 23a.

Table 23d.

	Rain	Snow	T(max)	T(min)	Q2	Variant	Köppen
sub-humid							
Shalahaddin	666		29.6	3.9	65.6	5	Csa
Bingol	910	229	26.3	−1.2	80	7	Csa
Mus	887	397	24.3	−5.9	76.1	8	Dsa
semi-arid – upper							
Malakiyen	991		31.1	5.9	51	5	Csa
Siverek	546	3	29.7	3.2	50.9	6	Csa
Hinis	593	243	21	−7.8	50.6	9	Dsb
semi-arid – lower							
Sinjar	402		32.9	6.6	38	4	BSsa
Diyarbakir	496	30	29.9	2.1	42	6	Csa
Urfa	440		31.1	5	39.7	5	Csa
Elazig	433	63	26.7	−1.1	39.2	7	Csa
Erzigan	374	73	23.2	−3.2	33.4	8	Dsa
Malazgirt	406	132	21.7	−6.8	33.1	9	Dsb
arid – upper							
Mashed-Soleyman	352		37.4	12.2	31.3	2	BWsa
Khanaqin	310		34.5	10.1	26.6	4	BSsa
Birecik	304		30.5	5.8	33	5	BSsa
Kunduz	350	13	30.9	2.4	29.2	6	BSsa
Arak	356	50	27.1	−0.8	29.9	7	BSsa
Erivan	317	62	25.6	−5.5	25.7	8	Dwa
arid – lower							
Basra	185		33.6	12.5	18.7	2	BWsa
Ahwaz	200		36.3	11.9	16.9	3	BWsa
Baghdad	140		33.8	9.8	12	4	BWsa
El Kom	135		29.4	6.1	12.5	5	BWsa
Mazari Sharif	160	6	32.1	3.6	16.1	6	BWsa
Sabzevar	166	13	29.5	2.3	14	7	BWsa
per-arid – upper							
Doha	83		34.9	17.2	9.6	1	BWsa
Ahmadi	122		36.8	12.9	11.2	2	BWsa
Hai	127		34.9	11.3	11.2	3	BWsa
El Meghaim	83		31.7	10.3	7.3	4	BWsc
T.3	115		29.8	7	10.5	5	BWsc
Semman	125	3	31.1	3.5	10.7	6	BWsa
Isfahan	108	9	28	1.9	9.1	7	BWsa
per-arid – mean							
Ras Tammurah	65		34.3	15.8	7.9	7	BWsa
Al Hufuf	46		33.9	14.2	4.5	3	BWsc
Rapha	32		33.3	12.2	2.9	3	BWsc
Tagounite	35		34.6	11.7	3	4	BWsc
Zabol	32		33.2	8.8	5.1	5	BWsc
per-arid – lower							
Reggane	9		38.1	14.7	0.8	2	BWsc
Aoulef El Ared	9		37.2	14.2	0.8	3	BWsc
Murzuq	8		31.6	11.3	0.7	4	BWsc
Awbari	18		32.6	10.6	1.5	5	BWsc

For legend, see Tab. 23a.

Table 23e.

	Rain	Snow	T(max)	T(min)	Q2	Variant	Köppen
arid – lower							
Ashad	227	15	28.9	−0.3	19.7	7	BWsa
per-arid – upper							
Najaf	99		36.1	10.5	8.5	3	BWsa
Kazalinsk	125	56	25.2	−11.9	9.5	9	BSsa
per-arid – lower							
Adrar	14		36.9	12.2	1.1	4	BWsc

For legend, see Tab. 23a.

Table 24. Climate stations in Greece. Compiled by M. Arianoutsou and S. Paraskevopoulos.

Climate station	Latitude (North)	Longitude (East)	Altitude (m)	Vegetation type
Agrinion	38° 37'	21° 24'	46	evergreen shrublands
Alexandroupolis	40° 51'	25° 57'	4	evergreen shrublands
Anoghia	35° 16'	24° 54'	776	phrygana
Argostolion	38° 10'	20° 30'	13	evergreen shrublands
Athens	37° 58'	23° 43'	107	phrygana
Corfu	39° 37'	19° 55'	20	evergreen shrublands and oak forest
Ierapetra	35° 00'	25° 45'	45	phrygana
Kalamata	37° 02'	22° 06'	32	evergreen shrublands and phrygana
Kavalla	40° 56'	24° 25'	27	evergreen shrublands and pine forest
Kimi	38° 38'	24° 06'	222	–
Konitsa	40° 03'	20° 45'	610	evergreen shrublands
Kozani	40° 20'	21° 47'	669	phrygana and deciduous forest
Lesvos	39° 06'	26° 03'	4	evergreen shrublands, pine forest and phrygana
Methoni	36° 50'	21° 43'	6	phrygana
Rhodes	36° 28'	28° 13'	42	evergreen shrublands
Siros	37° 27'	24° 57'	26	evergreen shrublands and phrygana
Thessaloniki	40° 38'	23° 01'	60	evergreen shrublands
Volos	39° 22'	22° 56'	2	evergreen shrublands

Reference

Karras, G. 1973. Climatic classification of Greece according to Thornthwaite. Ph. D. Thesis, University of Athens.

Table 25. Climatic data for Greece – Classification. Compiled by M. Arianoutsou and S. Paraskevopoulos.

Climate station	Mean annual precipitation (mm)	Mean temperature (°) hottest month max.	Mean temperature (°) coldest month min.	Emberger pluviothermic quotient (Q)	Climatic type Emberger	Climatic type Köppen
Agrinion	1,027	34.5	3.2	103	warm, sub-humid	Csa
Alexandroupolis	578	30.6	1.8	60	cool, semi-arid	Csa
Anoghia	1,098	27.6	4.3	187	warm, humid	Csa
Argostolion	1,005	30.8	7.2	146	hot, humid	Csa
Athens	416	33.0	6.2	53	warm, semi-arid	Csa
Corfu	1,309	31.3	6.3	158	warm, semi-arid	Csa
Ierapetra	526	32.2	7.7	70	hot, semi-arid	Csa
Kalamata	812	31.7	7.5	116	hot, sub-humid	Csa
Kavalla	550	29.8	1.6	60	cool, semi-arid	Csa
Konitsa	1,091	30.8	2.1	126	cool, humid	Csa
Kozani	594	29.3	−1.5	66	cold, sub-humid	Cfa
Lesvos	729	31.1	7.7	107	hot, sub-humid	Csa
Methoni	776	29.6	8.4	135	hot, sub-humid	Csa
Rhodes	748	32.3	7.5	87	hot, sub-humid	Csa
Siros	401	–	9.0	81	hot, semi-arid	Csa
Thessaloniki	443	31.6	1.9	55	cool, semi-arid	Csa
Volos	476	31.1	4.0	67	warm, semi-arid	Csa

Reference

Mavrommatis, G. 1980. Le Bioclimat de la Grèce. Relations entre le Climat et la Végétation Naturelle. Cartes Bioclimatiques Vol. 1 Annex. Institut de Recherches Forestieres d'Athènes.

Table 26. Climatic data for Greece – Growth indices. Compiled by M. Arianoutsou, L.B. Hutley, S. Paraskevopoulos and R.L. Specht.

Climatic station	Annual precip. (mm)	Annual* pan evap. (mm)	Evap. coeff. (k)	Smax (cm)	Moisture Index Annual	Moisture Index Summer++	Summer drought (mo)+	T.I. × M.I. (Ann. Total) Temp.	T.I. × M.I. (Ann. Total) Subtrop.
Agrinion	1,010	1,564	0.034	30+	0.69	0.13	1 (3)	1.82	1.34
Alexandroupolis	575	1,366	0.040	21	0.58	0.12	1 (3)	1.00	0.82
Anoghia (Crete)	1,109	1,318	0.046	30+	0.72	0.08	2 (3)	1.84	1.22
Argostolion (Kefallinia)	1,085	1,399	0.042	30+	0.73	0.14	1 (2)	2.96	1.90
Athens (Hellinikon)	359	1,689	0.033	14	0.32	0.04	3 (4)	1.03	0.66
Corfu	1,276	1,394	0.040	30+	0.77	0.20	1 (2)	2.39	1.76
Ierapetra (Crete)	548	1,601	0.035	27	0.51	0.03	3 (4)	3.08	1.03
Kalamata (Peloponnisos)	832	1,422	0.044	30+	0.68	0.10	1 (3)	2.75	1.68
Kavalla	570	1,316	0.044	18	0.58	0.12	1 (2)	1.03	0.78
Kimi	1,116	1,389	0.042	30+	0.73	0.16	1 (2)	1.86	1.21
Konitsa	1,081	1,354	0.041	30+	0.78	0.26	0 (1)	1.97	1.52
Kozani	589	1,277	0.043	15	0.56	0.21	0 (2)	1.02	0.92
Lesvos	725	1,597	0.035	30+	0.59	0.04	3 (4)	1.82	1.08
Methoni (Peloponnisos)	769	1,273	0.055	30+	0.71	0.09	2 (3)	2.88	2.05
Rhodes (Rodhos)	789	1,541	0.037	30+	0.60	0.04	3 (4)	2.30	1.42
Thessaloniki	473	1,387	0.040	14	0.44	0.12	1 (3)	0.70	0.85
Volos	446	1,449	0.039	14	0.41	0.10	1 (3)	1.06	0.78

* Estimated by the technique of Fitzpatrick, E.A. (1963) *J. Appl. Met.* 2: 780–792.
+ Months with M.I. < 0.10 and, in parentheses, M.I. < 0.20.
++ Driest (summer) months = July, August, September.

Table 27. Climate stations in Israel – Compiled by Ruhama Berliner and R.L. Specht.

Climate station	Latitude (North)	Longitude (East)	Altitude (m)	Vegetation type
Akko	32° 56'	35° 06'	10	hydrophilic salinas
Ashdot Yaaqov	32° 39'	35° 35'	200	hydrophilic (*Populus, Tamarix*)
Avedat	30° 48'	34° 46'	525	*Artemisia, Zygophyllum*
Beer Sheva	31° 14'	34° 47'	270	segetal (*Achillea*)
Bet Dagan	32° 00'	34° 50'	40	segetal (*Eragrostis*)
Bet Netofa	32° 51'	35° 20'	225	open forest and maquis (*Quercus, Styrax*) → segetal
Bet Qama	31° 27'	34° 46'	250	segetal (*Achillea*)
Elat	29° 33'	34° 57'	12	*Acacia* spp., *Hyphaene*
En Shemer	32° 28'	35° 01'	40	open forest (*Quercus, Styrax*) → batha & segetal
Erez	31° 34'	34° 34'	50	*Artemisia, Aristida, Retama*
Even Sapir	31° 46'	35° 08'	650	maquis (*Quercus, Pistacia*)
Gerar	31° 14'	34° 42'	225	segetal (*Achillea*)
Gilat	31° 19'	34° 39'	130	segetal (*Achillea*)
Hula Farm	33° 09'	35° 37'	70	hydrophilic (*Cyperus*)
Hulata	33° 03'	35° 36'	70	hydrophilic (*Cyperus*)
Kefar Yehezqel	32° 34'	35° 22'	30	batha
Kishon Reservoir	32° 38'	35° 08'	50	hydrophilic (*Phragmites*) → segetal + *Prosopis*
Lahav	31° 23'	34° 52'	440	batha & garrigue
Lod	32° 00'	34° 54'	40	segetal + *Prosopis*
Mevo Betar	31° 43'	35° 06'	760	maquis (*Quercus, Pistacia*)
Mishmar Ayyalon	31° 52'	34° 57'	150	(open maquis → batha & segetal + *Prosopis*)
Mizpe Ramon	31° 46'	35° 30'	− 385	very open forest (Pistacia) → *Artemisia, Zygophyllum*
Mishmar Ha Negev	31° 22'	34° 43'	200	*Artemisia, Asphodelus*
Nahal Shiqma	31° 30'	34° 39'	110	batha and garrigue
Nahariyya	33° 00'	35° 05'	5	maquis & garrigue → batha
Negba	31° 40'	34° 41'	85	coastal batha → segetal + *Prosopis*
Nir Yizhaq	31° 14'	34° 20'	70	*Artemisia, Lolium*
Ramat Yishay	32° 43'	35° 12'	125	open forest (*Quercus, Styrax*) → segetal
Saad	31° 28'	34° 32'	100	*Artemisia, Aristida*
Sasa	33° 02'	35° 24'	880	forests & maquis (*Quercus, Pistacia*)
Sede Moshe	31° 36'	34° 48'	130	batha → segetal + *Prosopis*
Sedom Pans	31° 02'	35° 23'	− 390	*Arthrocnemum, Suaeda, Tamarix*
Sheluhot	32° 28'	35° 29'	110	batha → segetal + *Prosopis*
Talme Yafe	31° 43'	34° 45'	80	batha & garrigue (*Prosopis*)
Tal Shahar	31° 48'	34° 55'	150	segetal + *Prosopis*
Tirat Zevi	32° 25'	35° 32'	220	batha + *Ziziphus*
Yavne	31° 54'	34° 46'	20	coastal batha → segetal + *Prosopis*
Zeelim	31° 12'	34° 32'	140	*Artemisia, Lolium*
Zora	31° 46'	34° 59'	350	open maquis & garrigue (*Ceratonia, Pistacia*) → batha

Table 28. Climatic data for Israel – Classification; R.L. Specht.

Climate station	Mean annual precipitation (mm)	Mean temperature (°)		Emberger pluviothermic quotient (Q)	Climatic type	
		hottest month max.	coldest month min.		Emberger	Köppen
Akko	549	31	9	85	warm, semi-arid	Csa
Ashdot Yaaqov	373	37	9	45	warm, semi-arid/arid	BShs
Avedat	91	34	6	11	temperate, very arid	BWhs
Beer Sheva	204	34	6	25	temperate, arid	BWhs
Bet Dagan	535	32	7	73	temperate/warm, semi-arid/sub-humid	Csa
Bet Netofa	576	33	8	79	warm, semi-arid/sub-humid	Csa
Bet Qama	315	33	9	45	warm, semi-arid/arid	BShs
Elat	25	40	10	3	warm/hot, very arid	BWhs
En Shemer	578	31	9	90	warm, semi-arid/sub-humid	Csa
Erez	426	32	9	63	warm, semi-arid	Csa
Even Sapir	600	33	9	85	warm, semi-arid/sub-humid	Csa
Gerar	174	34	6	21	temperate, arid	BWhs
Gilat	228	33	8	31	warm, arid	BShs
Hula Farm	499	36	6	57	temperate, semi-arid	Csa
Hulata	388	36	6	44	temperate, semi-arid	BShs
Kefar Yehezqel	403	34	8	53	warm, semi-arid	BShs
Kishon Reservoir	593	33	7	78	temperate/warm, sub-humid	Csa
Lahav	283	33	7	37	temperate/warm, arid	BShs
Lod	513	32	7	70	temperate/warm, semi-arid	Csa
Mevo Betar	537	33	9	76	warm, semi-arid	Csa
Mishmar Ayyalon	496	32	8	71	warm, semi-arid	Csa
Mishmar Ha Negev	244	33	7	32	temperate/warm, arid	BShs
Mizpe Ramon	88	29	10	16	warm/hot, very arid	BWks
Nahal Shiqma	328	33	8	45	warm, semi-arid	BShs
Nahariyya	601	31	9	93	warm, sub-humid	Csa
Negba	448	32	9	66	warm, semi-arid	Csa
Nir Yizhaq	198	33	8	27	warm, arid	BWhs
Ramat Yishay	578	31	8	86	warm, sub-humid	Csa
Saad	369	32	9	55	warm, semi-arid	BShs
Sasa	773	32	7	106	temperate/warm, sub-humid	Csa
Sede Moshe	350	33	9	50	warm, semi-arid	BShs
Sedom Pans	88	41	11	10	hot, very arid	BWhs
Sheluhot	309	36	8	37	warm, arid	BShs
Talme Yafe	455	32	8	65	warm, semi-arid	Csa
Tal Shahar	488	32	7	67	temperate/warm, semi-arid	Csa
Tirat Zevi	270	36	8	33	warm, arid	BShs
Yavne	489	31	8	73	warm, semi-arid	Csa
Zeelim	168	34	6	21	temperate, arid	BWhs
Zora	459	33	9	65	warm, semi-arid	Csa

Table 29. Climatic data for Israel – Growth Indices. Compiled by R. Berliner, R.L. Specht and L.B. Hutley.

Climatic station	Annual precip. (mm)	Annual pan evap. (mm)	Evap. coeff. (k)	Smax (cm)	Moisture Index Annual	Moisture Index Summer	Summer drought (mo)*	T.I. × M.I. (Ann. Total) Temp.	T.I. × M.I. (Ann. Total) Subtrop.
Akko	549	1,785	0.038	21	0.43	0.01	4 (5)	3.06	0.82
Ashdot Yaaqov	373	2,328	0.028	17	0.24	0.00	5 (6)	1.44	0.67
Avedat	91	2,912	0.023	5	0.04	0.00	10 (12)	0.19	0.04
Beer Sheva	204	2,558	0.029	9	0.12	0.00	7 (9)	0.40	0.13
Bet Dagan	535	1,918	0.038	25	0.43	0.00	5 (5)	2.42	0.56
Bet Netofa	576	2,285	0.032	24	0.38	0.00	5 (5)	0.85	0.42
Bet Qama	315	2,421	0.031	13	0.19	0.00	5 (7)	0.64	0.22
Elat	25	3,959	0.019	1	0.01	0.00	12 (12)	0.05	0.04
En Shemer	578	1,933	0.040	25	0.44	0.00	5 (5)	1.88	0.62
Erez	426	1,720	0.044	18	0.38	0.00	5 (5)	2.70	0.77
Even Sapir	600	2,263	0.033	25	0.38	0.00	5 (5)	1.00	0.36
Gerar	174	2,500	0.030	7	0.10	0.00	7 (10)	0.34	0.11
Gilat	228	2,424	0.031	10	0.14	0.00	7 (9)	0.47	0.14
Hula Farm	499	1,898	0.035	20	0.40	0.01	4 (6)	2.36	0.88
Hulata	388	2,200	0.031	19	0.27	0.01	5 (6)	1.32	0.50
Kefar Yehezqel	403	2,005	0.032	20	0.32	0.01	5 (6)	1.74	0.53
Kishon Reservoir	593	2,410	0.029	27	0.38	0.01	5 (5)	1.57	0.54
Lahav	283	2,226	0.034	12	0.18	0.00	5 (8)	0.74	0.20
Lod	513	2,180	0.033	22	0.36	0.00	5 (5)	1.97	0.48
Mevo Betar	537	2,312	0.030	21	0.33	0.00	5 (5)	0.85	0.43
Mishmar Ayyalon	496	1,830	0.043	17	0.39	0.01	4 (5)	2.12	0.58
Mishmar Ha Negev	244	2,245	0.034	10	0.16	0.00	5 (7)	0.54	0.17
Mizpe Ramon	88	2,541	0.025	5	0.05	0.00	10 (12)	0.14	0.05
Nahal Shiqma	328	2,398	0.031	12	0.19	0.00	5 (7)	1.00	0.32
Nahariyya	601	1,734	0.046	24	0.47	0.01	5 (5)	2.50	0.76
Negba	448	1,998	0.036	20	0.35	0.00	5 (5)	2.69	0.76
Nir Yizhaq	198	2,173	0.035	8	0.13	0.00	6 (8)	1.10	0.32
Ramat Yishay	578	1,965	0.035	29	0.46	0.01	4 (5)	1.73	0.68
Saad	369	1,933	0.039	15	0.28	0.00	5 (6)	1.67	0.46
Sasa	773	1,948	0.035	30+	0.50	0.01	4 (4)	0.94	0.88
Sede Moshe	350	2,205	0.033	15	0.23	0.00	5 (7)	1.31	0.36
Sedom Pans	88	3,890	0.019	4	0.03	0.00	12 (12)	0.18	0.13
Sheluhot	309	2,637	0.025	16	0.19	0.00	5 (7)	1.02	0.47
Talme Yafe	455	2,214	0.035	18	0.30	0.00	5 (5)	1.69	0.40
Tal Shahar	488	2,013	0.036	19	0.36	0.00	5 (5)	2.03	0.61
Tirat Zevi	270	2,551	0.026	14	0.17	0.00	6 (8)	0.90	0.41
Yavne	489	1,957	0.039	20	2.93	0.76	5 (5)	2.93	0.76
Zeelim	168	2,609	0.028	7	0.09	0.00	7 (10)	0.30	0.11
Zora	459	2,263	0.035	18	0.28	0.00	5 (6)	1.64	0.44

Table 30. Climate stations in Portugal; F.M. Catarino and A.I.D. Correia.

Climate station	Latitude (North)	Longitude (West)	Altitude (m)	Potential vegetation type
Alcobaça	39° 32'	08° 58'	75	Mixed sclerophyll and deciduous forest (*Quercus faginea*)
Alvalade	37° 57'	08° 24'	61	Open sclerophyll woodland (*Quercus suber, Q. rotundifolia*)
Beja	38° 01'	07° 52'	246	Open sclerophyll woodland (*Quercus rotundifolia*)
Cabo Carvoeiro	39° 21'	09° 24'	32	Heathland and coastal scrub
Caramulo	40° 34'	08° 10'	810	Deciduous woodland (*Quercus robur*) and scrub (*Rhododendron ponticum*)
Castelo Branco	39° 49'	07° 29'	390	Open sclerophyll woodland (*Quercus faginea, Q. suber, Q. rotundifolia*)
Faro/Aeroporto	37° 01'	07° 58'	8	Mixed sclerophyll forest (*Oleo-Ceratonion*)
Figueira de Castelo Rodrigo	40° 52'	06° 54'	635	Deciduous oakland (*Quercus pyrenaica*)
Lisboa/tapada	38° 42'	09° 11'	60	Mixed sclerophyll and deciduous forest (*Quercus faginea, Q. pyrenaica*)
Monchique	37° 19'	08° 33'	465	Mixed sclerophyll (*Myrica faya*) and deciduous forest (*Quercus canariensis*)
Montemór-o-Velho/Boiça	40° 11'	08° 43'	15	Mixed sclerophyll and deciduous woodland (*Quercus faginea*) and freshwater marsh
Monte Real/Base Aérea	39° 50'	08° 53'	58	Pinewood (*Pinus pinaster* + *Quercus faginea*)
Penhas douradas	40° 25'	07° 33'	1,380	Heathland and deciduous subalpine woodland (*Quercus pyrenaica*)
Porto/Serra P.	41° 08'	08° 36'	93	Mixed sclerophyll and deciduous forest (*Quercus robur*)
Sabugal	40° 21'	07° 06'	790	Deciduous oakland (*Quercus pyrenaica*)
Sesimbra/Maçã	38° 28'	09° 05'	120	Mixed sclerophyll forest (*Oleo-Ceratonion*)
Sintra/Pena	38° 47'	09° 23'	471	Mixed sclerophyll and deciduous forest (*Quercus faginea, Q. pyrenaica*) and pinewood (*Pinus pinaster*)
Tavira	37° 10'	07° 36'	25	Mixed sclerophyll forest (*Oleo-Ceratonion*)
Viana do Castelo	41° 42'	08° 48'	16	Pinewood (*Pinus pinaster*) and deciduous oak forest (*Q. robur*)
Vila Real	41° 19'	07° 44'	481	Mixed sclerophyll and deciduous forest (*Quercus robur, Q. suber, Q. rotundifolia*)
Vila Real de St. António	37° 11'	07° 25'	7	Mixed sclerophyll forest (*Oleo-Ceratonion*) and saltmarsh

Table 31. Climatic data for Portugal – Emberger pluviothermic quotient and moisture index; F.M. Catarino and A.I.D. Correia.

Climate station	Emberger pluviothermic quotient (Q)	Moisture index (Ea/Ep)		
		(Penman*)	(Thornthwaite+)	
		Annual	Annual	Summer
Alcobaça	177	0.58[a] 0.50[b]	0.68	0.31
Alvalade	70	0.40[a] 0.45[c]	0.50	0.13
Beja	100	0.43[a] 0.53[b]	0.50	0.13
Cabo Carvoeiro	171	0.57[a] 0.52[c]	0.63	0.22
Caramulo	372	0.85[a] 0.81[c]	0.83	0.64
Castelo Branco	103	0.48[a] 0.53[c]	0.56	0.21
Faro	83	0.38[a] 0.35[c]	0.46	0.09
Figueira de Castelo Rodrigo	72	0.56[a] 0.61[c]	0.60	0.35
Lisboa/tapada	127	0.46[a] 0.55[c]	0.57	0.18
Monchique	218	–	0.64	0.26
Montemor-o-Velho/Boiça	150	0.58[a] 0.61[c]	0.69	0.31
Monte Real/Base Aérea	103	–	0.68	0.28
Penhas douradas	283	0.71[a] 0.75[c]	0.84	0.67
Porto/Serra P.	208	0.65[a] 0.69[c]	0.78	0.49
Sabugal	112	–	0.61	0.28
Sesimbra	105	–	0.59	0.18
Sintra/Pena	228	–	0.70	0.32
Tavira	89	0.38[a] 0.43[c]	0.50	0.06
Viana do Castelo	236	0.78[a] 0.82[c]	0.85	0.65
Vila Real	149	0.55[a] 0.60[c]	0.64	0.30
Vila Real de St. António	67	0.38[a] 0.43[c]	0.46	0.09

* Potential evapotranspiration (Ep) calculated by Penman formula.
+ Potential evapotransporation (Ep) calculated by Thornthwaite formula.
[a] For a Water Holding Capacity of 100 mm.
[b] For a Water Holding Capacity of 200 mm.
[c] For a Water Holding Capacity of 150 mm.

References

Alcoforado, M.J., Alegria, M.F., Pereira, A.R. and Sirgado, C., 1982. Domínios bioclimáticos de Portugal definidos por comparação dos indíces de Gaussen e Emberger. Linha de Acção de Geografia Física. Rel. nº 14. Centro de Estudos Geográficos, INIC (Portuguese Ministry of Education).

Casimiro Mendes, J. and Bettencourt, M.L., 1980. Contribuição para o estudo do balanço climatológico de água no solo e classificação climática de Portugal continental. O Clima de Portugal. Fasc. XXIV. Instituto Nacional de Meteorologia e Geofísica, Lisboa.

Moreira, Tomaz J.S., 1981. Tése de Doutoramento. Instituto Superior de Agronomia, Lisboa.

Table 32. Climate stations in east Spain; C.A. Gracia.

Climate station	Latitude (North)	Longitude	Altitude (m)	Vegetation type
Viella	42° 42'	0° 44' E	974	beech and pine forest
Estany Gento	42° 30'	0° 56' E	2,174	mountain grassland
Capdella	42° 28'	0° 56' E	1,422	mountain grassland
Adrall	42° 20'	1° 20' E	639	pine forest
Oliana	42° 04'	1° 15' E	469	evergreen oak and pine forest
Solsona	42° 00'	1° 27' E	664	evergreen oak and pine forest
Puigcerda	42° 26'	1° 54' E	1,202	pine forest
Baget	42° 19'	2° 25' E	541	evergreen oak forest
Figueres	42° 16'	2° 54' E	39	evergreen oak forest
Olot	42° 11'	2° 26' E	443	beech forest
Gerona	41° 59'	2° 46' E	70	evergreen oak forest
S. Julián de Vilat	41° 55'	2° 15' E	685	evergreen oak forest
Montseny	41° 47'	2° 23' E	1,712	beech forest and *Calluna* heathland
Montserrat	41° 32'	1° 50' E	126	evergreen oak forest
Barcelona	41° 24'	2° 07' E	12	evergreen oak forest
Igualada	41° 35'	1° 33' E	312	evergreen oak forest
Lérida	41° 37'	0° 33' E	221	evergreen shrub communities
Tarragona	41° 07'	1° 11' E	20	evergreen shrub communities and evergreen forest
Flix	41° 14'	0° 29' E	42	evergreen shrub communities
San Mateo	40° 28'	0° 07' E	325	*Olea europaea* macchia
Castellfort	40° 30'	0° 15' W	1,181	macchia
Castellón	39° 59'	0° 06' W	47	macchia
Zucaina	40° 08'	0° 29' W	610	macchia and evergreen oak forest
Eslida	39° 53'	0° 22' W	370	macchia and evergreen oak forest
Gilet	39° 40'	0° 25' W	180	shrub communities
Chelva	39° 45'	1° 04' W	474	shrub communities
Requena	39° 29'	1° 10' W	692	shrub communities
Valencia	39° 28'	0° 26' W	15	*Pinus halepensis* forest
Cofrentes	39° 14'	1° 07' W	394	*Pinus halepensis* forest
Enguera	38° 59'	0° 45' W	318	*Pinus halepensis* forest and shrub communities
Beniatjer	38° 51'	0° 29' W	396	*Pinus halepensis* forest and macchia
Denia	38° 50'	0° 03' E	14	*Pinus halepensis* forest and macchia
Jijona	38° 32'	0° 34' W	516	*Pinus halepensis* forest
Alicante	38° 22'	0° 33' W	81	*Pinus halepensis* forest and macchia
Guardamar	38° 05'	0° 43' W	27	*Pinus* forest and *Tetraclinis*
Ibiza	38° 52'	1° 18' E	8	shrub communities
Lluch	39° 49'	2° 49' E	525	*Pinus halepensis* forest
Palma de Mallorca	39° 34'	2° 36' E	28	*Pinus* forest and evergreen shrub communities
Pollensa	39° 51'	2° 57' E	70	*Pinus* forest and evergreen oak forest
Mahón	39° 53'	4° 11' E	55	evergreen oak forest

Table 33. Climatic data for east Spain – Classification; C.A. Gracia.

Climate station	Mean annual precipitation (mm)	Mean temperature (°)		Emberger pluviothermic quotient (Q)	Climatic type	
		hottest month max.	coldest month min.		Emberger	Köppen
Viella	870.3	33.2	−10.9	69	very cold, sub-humid	Cfb
Estany Gento	1,259.4	23.0	−17.0	114	very cold, humid	Dfb/c
Capdella	1,264.9	29.4	−8.9	117	very cold, humid	Cfb
Adrall	634.7	30.9	−12.1	52	very cold, sub-humid	Csb
Oliana	741.2	36.9	−8.2	57	very cold, sub-humid	Csa
Solsona	713.0	34.7	−7.1	59	very cold, sub-humid	Csa/b
Puigcerda	997.1	28.9	−8.7	94	very cold, sub-humid/humid	Cfb
Baget	1,090.6	29.3	0.7	132	cool, humid	Cfa
Figueres	546.0	33.7	−3.0	52	very cold/cold, sub-humid	Csa
Olot	961.7	33.1	−7.3	83	very cold, sub-humid	Csb/Cfb
Gerona	745.8	35.0	−4.5	66	very cold, sub-humid	Csa
S. Julián de Vilat	759.5	33.3	−9.7	62	very cold, sub-humid	Csb
Montseny	952.5	24.7	−9.4	100	very cold, humid	Csb
Montserrat	708.6	31.2	−4.5	69	very cold, sub-humid	Csb
Barcelona	594.2	32.4	1.2	66	cool, sub-humid	Csa
Igualada	571.6	34.8	−5.2	50	very cold, sub-humid	Csa
Lérida	378.6	37.6	−5.9	30	very cold, semi-arid	Csa
Tarragona	502.7	30.8	−1.2	55	cold, sub-humid	Csa
Flix	366.3	38.5	−4.9	29	very cold, semi-arid	Csa
San Mateo	644.4	34.9	−4.3	57	very cold, sub-humid	Csa
Castellfort	687.0	31.9	−6.3	63	very cold, sub-humid	Csb
Castellón	418.4	33.1	0.6	44	cool, semi-arid	Csa
Zucaina	563.7	33.0	−3.5	54	very cold, sub-humid	Csa
Eslida	636.4	33.5	0.8	67	cool, sub-humid	Csa
Gilet	490.6	34.5	0.9	50	cool, semi-arid	Csa
Chelva	459.3	37.4	−3.1	39	very cold/cold, semi-arid	Csa
Requena	417.5	37.7	−6.1	33	very cold, semi-arid	Csa
Valencia	418.6	34.9	−0.1	41	cool, semi-arid	Csa
Cofrentes	450.0	41.3	−4.3	34	very cold, semi-arid	Csa
Enguera	676.8	41.3	−4.3	34	very cold, semi-arid	Csa
Beniatjar	797.9	38.3	−0.1	71	cold, sub-humid	Csa
Denia	664.5	36.2	4.0	70	temperate, sub-humid	Csa
Jijona	347.4	35.5	−2.2	32	cool, semi-arid	Csa
Alicante	335.0	36.6	0.7	32	cool, semi-arid	Csa
Guardamar	273.6	34.7	2.1	29	cool, arid	Csa
Ibiza	408.4	32.8	1.7	45	cool, semi-arid	Csa
Lluch	1,114.7	33.1	−2.8	108	cold, humid	Csb
Palma de Mallorca	466.7	34.7	1.7	49	cool, semi-arid	Csa
Pollensa	900.6	34.5	0.8	92	cool, sub-humid	Csa
Mahōn	607.3	33.5	2.9	68	cool, sub-humid	Csa

Table 34. Climatic data for east Spain – Growth Indices; C.A. Gracia.

Climatic station	Annual precip (mm)	Annual pan evap. (mm)	Evap. coeff. (k)	S_{max} (cm)	Moisture Index		Summer drought (mo)*	T.I. × M.I. (Ann. Total)	
					Annual	Summer		Temp.	Subtrop.
Viella	870.3	828.4	0.066	5	0.74	0.65	0	1.76	0.38
Estany Gento	1,259.5	638.8	0.093	5	1.00	1.00	0	0.30	0.12
Capdella	1,265.0	804.8	0.068	5	0.94	0.91	0	2.44	0.30
Adrall	634.7	812.4	0.066	5	0.60	0.57	0	1.44	0.25
Oliana	741.2	926.3	0.058	5	0.59	0.48	0	0.56	1.59
Solsona	713.9	895.9	0.059	5	0.59	0.50	0	1.01	0.77
Puigcerda	997.1	679.3	0.102	5	1.00	1.00	0	2.55	0.75
Baget	981.6	759.7	0.071	5	0.81	0.60	1	2.61	2.05
Figueras	546.0	789.8	0.068	5	0.55	0.31	3	1.07	1.13
Olot	961.7	731.9	0.074	5	0.85	0.81	0	1.98	1.40
Gerona	745.8	786.0	0.069	5	0.69	0.46	1	2.17	1.83
S. Julián de Vilat	759.5	776.1	0.069	5	0.70	0.56	0	1.53	0.81
Montseny	956.1	641.8	0.082	5	0.90	0.74	0	1.15	0.11
Montserrat	708.6	788.4	0.070	5	0.68	0.47	2	2.07	1.01
Barcelona	594.2	865.4	0.064	5	0.55	0.35	2	0.99	1.29
Igualada	571.6	828.8	0.066	5	0.56	0.36	2	1.55	1.38
Lérida	378.6	1,001.6	0.054	5	0.35	0.26	5	0.74	0.89
Tarragona	502.7	873.5	0.066	5	0.51	0.26	3	1.55	0.65
Flix	366.3	1,043.2	0.055	5	0.34	0.18	4	0.93	0.59
San Mateo	644.4	816.9	0.069	5	0.62	0.34	2	2.26	0.92
Castellfort	687.0	752.3	0.077	5	0.72	0.53	0	1.13	0.82
Castellón	418.4	851.6	0.067	5	0.45	0.18	4	1.69	1.05
Zucaina	563.7	803.0	0.068	5	0.57	0.37	2	1.50	1.38
Eslida	636.4	851.7	0.066	5	0.58	0.25	3	1.00	1.18
Gilet	490.6	924.0	0.064	5	0.48	0.21	3	1.76	0.98
Chelva	459.1	916.9	0.063	5	0.46	0.28	2	1.47	1.03
Requena	417.5	876.0	0.065	5	0.44	0.25	3	0.94	0.84
Valencia	418.6	843.5	0.064	5	0.44	0.23	5	1.52	1.10
Cofrentes	450.0	947.0	0.060	5	0.43	0.21	3	0.56	0.65
Enguera	676.8	905.2	0.063	5	0.58	0.29	3	1.38	0.97
Beniatjar	797.9	945.1	0.062	5	0.63	0.23	4	2.19	0.93
Denia	664.5	1,070.4	0.056	5	0.50	0.15	4	2.17	1.39
Jijona	340.0	798.0	0.069	5	0.42	0.20	5	1.26	0.78
Alicante	335.0	1,028.8	0.058	5	0.34	0.09	6	1.49	0.66
Guardamar	273.6	1,052.0	0.056	5	0.29	0.08	8	1.23	0.46
Ibiza	408.4	982.9	0.058	5	0.38	0.12	7	1.38	0.70
Lluch	1,114.7	895.9	0.062	5	0.71	0.22	3	1.31	0.75
Palma de Mallorca	466.7	972.1	0.059	5	0.43	0.13	3	0.93	0.62
Pollensa	900.6	959.6	0.060	5	0.63	0.19	3	2.00	1.18
Mahòn	607.3	995.2	0.063	5	0.53	0.13	3	2.10	1.10

Table 35. Climatic station in Cape Province, South Africa – Compiled by R.L. Specht & E.J. Moll.

Climate station	Latitude (South)	Longitude (East)	Altitude (m)	Vegetation type
Beaufort West	32° 19'	22° 38'	868	karroid broken veld; central lower karoo
Bird Island	33° 50'	26° 18'	4	Alexandria forest
Cape Agulhas	34° 50'	20° 01'	19	coastal fynbos
Cape Columbine	32° 50'	17° 51'	61	strandveld
Cape St Blaize	34° 11'	22° 09'	60	coastal fynbos
Cape St Francis	34° 12'	24° 50'	8	false macchia
Cape Town	33° 56'	18° 29'	12	fynbos
Citrusdal	32° 34'	18° 59'	250	fynbos (dry mountain)
Danger Point	34° 37'	19° 18'	28	coastal fynbos
Dasseneiland	33° 26'	18° 05'	5	strandveld
Deepwalls	33° 57'	23° 10'	519	Knysna rainforest
East London	33° 02'	27° 50'	125	coastal forest/thornveld; valley bushveld
Elgin	34° 08'	19° 02'	305	fynbos (mesic mountain)
George	33° 58'	22° 25'	221	coastal renosterveld; Knysna rainforest
Graaf Reinet	32° 15'	24° 32'	750	false karoo; succulent mountain scrub; karroid mountain veld
Grahamstown	33° 18'	26° 32'	539	false macchia
Great Fish Point	33° 32'	27° 06'	73	valley bushveld; Alexandria forest
Groot Drakenstein	33° 52'	19° 00'	146	fynbos (mesic mountain)
Hermitage	33° 31'	25° 40'	122	valley bushveld
Jansenville	32° 56'	24° 40'	442	Noorsveld
Jonkershoek	33° 58'	18° 56'	274	fynbos (mesic mountain)
King William's Town	32° 53'	27° 23'	375	valley bushveld; Eastern Province thornveld
Langgewens	33° 17'	18° 42'	91	coastal renosterveld
Lovedale	32° 46'	26° 50'	536	valley bushveld; false thornveld
Matroosberg	33° 26'	19° 49'	963	fynbos; mountain renosterveld
Montagu	33° 47'	20° 07'	223	fynbos; false fynbos; karroid broken veld
Oudtshoorn	33° 35'	22° 12'	335	karroid broken veld; succulent mountain scrub
Port Elizabeth	33° 59'	25° 36'	58	Alexandria forest; valley bushveld
Riversdale	34° 06'	21° 16'	105	coastal renosterveld
Somerset East	32° 43'	25° 35'	754	false central lower karoo; karroid mountain veld
Spes Bona	33° 06'	19° 48'	604	succulent karoo
Sutherland	32° 23'	20° 40'	1,456	mountain renosterveld; western mountain karoo
Table Mountain	33° 59'	18° 24'	761	fynbos (+ closed forests)
Wingfield	33° 54'	18° 32'	17	coastal fynbos

Table 36. Climatic data for Cape Province, South Africa – Classification; R.L. Specht & E.J. Moll.

Climate station	Mean annual precipitation (mm)	Mean temperature (°)		Emberger pluviothermic quotient (Q)	Climatic type	
		hottest month max.	coldest month min.		Emberger	Köppen
Beaufort West	223	32.5	4.9	28	temperate, arid	BWh
Bird Island	461	23.7	13.2	151	hot, sub-humid	BSh
Cape Agulhas	445	23.5	10.0	114	hot/warm, sub-humid	C(s)b
Cape Columbine	229	21.1	10.0	72	hot/warm, semi-arid	BSks
Cape St. Blaize	417	24.6	10.8	104	hot, sub-humid	BSh/k
Cape St. Francis	666	22.9	10.0	178	hot/warm, humid/sub-humid	Cfb
Cape Town	627	27.1	8.5	116	warm, sub-humid	Csb
Citrusdal	334	32.4	5.6	43	temperate, semi-arid	BSh/k
Danger Point	544	22.1	10.1	157	hot/warm, sub-humid	Csb
Dasseneiland	346	19.9	10.6	129	hot, sub-humid	Csb
Deepwalls	1,214	23.4	8.1	275	warm, per-humid	Cfb
East London	808	25.6	10.2	180	hot/warm, humid	Cfb
Elgin	919	25.6	4.8	153	temperate, humid	Csb
George	860	24.5	7.4	174	warm, humid	Cfb
Graaf Reinet	346	32.2	6.4	46	temperate, semi-arid	BSh
Grahamstown	697	26.8	4.3	107	temperate, sub-humid	Cfb
Great Fish Point	559	24.3	10.2	137	hot/warm, sub-humid	Cfb
Groot Drakenstein	910	30.1	5.9	129	temperate, sub-humid	Csa
Hermitage	360	29.8	5.0	50	temperate, semi-arid	BSh
Jansenville	268	32.9	4.2	32	temperate, arid	BWh
Jonkershoek	1,068	28.4	6.9	171	temperate/warm, humid	Csb
King William's Town	618	28.7	5.5	92	temperate, sub-humid	Cfa
Langgewens	368	30.7	8.0	56	warm, semi-arid	Csa
Lovedale	574	30.5	4.9	77	temperate, sub-humid	Cfa
Matroosberg	231	28.1	2.6	31	cool, arid	BSh(s)
Montagu	312	31.4	3.7	39	temperate, semi-arid/arid	BSk(s)
Oudtshoorn	232	32.3	3.3	28	temperate, arid	BWk
Port Elizabeth	576	25.5	7.1	108	warm/temperate, sub-humid	Cfb
Riversdale	445	29.1	4.5	62	temperate, semi-arid	BSk
Somerset East	603	29.3	5.9	89	temperate, sub-humid	Cfa/b
Spes Bona	110	32.2	3.3	13	temperate, arid/very arid	BWk
Sutherland	229	27.9	−2.0	27	cold, semi-arid	BSk
Table Mountain	1,780	20.9	6.2	423	temperate, per-humid	Csb
Wingfield	508	26.2	7.2	92	warm, sub-humid	Csb

Table 37. Climatic data for Cape Province, South Africa – Growth Indices; R.L. Specht & E.J. Moll.

Climatic station	Annual precip. (mm)	Annual pan evap. (mm)	Evap. coeff. (k)	Smax (cm)	Moisture Index		Summer drought (mo)*	T.I. × M.I. (Ann. Total)	
					Annual	Summer		Temp.	Subtrop.
Beaufort West	223	3,203	0.024	2	0.07	0.07	12 (12)	0.23	0.27
Bird Island	461	1,915	0.037	7	0.27	0.13	2 (3)	1.88	1.13
Cape Agulhas	445	1,661	0.041	10	0.33	0.11	2 (5)	2.14	0.52
Cape Columbine	229	2,058	0.037	6	0.14	0.02	6 (8)	0.95	0.04
Cape St. Blaize	417	1,319	0.057	4	0.33	0.21	0 (1)	2.52	1.33
Cape St. Francis	666	1,611	0.044	12	0.48	0.17	0 (2)	4.26	0.78
Cape Town	627	2,049	0.032	21	0.43	0.06	3 (5)	2.90	0.63
Citrusdal	334	2,725	0.024	15	0.19	0.02	5 (7)	0.80	0.38
Danger Point	544	1,661	0.041	13	0.41	0.11	2 (5)	2.92	0.33
Dasseneiland	346	2,058	0.037	8	0.21	0.04	5 (6)	1.52	0.06
Deepwalls	1,214	1,197	0.098	20	1.00	1.00	0 (0)	6.16	1.66
East London	808	1,639	0.049	6	0.49	0.44	0 (0)	2.82	2.67
Elgin	919	1,484	0.046	30+	0.66	0.16	0 (3)	3.00	0.57
George	860	1,319	0.057	7	0.66	0.53	0 (0)	4.52	1.60
Graaf Reinet	346	2,147	0.032	4	0.16	0.16	1 (11)	0.59	0.73
Grahamstown	697	1,639	0.049	6	0.43	0.37	0 (0)	2.62	1.40
Great Fish Point	559	1,639	0.049	5	0.35	0.27	0 (1)	2.52	1.54
Groot Drakenstein	910	1,530	0.044	30+	0.66	0.12	1 (3)	3.12	1.79
Hermitage	360	1,910	0.039	4	0.20	0.13	1 (7)	1.00	0.89
Jansenville	268	2,076	0.033	4	0.13	0.12	2 (12)	0.47	0.54
Jonkershoek	1,068	2,065	0.032	30+	0.59	0.11	1 (4)	2.82	0.74
King William's Town	618	1,690	0.048	4	0.36	0.41	0 (0)	1.46	2.09
Langgewens	368	2,763	0.024	14	0.19	0.03	5 (7)	1.13	0.33
Lovedale	574	1,690	0.048	4	0.33	0.38	0 (0)	1.25	1.83
Matroosberg	231	2,284	0.030	6	0.13	0.03	5 (9)	0.38	0.09
Montagu	312	2,166	0.030	7	0.18	0.05	4 (6)	0.70	0.46
Oudtshoorn	232	2,083	0.032	4	0.12	0.06	4 (12)	0.48	0.45
Port Elizabeth	576	1,920	0.037	9	0.33	0.15	0 (2)	2.19	0.90
Riversdale	445	1,669	0.041	7	0.29	0.16	1 (3)	1.51	0.98
Somerset East	603	2,035	0.035	6	0.30	0.30	0 (0)	1.35	1.51
Spes Bona	110	2,202	0.030	3	0.06	0.02	10 (12)	0.19	0.14
Sutherland	229	2,200	0.030	6	0.13	0.06	6 (9)	0.29	0.08
Table Mountain	1,780	1,232	0.071	30+	0.94	0.76	0 (0)	4.40	0.30
Wingfield	508	2,049	0.032	18	0.36	0.05	4 (5)	1.95	0.34

* Months with M.I. <0.10 and, in parentheses, M.I. <0.20.

CHAPTER 3

Vegetation, nutrition and climate – examples of integration

Vegetation, nutrition and climate – examples of integration

(1) Mediterranean bioclimate and its variation in the palaearctic region

Ph. Daget, L. Ahdali and P. David
(with technical assistance of F. Taillole)

Contents

Introduction	139
Generalizations on the Mediterranean climate	139
Climatic typology	139
Analysis of climate stations	141
Network of climate stations	142
Results	143
Discussion	143
Conclusion	147
References	147

1. Introduction

Numerous definitions of the mediterranean zone have been given; they originate in several disciplines. However, numerous biogeographical works converge on an essential principle: 'Mediterraneaneity' is a climatic phenomenon which must be analysed in physical terms. We have chosen to use the method of the late Professor Emberger. He proposed the basis of a new approach to the mediterranean climate; we pursue this project and we present here, not the final result of our work on mediterranean climates, but a stage in this research; a stage now well developed, which collates our own works with results acquired in other laboratories (U.R.B.T. of Algeria, Biological Institute of Ankara, the Faculties of Science of the Universities of Damascus, Aleppo (Nahal 1981) and Thessaloniki.

2. Generalizations on the Mediterranean climate

Several previous works have allowed us to summarise the comprehensive meaning of the mediterranean bioclimate as Emberger expressed it (Daget 1977a, 1977b, 1980; Daget and David 1982; Ahdali et al., 1976). In order that a climate be retained as mediterranean, it must, and is sufficient that, it should satisfy the two following conditions:
(1) Summer is the season of least precipitation.
(2) Summer is dry.
The summer is defined as the three months of the year when the daylengths are long and the days hottest *and* driest.

This last definition is a little different from that which was presented previously (Akman and Daget 1981; Daget 1980) but it permits elimination of most of the anomalies which were particularised during a recent conference (Daget 1984).

2.1 Climatic typology

Emberger (1930, 1955, 1971a, b) defined subclasses in the mediterranean bioclimate on the basis of two criteria:

global humidity of the climate
the severity of its winter

the first being characterised by the pluviothermic quotient (Q2):

$$Q2 = \frac{2000P}{(M+m)(M-m)} \quad (1)$$

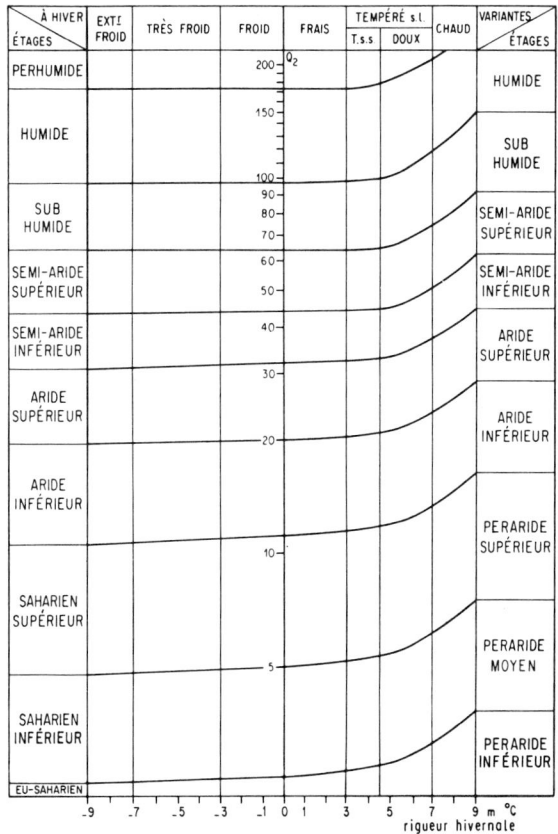

Fig. 1. Climagram of Emberger for the characterisation of mediterranean climates.

where P = mean annual precipitation (mm)
M = the mean of the maxima of the hottest month (°K)
m = the mean of the minima of the coldest month (°K)

A two-dimensional grid, the 'climagram of Emberger' enabled the mediterranean bioclimates to be classified. The last version of this grid (Akman and Daget 1981; Ahdali *et al.* 1976–1981) presented ten increasing levels of humidity: the bioclimatic zones and nine levels of 'winter severity': the thermal variants (Fig. 1).

Elsewhere, we have characterized the seasonal 'contrast' of diverse extratropical climates (including the mediterranean bioclimate) by the value of the index K^1 of thermal continentality, modified from Gorczinski (Daget 1968, 1977a). The index K^1 of thermal continentality may be calculated as follows:

$$K^1 = \frac{1.7 A}{\sin (L + 10 + 9h)} - 14 \qquad (2)$$

where A = mean annual amplitude of temperature, i.e. the difference between the maximal and minimal values of the mean monthly temperatures
L = latitude of the locality (degrees of arc)
h = altitude of the locality (km)

From these studies, five contrasting classes of the mediterranean climate (*sensu stricto*) and one sub-mediterranean class have been retained:

– *sub-mediterranean*
  Summer is only moderately dry; the Emberger-Giacobbe index of summer drought S lies between 7 and 5.
– *mediterranean sensu stricto*
  Summer is distinctly dry; the index of summer drought S is less than 5; one can recognise the following groups:
  – *littoral mediterranean*
  Thermal continentality is very reduced here: $K^1 < 25$.
  – *semi-continental mediterranean*
  Thermal contrasts are distinct: $K^1 > 25$.
  – *weak semi-continental mediterranean*
  $25 < K^1 < 33$
  – *moderate semi-continental mediterranean*
  $33 < K^1 < 50$
  – *contrasted semi-continental mediterranean*
  $50 < K^1$

There are then, finally, three criteria of classification:
– global humidity, with 10 classes (groups)
– winter severity, with 9 classes
– seasonal contrast, with 5 classes

The simultaneous use of these three criteria enables a classification of the climates in a system with three dimensions (Fig. 2) in which each climatic type corresponds to a three-dimensional 'cell', with a total of 450 (10 × 9 × 5) 'cells' possible.

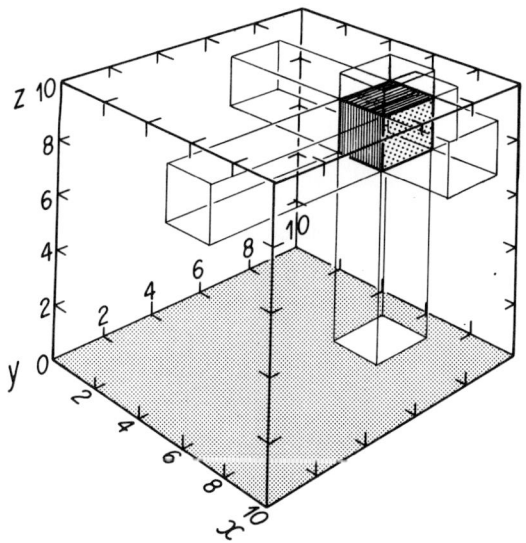

Fig. 2. Three-dimensional relationships for a modern study of mediterranean climates.
– Example of a three-dimensional 'cell'
x: Global humidity
y: Winter severity
z: Thermal contrasts expressing the degree of continentality.

2.1 Analysis of climate stations

All the stations of the data-bank have been submitted to an objective analysis carried out by computer with the help of a special algorithm called FERULA. The program constructs for each station whose climate is mediterranean, a card assembling the usual main climatic parameters.

The card for each station is similar to that which is given in Fig. 3. They carry the following elements. At the top left, outside the frame, a reference number (here 40102). This is the code of the station in the data base; the numbers begin with a 3 corresponding to the stations in Europe, with a 4 in Africa, with a 5 in Asia.

In the frame, at the top left, the name of the station (here: TLEMCEM), then that of the country where it is found (here: ALGERIA).

The following part comprises the complete diagnosis of climate in the three-dimensional system described previously (here: MEDITERRANEAN, LITTORAL, ...). Only stations of which the climate is mediterranean or sub-mediterranean are concerned in this diagnosis.

```
40102
┌─────────────────────────────────────────────────────┐
│         TLEMCEM                    ALGERIA          │
├─────────────────────────────────────────────────────┤
│ LITTORAL SUB-HUMID   WITH WARM WINTER               │
│ PM = 5.6 GM = 31.7 TMIN = 8.9 TMAX = 26.1 A = 17.2 K¹ = 23.1 │
│ P = 546 P.E. = 21.000   >100: 0   <30: 4   <1: 0   <0.1: 0 │
│ Q2 = 71.7 S = 0.66 G = 4 I = 1.31                   │
│ E.G. = 358 ST = 2671                                │
│ P(MIN) : I = 7 8  P = 3.0                           │
│ P(MAX) : I = 12  P = 76.0                           │
└─────────────────────────────────────────────────────┘
```

Fig. 3. Example of a computer card printed for each climate station by the programme FERULA.

The third part of the card presents an important number of climatic parameters which can be of use in the diagnosis (PM, Q2, etc.), or in the calculation of climatic indices. These parameters are the following:

On the first line:

PM – Mean value of the minima of the coldest month, designated normally by m (here m = 5.6° C)

GM – Mean value of the maxima of the hottest month, designated usually by M (here M = 31.7° C)

TMIN – This is the lowest value of the monthly thermal means (here, it is 8.9° C)

TMAX – This is the highest value of the monthly thermal means (here 26.1° C)

A – Represents the mean annual thermal amplitude A = TMAX-TMIN (which is at TLEMCEM 17.2° C)

K^1 – Is the degree of thermal continentality, modified from Gorczinski (here, K^1 = 23.1)

On the second line:

P – Represents the mean annual precipitation, or sum of mean monthly precipitations; it is given in millimetres

P.E. – Is the mean summer precipitation; at TLEMCEM, 546 mm falls per year and 21 mm in summer

>100 – Indicates the number of months with mean precipitation above 100 mm; equivalent to the number of humid months according to Aubréville (1949), Walter (1957) and Water and Lieth (1960).

<30 – Indicates the number of months with mean precipitation below 30 mm; equivalent to the dry months according to Köppen (1918) and Aubréville (1949)

<1 – The number of months with mean precipitation at the most equal to 1 mm
⩽0,1 – The number of months with mean precipitation at the most equal to 0.1 mm

One may consider these months as arid or very arid. The preceding card shows that at TLEMCEM, there is not any humid month, nor arid nor very arid and 4 dry months (<30).

On the third line:

Q_2 – Corresponds to the pluviothermic quotient of Emberger which is equal to 71.7 at TLEMCEM

S – Is the Emberger-Giacobbe index of summer drought: S = P.E./GM; it is always less than 7 because all the cards correspond to stations of the mediterranean type (here, S = 0.66)

G – Indicates the number of dry months in the sense of Gaussen (1954a, 1954b, 1963) and Walter (1957), that is to say such as $p(i) < 2xt(i)$; there are 4 months at TLEMCEM

I – Is the index of aridity of Budyko (1964). The computation is complex and to summarise it would be beyond the limits of this paper. It is necessary to specify that this index varies from 0 to infinity, but we have fixed an upper limit at 999 (in the present case, I = 1.31)

The value of I shows that, in the system of Budyko, TLEMCEM experiences a climate 'chaud, insuffisament humide' under a vegetation of 'forêt-steppe'.

On the fourth line:

E.G. – Represents the effective global evapotranspiration calculated according to the formula of Turc (1961); in the case cited, it is 358 mm per year

ST – Is the sum of temperatures above 10°C, calculated according to the method of Budyko (1964), which differs from that which is used in France, at the I.N.R.A., for example (in the present case, ST = 2671°C J^{-1}).

On the two last lines:

P(MIN) – the lowest monthly precipitation presents successively the months where it occurs (here: I = 7–8, therefore July and August) and the value of this minimum (here: P = 3, therefore 3 mm). The same data are given for the highest mean monthly precipitation P(MAX) in the next line.

At the same time, another algorithm, PEGANUM, has allowed all the stations, whose images are found in the same cell of the three-dimensional space of Fig. 2, to be united. These stations have therefore the same type of climate.

3. Network of climate stations

A climatic data-bank for the countries around the Mediterranean is in the course of construction at Montpellier. It encompasses a territory sufficiently vast to include most of the Mediterranean Isoclimatic Area. An important part of the data published by Ahdali et al. (1976–81) is included in this data-bank, but it allows for data originating from other sources.

In summary, the data-bank includes 1363 stations distributed in Europe, Africa and Western and Central Asia, between the 14th and 52nd degrees of latitude north and between the 15th degree of longitude west and the 82nd degree east. The map shown in Fig. 4 gives the distribution of stations utilized in a network of 1782 grid squares (27 × 66 of 1.5° latitude and longitude). The station present in the grid-square is localized in its centre; but a single set of values is represented if the grid-square includes one or several stations; following this convention, one notes that:

366 grid-squares contain 1 station
130 grid-squares contain 2 stations
81 grid-squares contain 3 stations
41 grid-squares contain 4 stations
19 grid-squares contain 5 stations
11 grid-squares contain 6 stations
9 grid-squares contain 7 stations
6 grid-squares contain 8 stations
2 grid-squares contain 10 stations
1 grid-squares contains 11 stations
1 grid-squares contains 12 stations
1 grid-squares contains 15 stations

There results over-sampling in certain areas, most often localized in mountainous regions; there is, by contrast, an important under-sampling in all the north-east quarter of the territory studied and some reduction in Spain, Italy, Greece, Pakistan and India.

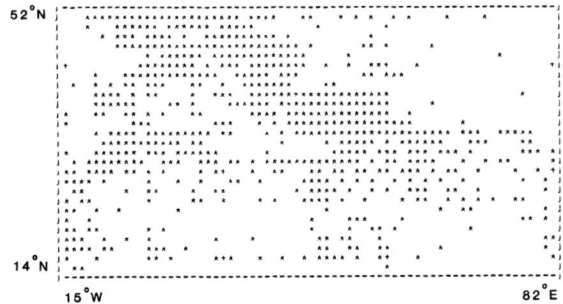

Fig. 4. Distribution of climate stations (between latitude 14° N and 52° N and longitude 15° W and 82° E).

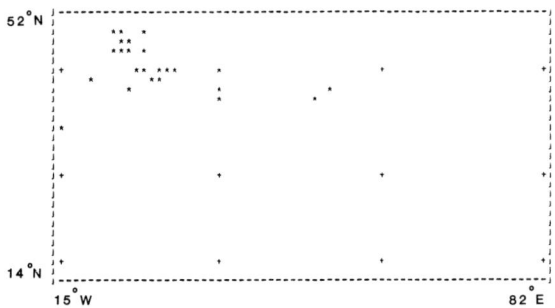

Fig. 5. Distribution map of stations with a sub-mediterranean climate.

4. Results

For each climatic type, a synthesis card comprises firstly one card, constructed automatically. The computer has printed on it the stations which show this type of climate. 'Official' coordinates (published in the sources used) have been utilized; as it is clear that several among these coordinates were false, the errors detected have been corrected, but some may be substituted.

After the first card, the climatic index card of one representative station is given. The station type has been determined in the following manner:
– when there is only one station in a climatic unit, it is evidently taken as type;
– when there are two, the one nearest to the centre of the cell is retained;
– when there are more than two stations, the type-station is that which is nearest to the barycentre of the stations of the cell.

Global results

Of 1363 stations studied in the Palaearctic Region, 32 of them have been recognised as sub-mediterranean; 725 stations belong to the mediterranean climate *sensu stricto*; these last stations are distributed in 160 of 450 possible three-dimensional cells, represented by one station at least in the following manner:

42 climatic types are represented by 1 station
26 climatic types are represented by 2 stations
19 climatic types are represented by 3 stations
16 climatic types are represented by 4 stations
14 climatic types are represented by 5 stations
 6 climatic types are represented by 6 stations
 7 climatic types are represented by 7 stations
 3 climatic types are represented by 8 stations
 7 climatic types are represented by 9 stations
 2 climatic types are represented by 10 stations
 3 climatic types are represented by 11 stations
 1 climatic types are represented by 12 stations
 4 climatic types are represented by 13 stations
 2 climatic types are represented by 14 stations
 2 climatic types are represented by 15 stations
 1 climatic type is represented by   17 stations
 1 climatic type is represented by   18 stations
 1 climatic type is represented by   19 stations
 1 climatic type is represented by   21 stations
 1 climatic type is represented by   24 stations
 1 climatic type is represented by   26 stations

5. Discussion

The following maps of the study-area (Fig. 4) show all the stations recognised as sub-mediterranean (Fig. 5), then (Fig. 6) those stations whose climate is mediterranean littoral, and lastly, on Figs. 7, 8 and 9, those stations with a semi-continental mediterranean climate successively weak, moderate and seasonally contrasted.

It will be noted that the sub-mediterranean climate stations (Fig. 5) are almost all situated in France and on the North-Atlantic boundary of the territory studied. These, therefore, are the stations presenting a transitional climate between oceanic and mediterranean climates, sometimes difficult to discriminate (Mounier 1979). Two stations are situated in Yugoslavia and constitute transitions towards continental climates; there summer remains 'sub-dry' whereas the thermal contrasts are sufficiently pronounced, much more in any case than in the first group. Fig. 6 shows that the major part of stations classed as the littoral mediterranean climatic form are situated on the edge of the sea; but not only there, since two of them are located, far from all sea, in Afghanistan, at a very high altitude.

Tables 1 and 2 summarise the climatic types, zones and variants, respectively, of sub-mediterranean and littoral mediterranean. The analysis of the values presented was very difficult for they

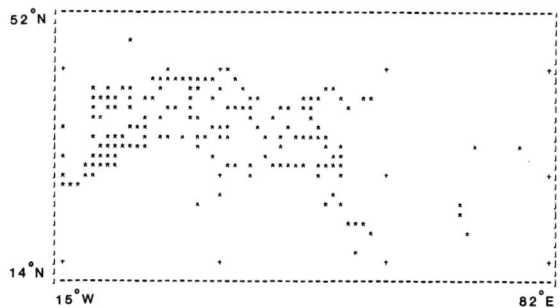

Fig. 6. Distribution map of stations with a littoral mediterranean climate.

Table 1. Number of the different forms of sub-mediterranean climate.

Zones →	Humid s.l.			Arid s.l.				Per-arid		
Winter variants* ↓	PH	H	½H	$\frac{AS}{2}$	$\frac{AI}{2}$	AS	AI	PAS	PAM	PAI
frozen										
extremely cold										
very cold			1							
cold		1	1							
cool		13	4							
temperate		4								
warm	1	5								
hot										
very hot										
Zonal totals	1	24	6							
Global totals	31									

* See Fig. 1, p. 140.

Table 2. Number of the different forms of littoral mediterranean climate.

Zones →	Humid s.l.			Arid s.l.				Per-arid		
Winter variants* ↓	PH	H	½H	$\frac{AS}{2}$	$\frac{AI}{2}$	AS	AI	PAS	PAM	PAI
frozen		2								
extremely cold										
very cold		2	2	2		2				
cold	2	3	11	9	1		1			
cool	4	5	23	4	2	1	2			
temperate	4	14	12	5	6	1				
warm		10	24	13	9	8	3			2
hot		15	21	17	5	4	9	1		1
very hot		1	5	5		3	3	6	2	4
Zonal totals	10	52	98	55	23	19	18	7	2	7
Global totals	160			115				16		

* See Fig. 1, p. 140.

reflect narrowly the spatial irregularities of the grid used.

The weakly semi-continental mediterranean climate is largely represented (Fig. 7) in Northern Africa and in the Eastern Mediterranean: Balkans, Turkey, Syria, Egypt. The distribution (Fig. 8) of the moderate semi-continental mediterranean climate is more concentrated, on the one hand in Northern Africa and on the other, in the Middle-East: Iraq, Iran, Afghanistan, Turkestan and the borders of Pakistan. Lastly, the contrasted forms are very sparse, at the west of the Sahara and in Central Asia (Fig. 9). It is certain that these two last forms are inderestimated, both in their diversity

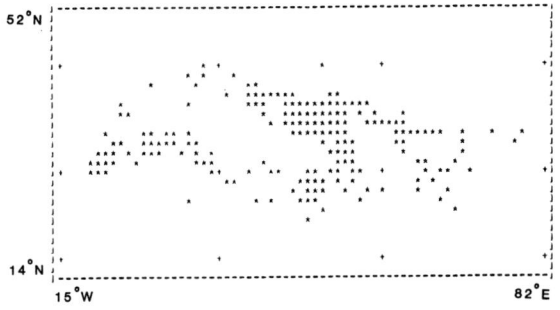

Fig. 7. Distribution map of stations experiencing a mediterranean climate with weak semi-continentality.

Table 3. Number of the different forms of mediterranean climate, with weak semi-continentality.

Zones →	Humid s.l.			Arid s.l.				Per-arid		
Winter variants* ↓	PH	H	½H	$\frac{AS}{2}$	$\frac{AI}{2}$	AS	AI	PAS	PAM	PAI
frozen			2	2						
extremely cold			1	3	7	5				
very cold		1	5	11	18	1	4	1		
cold		4	9	15	13	3.	2	3	1	
cool	1	7		14	7	10	13	7	4	
temperate		3	6	6	1	5	2	8	2	5
warm		4	9	5	3	2	3	4		11
hot		1			1	2			4	5
very hot						1		4	2	1
Zonal totals	1	20	32	50	50	29	24	27	13	22
Global totals	53			153				62		

* See Fig. 1, p. 140.

Table 4. Number of the different forms of mediterranean climate, with moderate semi-continentality.

Zones →	Humid s.l.			Arid s.l.				Per-arid		
Winter variants* ↓	PH	H	½H	$\frac{AS}{2}$	$\frac{AI}{2}$	AS	AI	PAS	PAM	PAI
frozen				2	3					
extremely cold			2	2	2	1		1		
very cold			1		3	3	2	1		1
cold				1	1	2	7	4		
cool			3	1	3	6	10	9	7	1
temperate					1	2	1	8	5	4
warm							4	10	4	3
hot						1	1	5	1	1
very hot								1	4	
Zonal totals			6	4	13	15	25	38	21	10
Global totals	6			57				69		

* See Fig. 1, p. 140.

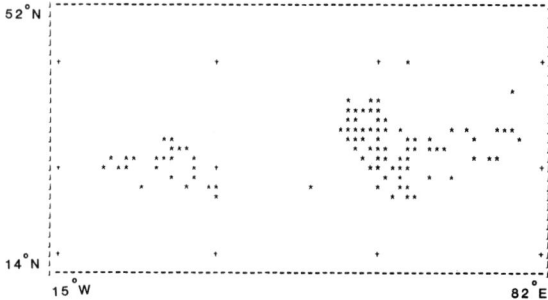

Fig. 8. Distribution map of stations experiencing a mediterranean climate with moderate semi-continentality.

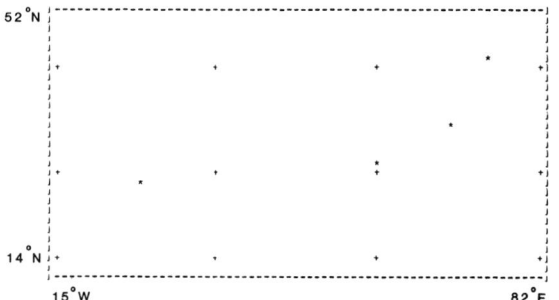

Fig. 9. Distribution map of stations experiencing a mediterranean climate with contrasted semi-continentality.

and in their territorial extension, by the undersampling of the grid utilized in the territories of the Soviet Union.

The comparison of different states presented on

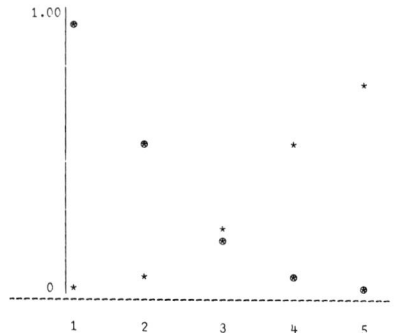

Fig. 10. Relationships between continentality, humidity and aridity.

ⓧ Proportion of humid climatic forms.
* Proportion of per-arid climatic forms.
1. Stations with a sub-mediterranean climate.
2. Stations with a littoral mediterranean climate.
3. Stations experiencing a mediterranean climate, with weak semi-continentality.
4. Stations experiencing a mediterranean climate, with moderate semi-continentality.
5. Stations experiencing a mediterranean climate, with contrasted semi-continentality.

Tables 1, 2, 3, 4 and 5 suggests a relationship between humidity and the continentality of the climate. To be specific, it is noticed from the range of 'global data' that the proportion of 'humid' stations diminishes regularly when the continentality increases – while that of 'perarid' stations increases (Fig. 10). These relationships (in Fig. 10) are both highly significant (with respectively $Z = 13$ and $Z = 11$: Test of a linear trend in proportions in Snedecor and Cochran 1968). This is similar to the

Table 5. Number of the different forms of mediterranean climate, with contrasted semi-continentality.

Zones → Winter variants* ↓	Humid s.l.			Arid s.l.				Per-arid		
	PH	H	½H	$\frac{AS}{2}$	$\frac{AI}{2}$	AS	AI	PAS	PAM	PAI
frozen								1		
extremely cold										
very cold							1			
cold										
cool										
temperate										
warm										1
hot								1		
very hot										
Zonal totals							1	2		1
Global totals				1				3		

* See Fig. 1, p. 140.

known phenomenon of rainfall decreasing with distance from the sea in general, and above all, from the Atlantic Ocean.

Conclusion

The method of description of mediterranean climates reported here allows a fine analysis of concrete situations; but it is not considered as an end in itself. The units described must serve as a basis for more developed studies. Some studies are orientated towards agronomy, horticulture and the appreciation of local or regional potentialities (Ahdali 1976–81). Others are directed towards determining the distribution of species (Djellouli 1981; Al-Hakim 1983). Others again have regional studies as their objective (Abi Saleh 1978; Bortoli *et al.* 1969; Floret and Pontanier 1982; Le Houérou 1959, Le Houérou *et al.* 1979; Long 1957; Sauvage 1963; Daget and Maheras, in prep.). The last are specific monographs (Mouffaddal 1983; M'Hirit 1982). Finally, certain studies combine the descriptive approach utilized here with a genetic analysis of climatic forms described (Daget and Reyes, in prep.). Many Mediterranean countries still present some problems both from the point of view of climate and the matter of management for which the climatic analyses according to Emberger's Method may serve as a framework.

References

Abi Saleh, B., 1978. Etude Phytoécologique des Peuplements Sylvatiques du Liban. Thèse, Univ. St Jérôme, Marseille, 300 p.

Ahdali, L. *et al.*, 1976–1982. Atlas Agroclimatique des Pays Arabes. O.A.D.A., Khartoum, 18 vol.

Ahdali, L. *et al.*, 1976. Etude Agroclimatologique des Pays Arabes. 1st Part. O.A.D.A. Khartoum, 23 vol. (in Arabic).

Ahdali, L. *et al.*, 1977. Etude Agroclimatologique des Pays Arabes. 2nd Part. O.A.D.A., Khartoum, 23 vol. (in Arabic).

Ahdali, L. *et al.*, 1980. Etude Agroclimatologique des Pays Arabes. 3rd Part. O.A.D.A., Khartoum, 23 vol. (in Arabic).

Ahdali, L. *et al.*, 1981. Agroecologic Map of the Arab Countries. O.A.D.A., Khartoum, 200 p.

Akman, Y. and Daget, Ph., 1981. Problèmes posés par la détermination des climats méditerranéens. Comm. Fac. Sci. Ankara, C2, 24: 15-27.

Al Hakim, W., 1983. L'étude agroclimatique des Pays Arabes. Mém. D.E.A., Univ. Sci. Tech. Lang., Montpellier, 37 p (+ annexes).

Aubréville, A., 1949. Climats et Désertification de l'Afrique Tropicale. Soc. Edit. Geogr. & Marit. Paris, 351 p.

Bortoli, L., Gounot, M. and Jacquinet, J., 1969. Climatologie et bioclimatologie de la Tunisie septentrionale. Ann. Inst. Nat. Rech. Agron. Tunisie, 42, 1: 1–235 (annexes).

Budyko, M., 1964. Climate and Life. Academic Press, N.Y., 508 p.

Daget, Ph., 1968. Quelques remarques sur le degré de continentalité des climats de la région holarctique. C.N.R.S.-C.E.P.E., Montpellier, 12 p.

Daget, Ph., 1977a. Le bioclimat méditerranéen, caractères généraux, modes de caractérisation. Vegetatio 34, 1–20.

Daget, Ph., 1977b. Le bioclimat méditerranéen, Analyse des formes climatiques par le système d'Emberger. Vegetatio 34, 87–103.

Daget, Ph., 1980. Un élément actuel de la caractérisation du monde méditerranéen: le climat. Comm. I° Colloque Emberger, Montpellier, Nat. Monsp., H.S.: 101–126.

Daget, Ph., 1984. Introduction à une théorie générale de la méditerranéité. Bull. Soc. Bot. Fr. 131, Act. Bot. 31–36.

Daget, Ph. and David, P., 1982. Essai de comparaison de diverses approaches climatiques de la Méditerranéité. Ecologia Mediterranea 8(1/2): 33–48.

Daget, Ph. and Maheras, T. In prep. Les bioclimats méditerranéens en Grèce.

Daget, Ph. and Reyes, S. In prep. Les influences tropicales sur les précipitations dans la Basse Californie du Nord (Mexique).

Djellouli, Y., 1981. Etude Climatique et Bioclimatique des Hautes Plaines du Sur-oranais Wilaya de Sai da Comportement des Espèces vis-à-vis du Climat. Thèse, Univ. H. Boumedienne, Alger. 250 p. (+ annexes).

Emberger, L., 1930. La végétation de la région méditerranéenne, essai d'une classification des groupements végétaux. Rev. Gén. Bot. 42: 641–662, 705–721.

Emberger, L., 1955. Une classification biogéographique des climats. Rev. Trav. Lab. Bot. Fac. Sci., Montpellier 7: 3–43.

Emberger, L., 1971a. Considérations complémentaires au sujet des recherches bioclimatologiques et phytogéographiques-écologiques. In: Emberger, 1971b: 291–301.

Emberger, L., 1971b. Travaux de Botanique et d'Ecologie. Masson, Paris, 520 p.

Floret, Ch. and Pontanier, R., 1982. L'Aridité en Tunisie Présaharienne. O.R.S.T.O.M., Paris, 544 p.

Gaussen, H., 1954a. Les limites des climats méditerranéens. C.R. VIII° Congr. Int. Bot., Paris, Sect. 27: 161–164.

Gaussen, H., 1954b. Théorie et classification des climats et microclimats. C.R. VIII° Congr. Int. Bot., Paris, Sect. 27: 125–130.

Gaussen, H., 1963. Carte Bioclimatique de la Zone Méditerranéenne. Notice Explicative. UNESCO, Paris, 21: 1–60.

Köppen, W., 1918. Une nouvelle classification générale des climats. Rev. Gén. Sci. 30: 550–554.

Le Houérou, H., 1959. Recherches Ecologiques et Floristiques sur la Végétation de la Tunisie Méridionale. Univ. Alger, Inst. Rech. Sahariennes, Mémoire No. 6, 281 p. & 229 p. (annexes).

Le Houérou, H., Claudin, J. and Pouget, M., 1979. Etude bioclimatique des steppes algériennes. Bull. Soc. Hist. Nat. Afr. Nord 8, 3–4: 33–74.

Long, G., 1957. The Bioclimatology and Vegetation of Eastern Jordan. FAO 52/2/2/109, 97 p.

M'Hirit, O., 1982. Etude écologique et forestière des cèdraies du Rif marocain. Ann. Rech. Forest. Maroc. 22: 1–502.

Mouffaddal, M., 1983. Contribution à l'Etude des Possibilités d'Extension du Cédre de l'Atlas en dehors de son Aire Actuelle. Thèse, Univ. Sci. Techn. Lang., Montpellier, 230 p.

Mounier, J., 1979. Les Climats Océaniques de Régions Atlantiques de l'Espagne et du Portugal. 2 vols. H. Champion, Paris.

Nahal, I., 1981. The mediterranean climate from a biological viewpoint. In: di Castri, F., Goodall, D.W. and Specht, R.L. (eds.) Ecosystems of the World. Mediterranean-type Shrublands, Vol. 11. Elsevier, Amsterdam, pp. 63–86.

Sauvage, Ch., 1963. Etages Bioclimatiques. In: Atlas du Maroc. Com. Nat. Géoge. Maroc, Rabat, 44 p. (annexes).

Snedecor G. and Cochran, W., 1968. Statistical Methods. Iowa State Univ. Press, Ames, 250 p.

Turc, L., 1961. Evaluation des besoins en eau d'irrigation. Ann. Agron., 12: 13–49.

Walter, H., 1957. Wie kann man den Klimatypes anschaulich darstellen? Die Umshau Wissenchaft u. Technik 24: 751–753.

Walter, H. and Lieth, H., 1960. Klimadiagramm-Weltatlas. Fischer, Jena.

Vegetation, nutrition and climate – examples of integration

(2) Climatic control of ecomorphological characters and species richness in mediterranean ecosystems of Australia

R.L. Specht

Contents

Introduction	149
Canopy structure: Attributes	150
Canopy structure: Seasonality	151
Species richness: Overstorey	153
Species richness: Understorey	153
Species richness: Small mammals	153
Conclusions	154
References	155

Introduction

The canopy of almost all mediterranean-type plant communities is composed of evergreen foliage which is renewed annually by seasonal shoot growth (Specht *et al.* 1983). The evaporative power of the atmosphere in the boundary layer above the canopy influences the fluxes of water vapour out of the plant community into the atmosphere. The evaporative balance with the annual evaporative cycle, expressed by the evaporative coefficient (k), is the result of the evolution of spatial distribution and ecomorphological attributes of the foliage of the canopy.

The evaporative coefficient (k) for a climate station can be computed using the iterative computing technique outlined by Specht (1972). This is based on the empirical relationship between the monthly value of the Moisture Index (Actual Evapotranspiration (Ea)/Pan Evaporation (Eo)) and the amount of Water Available to the plant community during the month.

$$\text{Moisture Index (M.I.)} = E_a/E_o = k(P - R - D + S_{ext}) \quad (1)$$

where
E_a = actual evapotranspiration (cm per month)
E_o = pan evaporation (cm per month)
P = precipitation (cm per month)
R = runoff (cm per month)
D = drainage (cm per month)
S_{ext} = extractable soil water (cm at beginning of the month)
k = evaporative coefficient

Evergreen plant communities, finely-tuned to the annual evaporative cycle, 'conserve' excess water deep in the soil during the humid season for use, albeit at a low rate, during the dry summer season (Specht 1957, 1972, 1987). This long-term, community-physiological equilibrium of monthly values of relative evapotranspiration (E_a/E_o) to Available Water ($W = P - R - D + S_{ext}$) is termed the evaporative coefficient (k).

For most mediterranean-type ecosystems, monthly values of Available Water (W) are limiting, with a Moisture Index (E_a/E_o) less than unity. Towards the perhumid section of the mediterranean-type climate, monthly values of Available Water (W) are no longer limiting except during the drier mediterranean-summer months. Then the Moisture Index equation departs from linearity (Eq. 1) to asymptot towards unity, and excess water is lost as runoff and drainage.

$$E_a/E_o \rightarrow 1.0 \quad (2)$$

R.L. Specht (ed.) Mediterranean-type Ecosystems. ISBN 90-6193-652-7.
© Kluwer Academic Publishers.

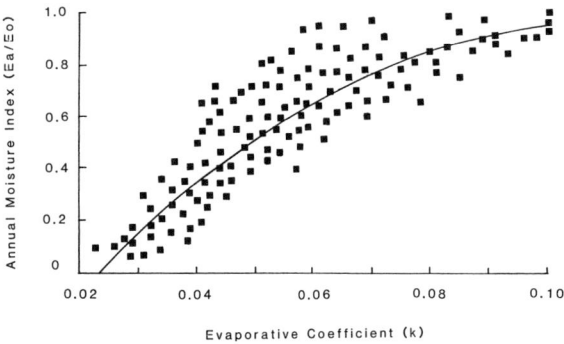

Fig. 1. The relationship between the mean annual value of the Moisture Index (Actual Evapotranspiration (Ea)/Pan Evaporation (Eo)) and the Evaporative Coefficient (k) of evergreen plant communities in perhumid to arid climatic zones of Australia – Specht 1972, Specht and Specht in prep.

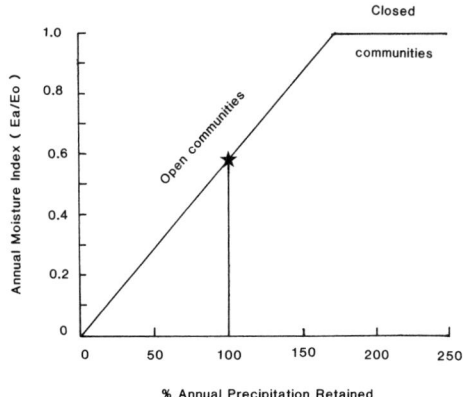

Fig. 2. Annual Moisture Index of a mediterranean sub-humid climate station when Available Water is reduced (by runoff or drainage) or increased (by runon, seepage or increased Smax) – Specht 1972, 1987.

Typical relationships between the Moisture Index and Available Water (W) are shown for perhumid to arid mediterranean-climates in Chapter 1, Fig. 2 of the Volume.

If monthly values for Precipitation (P), Pan Evaporation (Eo), Runoff (R) and Smax (maximum extractable soil water) are available or can be estimated, then the evaporative coefficient (k) for evergreen plant communities can be computed, together with monthly (and hence annual) estimates of the Moisture Index (Ea/Eo). Computations for over 400 climate stations in Australia show (Fig. 1) that there is a general relationship between the median values of the Annual Moisture Index and the evaporative coefficient (k) for the climate station. In any evaporative climate, some stations receive more precipitation, some less; micro-habitats around each climate station may receive more Available Water (by runon, or seepage or increased Smax) or less Available Water (by runoff or drainage) – a gradient of microhabitats with an Annual Moisture Index from zero to unity may result (Fig. 2).

The two parameters (k and Annual M.I.) have a marked influence on canopy structure and species richness of Australian mediterranean ecosystems.

Canopy structure: Attributes

Five major attributes of the evergreen canopy of mediterranean plant communities can be measured: (1) horizontal coverage (Foliage Projective Cover %), (2) size of leaf (Leaf Area), (3) degree of sclerophylly (Leaf Specific Area), (4) annual growth of foliage shoots (Shoot Length), and (5) vertical stratification of leaves (Leaf Area Index).

The annual evaporative cycle (as expressed by the evaporative coefficient k – a macro-climatic parameter) at a particular climate station directly influences the development (Fig. 3) of Foliage Projective Cover, Leaf Area and Leaf Specific Area, all of which control the fluxes of water vapour from the overstorey canopy into the atmosphere.

Within any particular evaporative macro-climate, a median value of the Annual Moisture Index may be expected (Fig. 1). Considerable variation in the intensity of seasonal rainfall and the subsequent distribution (by runoff, runon, seepage, drainage, etc.) of that rainfall will induce a wide range of values of the Annual Moisture Index around the median value. Foliage shoot production (mean Shoot Length per annum) and Leaf Area Index of 'climax' plant communities (both parameters related to photosynthetic production, not evapotranspiration) will show a median relation with the evaporative coefficient (Fig. 4); but con-

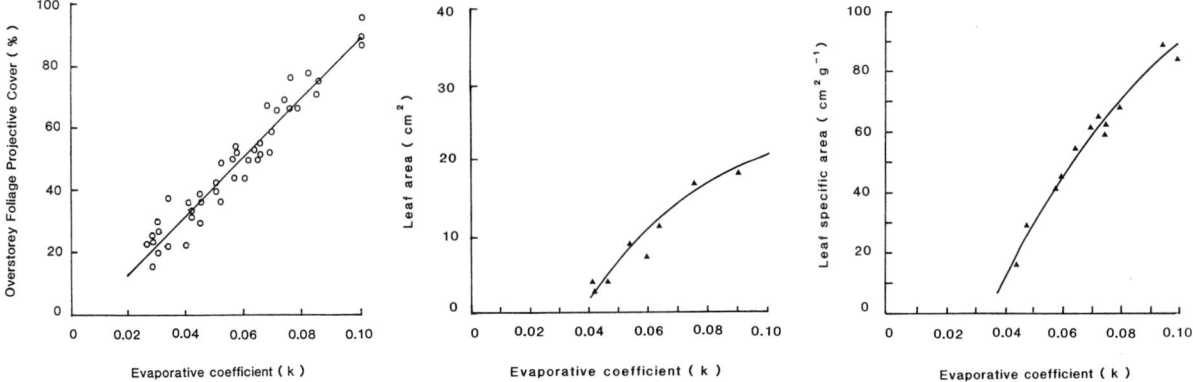

Fig. 3. Mean values of Foliage Projective Cover (%), Leaf Area (cm^2), and Leaf Specific Area (cm^2 g^{-1} dry weight of lamina) of overstorey species from evergreen plant communities in the mediterranean-climate of Australia, plotted against the evaporative coefficient (k) – Specht 1983, Specht and Specht in prep.

siderable variation will be found in microhabitats where the Annual Moisture Index departs from the median value. (Similar variation may be expected in nutrient-poor to nutrient-rich microhabitats, where the Soil Fertility Index influences photosynthetic production – Specht and Moll 1983, Specht in press).

Canopy structure: Seasonality

Photosynthetic production in evergreen mediterranean-type ecosystems will be markedly influenced by the seasonality of the Moisture Index throughout the year. Other environmental factors may modify photosynthetic production to a lesser extent. The following multiplicative equation (using indices of each community-physiological response – Specht 1981a, b) enables the plant community response to be estimated seasonally on a relative scale:

$$\text{N.P.I.} = \text{F.P.C.} \times \text{M.I.} \times \text{L.I.} \times \text{T.I.} \times \text{S.F.I.} \quad (3)$$

where N.P.I. = Net Photosynthetic Index
F.P.C. = Foliage Projective Cover (related to the evaporative coefficient k)
M.I. = Moisture Index = Evapotranspiration/Pan Evaporation
L.I. = Light Index
T.I. = Thermal Index
S.F.I. = Soil Fertility Index (see Specht in press)

In this multiplicative equation, community photosynthesis shows relatively minimal, seasonal variation with incident solar radiation and leaf temperature in mediterranean-regions (Mooney 1983).

Although photosynthetic production of evergreen plant communities may be little influenced by incident solar radiation (and associated changes in temperature), the production of growth hormones which activate foliage shoot growth, flowering, cambial activity and root growth appears to be strongly dependent on temperature. Seasonal patterns of translocation of photosynthates to different organs in the woody plant appear to be strongly temperature-controlled.

The growth of tagged shoots of overstorey species show the relative growth responses to ambient temperature observed in Australia (Fig. 5).

In the mediterranean-climate of southern Australia, overstorey species have been shown to have a late spring to summer shoot-growth rhythm (mesotherm plants with a peak of shoot growth at 20–21°C), while understorey species tend to be microtherms (peak at 15–16°C), or even nanotherms (peak at 10–11°C). In post-fire succession, the shoot growth of understorey species (more attuned to the mediterranean-winter/spring seasons)

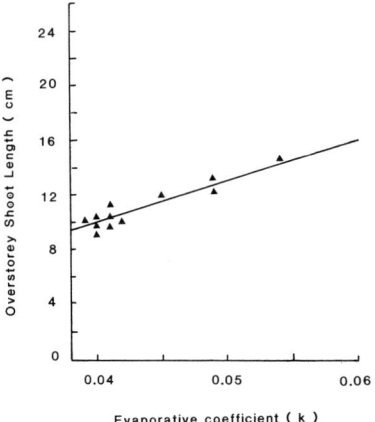 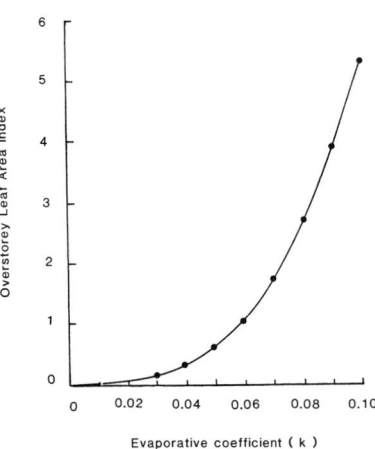

Fig. 4. Mean values of Shoot Length (cm) growth per annum and Leaf Area Index (cm^2 leaf cm^{-2} land) of overstorey species from evergreen plant communities in the mediterranean-climate of Australia, plotted against the evaporative coefficient (k) – Specht and Specht in prep. (N.B. Considerable variation around the mean values will result in microhabitats of differing Available Water and Soil Fertility).

commences rapidly, while the summer shoot growth of overstorey species develops slowly utilizing photosynthates stored during the more favourable seasons.

One interesting facet of this distinction between the shoot growth of the dicotyledonous overstorey and understorey species in the mediterranean-climate of southern Australia is the length of the internodes of species in the two strata: long internodes are characteristic of the overstorey, short internodes of the understorey. The temperatures (Fig. 6) experienced at the stem-air interface (where axillary buds are located) on a sunny, windless day may be 5°C higher in understorey plants (with short internodes) than in the overstorey species (where the temperatures are almost that of the ambient air temperatures). The foliar-shoot attributes of overstorey and understorey appear to be complementary (Specht *et al.* (in press)); they optimize the use of resources in the mediterranean climate of Australia.

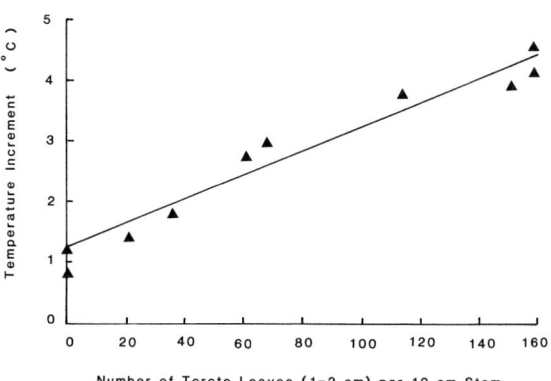

Fig. 5. Relative shoot growth of overstorey species plotted against mean daily temperature (Specht 1981a, 1986). Meg = Megatherm, Mac = Macrotherm, Mes = Mesotherm, Mic = Microtherm, Nan = Nanotherm.

Fig. 6. Increment in temperature within the stem-air boundary layer surrounding leafy stems when exposed to solar radiation on calm days – Specht and Yates in prep.

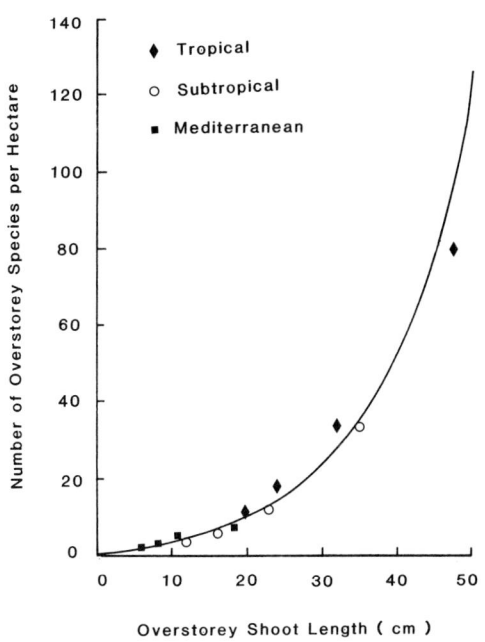

Fig. 7. Species richness of species of overstorey trees and tall shrubs (per hectare) plotted against mean values of Shoot Length (cm) growth per annum of overstorey species in evergreen plant communities of tropical, subtropical and temperate Australia – Specht and Specht 1988b.

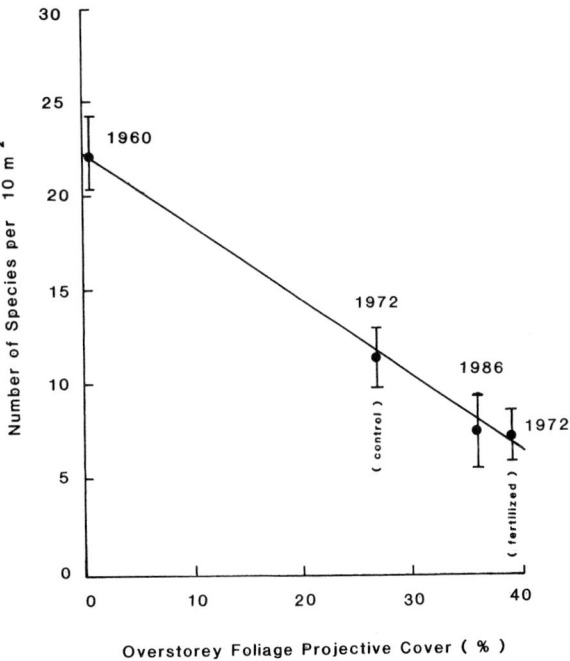

Fig. 8. Decrease in species richness of the understorey as Foliage Projective Cover of the overstorey increases in post-fire succession – Specht and Morgan 1981, Specht and Specht 1988a.

Species richness: Overstorey

The vigour of foliar shoot-growth of overstorey species (observed on tagged shoots, or estimated using Equation (3) – with substitution of the shoot-growth temperature response of Fig. 5 for the relatively-insignificant photosynthetic temperature response (Mooney 1983) – is closely correlated with species richness (number of species of trees/tall shrubs per hectare) of that stratum (Fig. 7). Shoot growth can be markedly influenced by variation in the Annual Moisture Index (Fig. 2) and the Soil Fertility Index (Specht *et al.* in press); species-richness of the overstorey parallels these variations.

Species richness: Understorey

In the pyric succession following fire which frequently razes mediterranean-type ecosystems, the Foliage Projective Cover of the overstorey regenerates slowly while the understorey develops rapidly (see above). Initially, species richness of the understorey is high, but as the Foliage Projective Cover of the overstorey gradually increases towards the 'climax' equilibrium, species richness of the understorey decreases (Fig. 8). Any community-physiological process which may speed or slow the development of overstorey cover will have a similar effect on species richness of the understorey (Specht and Specht 1988a).

Species richness of mediterranean plant communities (overstorey + understorey) in southeastern Australia, early in post-fire succession, increases with increasing values of the Annual Moisture Index for major climatic zones with similar evaporative coefficient (Fig. 9).

Species richness: Small mammals

It would appear there is a strong correlation between the Number of Species of Small Mammals

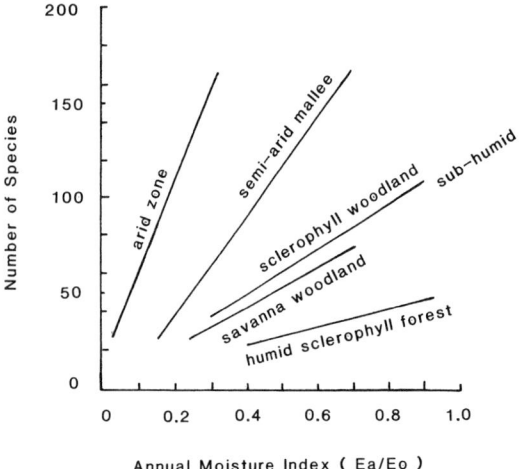

Fig. 9. Total species richness of plant communities in arid, semi-arid, subhumid and humid mediterranean climatic zones of southeastern Australia plotted against Annual Moisture Index – Specht 1984.

and the Number of Plant Species recorded in Australian ecosystems. Braithwaite *et al.* (1985) have demonstrated this relationship for tropical Australia (Fig. 10a); a similar correlation appears to hold for the mediterranean-ecosystems of Australia (as shown by data assembled by Westman, Chapter 2(3) and Catling, Chapter 4 of this Volume – Fig. 10b).

Conclusions

The nature of the horizontal cover (Foliage Projective Cover) and ecomorphological attributes of the foliage of the evergreen overstorey of mediterranean-climate plant communities have together achieved an evaporative balance with the annual evaporative cycle, expressed by the evaporative coefficient (k).

Although photosynthetic production is little affected by seasonal changes in air/leaf temperatures, growth of foliar shoots is strongly controlled by temperature. The temperatures experienced by axillary buds at the stem-air interface of overstorey species (with long internodes) is little different from ambient air temperature. Shoots of dicotyledonous understorey species possess short internodes, between which the temperature of the stem-air interface increases up to 5°C above ambient air temperature during periods of sunny, calm weather. These understorey species shown microtherm shoot-growth rhythms (with growth peaks at 15–16°C), while the overstorey species show mesotherm shoot-growth rhythms (with growth peaks at 20–21°C). Such plants produce shoot growth during mediterranean summer, using (a) photosynthates produced during spring and (b) water stored at depth in the soil during the humid season of the year.

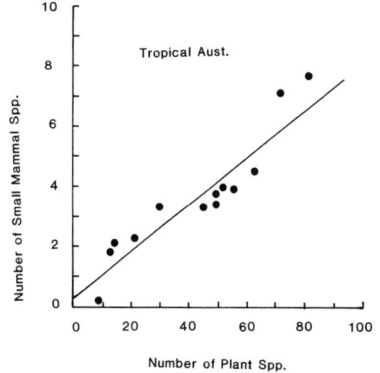
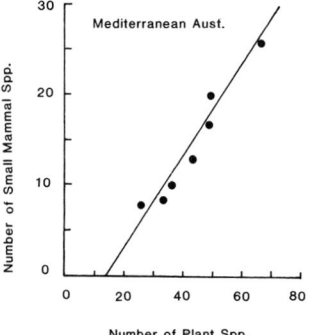

Fig. 10. Number of small mammals (M) recorded in plant communities in tropical Australia (Braithwaite *et al.* 1985) and in mediterranean southern Australia (Catling Ch. 4) plotted against number of plant species (P) recorded in the same ecosystems.
Tropics $M = 0.079P + 0.26$ *Mediterranean* $M = 0.50P - 6.81$
  ($n = 13$, $r^2 = 0.85$)   ($n = 6$, $r^2 = 0.90$)

Species richness of the trees/tall shrubs in the overstorey is closely associated with the vigour of foliar shoot growth of overstorey species. The development of Foliage Projective Cover in post-fire succession suppresses the species richness of the understorey. The balance between the species richness of both overstorey and understorey strata is closely influenced by foliar attributes via evapotranspiration, photosynthetic production and seasonal translocation to new foliage shoots.

Species richness of small mammals (and probably other faunal groups) is closely correlated with species richness of the plant species of the same ecosystem.

References

Braithwaite, R.W., Winter, J.W., Taylor, J.A. and Parker, B.S., 1985. Patterns of diversity and structure of mammalian assemblages in the Australian tropics. Aust. Mamm. 8: 171–186.

Catling, P.C. (co-ordinator), 1988. Vertebrates. Chapter 4. In: Mediterranean-Type Ecosystems. A Data Source Book, this volume.

Mooney, H.A., 1983. Carbon-gaining capacity and allocation patterns of mediterranean-climate plants. pp. 103–119. In: Kruger, F.J., Mitchell, D.T. and Jarvis, J.U.M. (eds.) Mediterranean-Type Ecosystems. The Role of Nutrients. Springer-Verlag, Berlin.

Specht, R.L., 1957. Dark Island heath (Ninety-Mile Plain, South Australia). 5. The water relationships in heath vegetation and pastures on the Makin Sand. Aust. J. Bot. 5: 151–172.

Specht, R.L., 1972. Water use by perennial evergreen plant communities in Australia and Papua New Guinea. Aust. J. Bot. 20: 273–299.

Specht, R.L., 1981a. Growth indices – their role in understanding the growth, structure and distribution of Australian vegetation. Oecologia (Berl.) 50: 347–356.

Specht, R.L., 1981b. Primary production in mediterranean-climate ecosystems regenerating after fire. pp. 257–267. In: di Castri, F., Goodall D.W. and Specht, R.L. (eds.) Ecosystems of the World. 11. Mediterranean-Type Shrublands. Elsevier, Amsterdam.

Specht, R.L., 1983. Foliage projective covers of overstorey and understorey strata of mature vegetation in Australia. Aust. J. Ecol. 8: 433–439.

Specht, R.L., 1984. Species-richness of plant communities in the mediterranean climate of southeastern Australia. pp. 143–144. In: Dell, B. (ed.) MEDECOS IV. Proceedings of the 4th International Conference on Mediterranean Ecosystems. Botany Dept., University of Western Australia, Nedlands, W. Aust.

Specht, R.L., 1986. Functioning of tropical plant communities. 6. Phenology. pp. 78–90. In: Clifford, H.T. and Specht, R.L. (eds.) Tropical Plant Communities. Their Resilience, Functioning and Management in Northern Australia. Botany Dept., University of Queensland, St Lucia, Qld.

Specht, R.L., 1987. The effect of summer drought on vegetation structure in the mediterranean climate region of Australia. pp. 625–639. In: Tenhunen, J.D., Catarino, F.M., Lange, O.L. and Oechel, W.C. (eds.) Plant Response to Stress. Functional Analysis in Mediterranean Ecosystems. Springer-Verlag, Berlin.

Specht, R.L., in press. Nutrient concentration in leaves of overstorey species as an assessment of the nutrient status of Australian ecosystems. Aust. J. Bot. 36.

Specht, R.L. and Moll, E.J., 1983. Mediterranean-type heathlands and sclerophyllous shrublands of the world: an overview. pp. 41–65. In: Kruger, F.J., Mitchell, D.T. and Jarvis, J.U.M. (eds.) Mediterranean-Type Ecosystems. The Role of Nutrients. Springer-Verlag, Berlin.

Specht, R.L. and Morgan, D.G., 1981. The balance between the foliage projective covers of overstorey and understorey strata in Australian vegetation. Aust. J. Ecol. 6: 193–202.

Specht, R.L. and Specht, A., 1988a. Species richness of sclerophyll (heathy) plant communities in Australia – The influence of overstorey cover. Aust. J. Bot. 36.

Specht, R.L. and Specht, A., 1988b. Species richness of overstorey strata in Australian plant communities – The influence of overstorey growth rates. Aust. J. Bot. 36.

Specht, R.L. and Yates, D.J., in prep. The structure and location of foliage shoots in Australian heathlands and their effect on the temperature experienced at growth apices.

Specht, R.L., Moll, E.J., Pressinger, F. and Sommerville, J., 1983. Moisture regime and nutrient control of seasonal growth in mediterranean ecosystems. pp. 120–132. In: Kruger, F.J., Mitchell, D.T. and Jarvis, J.U.M. (eds.) Mediterranean-Type Ecosystems. The Role of Nutrients. Springer-Verlag, Berlin.

Specht, R.L., Arianoutsou, M., Bolton, M.P., Forster, P.I., Grundy, R., Hegarty, E.E. and Specht, A., in press. Foliar attributes and species richness of plant communities in subtropical Australia: Changes induced by climate variation. Aust. J. Bot.

Westman, W.E. (co-ordinator) 1988. Species richness. Chapter 2(3). In: Mediterranean-Type Ecosystems. A Data Source Book. This volume.

Vegetation, nutrition and climate – examples of integration

(3) Leaf structure and nutrition in mediterranean-climate sclerophylls

P.W. Rundel

Contents

Introduction	157
Statistical analyses	159
Leaves from mediterranean-type and non-mediterranean-type ecosystems	160
Relationships between leaf structure and nutrition	162
Inter- and intra-regional patterns of leaf structure and nutrition	163
References	165

Introduction

Evergreen, sclerophyllous shrubs and trees are the dominant community elements in all of the five mediterranean-type ecosystems of the world. The remarkable convergence of leaf structure and morphology among unrelated taxa in these regions has led to a major interest in leaf sclerophylly as an adaptation in these ecosystems (Mooney and Dunn 1970). The major focus of this interest in mediterranean-type ecosystems has been on sclerophylly as a drought adaptation (see Seddon 1974 for a historical discussion of concepts of sclerophylly and xeromorphy), but there has also been considerable attention given recently to the significance of evergreen leaves as a physiological strategy to increase nutrient-use efficiency, and to sclerophylly as an adaptation to reduce herbivory.

A fundamental physiological difference between evergreen and deciduous leaves is, of course, longevity of the photosynthetic tissues (Chabot and Hicks 1982). Evergreen leaves are retained from one to many years on the plant, most commonly two years in mediterranean sclerophylls. While evergreen leaves are commonly more expensive to produce than deciduous leaves (Miller and Stoner 1979), they are able to amortize their cost of production over an extended period. Relatively low rates of photosynthetic capacity in evergreen sclerophylls are thus balanced by the potential ability to photosynthesize throughout the year. In the highly unpredictable environments characteristic of mediterranean-type ecosystems, this is an important adaptation (Mooney and Dunn 1970, Harrison *et al.* 1971). Deciduous leaves in contrast must have a relatively high photosynthetic capacity to amortize the costs of leaf production over a much shorter period.

Selective factors favoring differing levels of leaf longevity can be shown in a generalized relationship between leaf duration and environmental stress, both physical and biotic (Fig. 1). Integrated resource competition increases with greater longevity, at the expense of increased carbon cost for leaf production and maintenance as well as increased probability of encountering predators (Mooney and Gulmon 1982). Sclerophyllous leaf structures, among many possible functions, clearly may limit the latter expense. Greater leaf duration is advantageous where water or nutrient resources are continuously available in low amounts, but disadvantageous where they are predictably present in brief flushes. However, water and nutrient availability may not co-vary together, leading to the potential for intermediate strategies of adaptation (Mooney and Gulmon 1982).

Comparative anatomical studies of sclerophylls certainly support the importance of xerophytic ad-

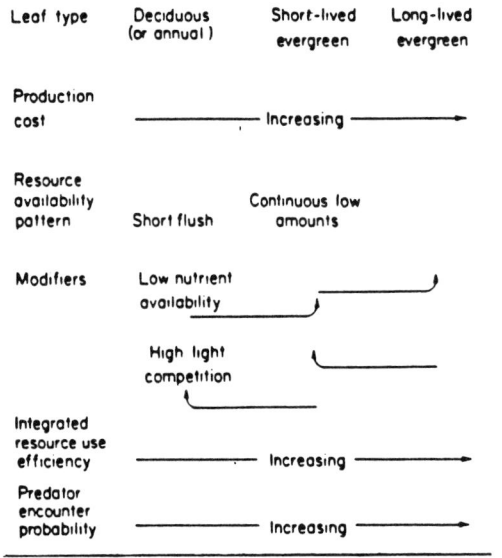

Fig. 1. Relationship between leaf longevity and resource availability (from Mooney and Gulmon 1982).

aptations such as small cell size, thick cuticles and high fibre content minimizing water loss (Kummerow 1973). Physiological studies support this importance (Dunn *et al.* 1977). The majority of the paleoecological discussions of the origins of mediterranean-type ecosystems explicitly hypothesize increasing drought as the primary selective factor in the evolution of community dominance by evergreen sclerophylls (e.g. Axelrod 1975, 1977, Beard 1977). Finally, modelling efforts combining elements of both leaf structure and function have been used to demonstrate how seasonal patterns of temperature and water availability may control the relative dominance of evergreen sclerophylls and deciduous species along altitudinal or latitudinal gradients (Miller 1981).

Despite this major focus on sclerophylly as an adaptation to water stress, it has long been known that sclerophylls are characteristic of many non-mediterranean ecosystems. Sclerophyllous heathlands with both structural and floristic affinities to mediterranean-type ecosystems extend into regions of summer rainfall along the east coast of Australia (Specht 1979) and similarly into the mountains of East Africa. Floristic elements of California chaparral extend into biseasonal rainfall areas of Arizona and eastward into summer rainfall regions of Mexico with little obvious change in leaf structure (Muller 1939, 1947). These biogeographical observations have led to a realization of the importance of nutrient availability as a selective factor in the evolution of sclerophyllous leaf morphologies.

There are a number of advantages which evergreen leaves may have over deciduous leaves in environments with low nutrient availability (Chapin 1980). Longer leaf life-spans allow leaves to utilize limited resources of nitrogen over more extended periods of time and thus allow a higher nutrient use efficiency (Small 1972, Schlesinger and Chabot 1977, Rundel 1982). Furthermore, the sclerophyllous nature of evergreen leaves promotes gradual litterfall and slow rates of decomposition (Schlesinger and Hasey 1981), thus moderating pulses of nutrients back into the soil and promoting tighter cycles of nutrient flow.

Nitrogen-use efficiency has been studied carefully in California chaparral ecosystems where this element is commonly a limiting resource. Nitrogen is of critical importance in foliar nutrition because of the central role of this element in the biochemistry and photobiology of photosynthesis. In a typical leaf, 70–80% of the organic nitrogen lies in proteins, 10% in nucleic acids and 5–10% in chlorophyll and lipoproteins, with the remainder largely in the form of free amino acids (Chapin and Kendrowski 1983). A good approximation may be that approximately 75% of the total nitrogen in a typical leaf may well be directly related to the process of photosynthesis (Field and Mooney 1986). As a result it is not surprising to find that leaf nitrogen content is strongly correlated with photosynthetic capacity. This relationship is shown very well even with leaves collected across a range of ecosystem types (Fig. 2).

Studies in California chaparral have demonstrated that leaf nitrogen content is a major determinant of photosynthetic capacity. Under controlled conditions, it is significant to note that there is an inverse relationship between nitrogen use efficiency and water use efficiency (Field *et al.* 1983). These data suggest that complex models of leaf structure and function will be necessary to properly

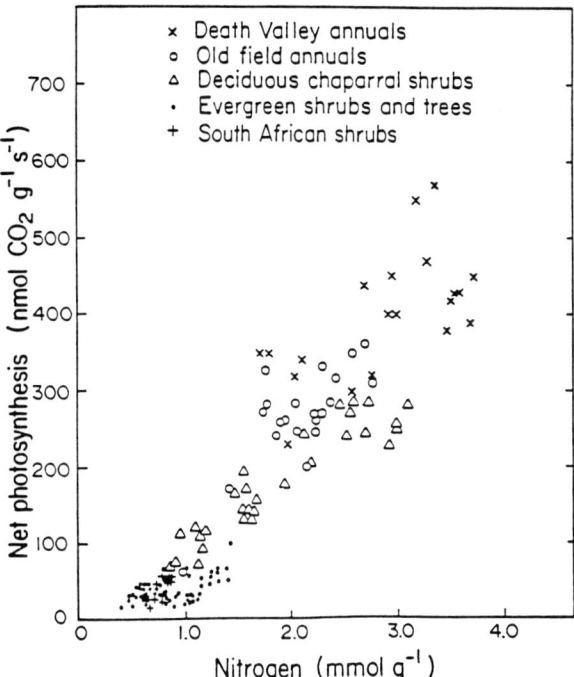

Fig. 2. Relationship between leaf nitrogen content (mmol g^{-1}) and net photosynthesis in a variety of wild plant species covering a range of growth-forms and ecosystems (from Field and Mooney 1986).

separate the significance of these two limiting factors in the evolution of sclerophyllous leaves.

Statistical analyses

In order to develop conceptual models of the leaf structural and chemical traits which characterize mediterranean-type ecosystems, I have analyzed mature leaf tissues from a range of plant communities in a variety of mediterranean climate regions. The community samples are outlined in Table 1. From California they include a mixed chaparral community from San Diego County (see Miller 1981 for details of this site) and a maritime chaparral community on nutrient poor substrates in Monterey Country (Griffin 1978). In Chile I collected samples from a geographical gradient of four vegetation types. These include matorral from the Coast Range (Miller 1981), semi-deciduous coastal matorral, montane matorral from about 2200 m in the foothills of the Andes, and hygrophilous forest from the Coast Range (see Rundel (1981) for a discussion of ecological relationships of these Chilean community-types). Dominant maquis species were collected near Marseille in France. Finally a collection of South African samples of mountain fynbos species were collected at two sites near Cape Town (Mooney *et al.* 1983). One group came from Jonkershoek Experimental Forest near Stellenbosch (Kruger 1979) and the other from the Kogelberg (Boucher 1978). Because of very different suites of structural and nutritional traits, the South African species were subdivided into three groups representing the dominant groups in fynbos. These are proteoid group (comprised of Proteaceae), the ericoid group (needle-leaved subshrubs principally in the Ericaceae, and Bruniaceae), and the restioid group (sedgelike species, largely in the Restionaceae). Summaries of some of these data were previously published by Specht and Moll (1983).

For comparison, parallel sets of data were collected for four communities of sclerophyllous-leaved species in tropical and subtropical ecosystems characterized by low nutrient availability. These included a white sand scrub community (campina) of low trees near Manaus in Brazil (Anderson 1981), a sand pine scrub community from central Florida, and two communities from the Hawaiian Island – a subalpine scrub of sclerophyllous-leaved shrubs from Haleakala on Maui and a low wet forest in the Koolau Range on Oahu, Sclerophyllous leaves of tropical 'bana' communities in the Amazon Basin have been described by Sabrado and Medina (1980).

I have focused on collecting data on ten basic parameters of foliar structure and nutrition. These are the following, listed with the method of analysis:

1) nitrogen (N) – Kjehldahl digestion and subsequent colorimetric analysis by Technicon Autoanalyzer II
2) phosphorus (P) – acid digestion and colorimetric analysis by Technicon Autoanalyzer II
3) potassium (K) – acid digestion and analysis by atomic absorption spectrophotometer
4) calcium (Ca) – as for potassium
5) magnesium (Mg) – as for potassium

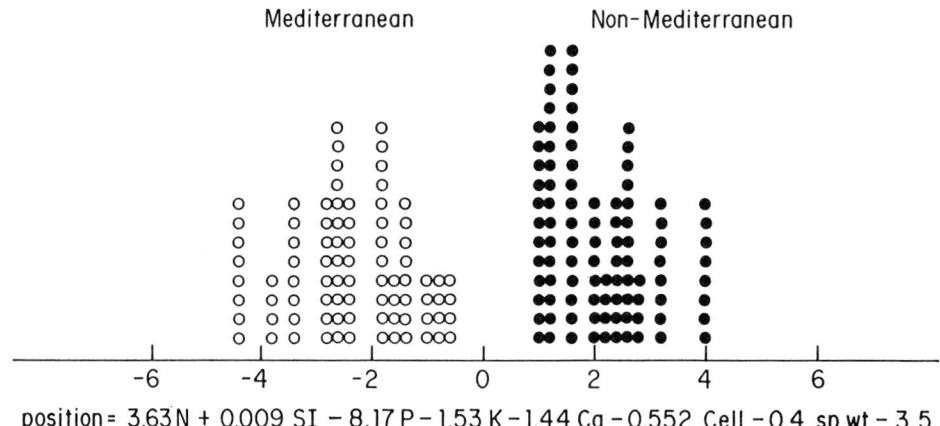

Fig. 3. Canonical discriminant function to allow the separation of mediterranean and non-mediterranean sclerophyllous leaves. See text for abbreviations of the parameters used.

6) lignin (Lig) – van Soest procedure
7) cellulose (Cell) – as for lignin
8) ether extractives (Ether) – ether soluble fraction including facts, waxes and oils (Randall 1974)
9) sclerophyll index (SI) – ratio of total fibre (lignin and cellulose) to protein concentration (nitrogen × 6.25) multiplied by 100 (Loveless 1961)
10) specific leaf weight (spwt) – leaf dry weight per unit area, defined as mg cm^{-2}

Details of these analytical procedures are given in Allen *et al.* (1974).

Leaves from mediterranean-type and non-mediterranean-type ecosystems

There is a variety of criteria by which one might assess the relative similarities and dissimilarities among sclerophyllous leaves from mediterranean and non-mediterranean-type ecosystems. One approach is to utilize the individual species values for data collected from the samples described in Table 1 (see Section 2 of this Volume for data on individual species). From these data, a canonical discriminant function can be developed (Gauch and Wentworth 1976) which will allow such a separation.

Complete separation of the two groups of species

Table 1. Geographical origin of leaf analysis data used in statistical analyses.

Region	Vegetation type	No. of species	Location
Mediterranean-type ecosystems			
California	Chaparral	4	Descanso, San Diego County
California	Maritime chaparral	9	Monterey Peninsula
Chile	Matorral	7	Fundo Santa Laura, Cordillera de la Costa
Chile	Montane matorral	4	Cordillera de los Andes
Chile	Coastal matorral	5	Pichidangui
Chile	Hygrophilous forest	11	Cerro La Campana
France	Maquis	10	Marseille
South Africa	Mountain fynbos	18	Jonkershoek and Kogelberg, Cape Province
Non-mediterranean sclerophyll			
Brazil	Campina	15	Manaus
Florida	Sand pine scrub	9	near Orlando
Hawaiian Islands	Subalpine scrub	7	Haleakala National Park, Maui
Hawaiian Islands	Wet forest	7	Koolau Mountains, Oahu

(mediterranean and non-mediterranean) can be achieved utilizing seven variables, as shown in Fig. 3. The position along the axis, with negative values indicative of mediterranean species and positive values of non-mediterranean species, can be calculated as:

$$\text{position} = 3.63\,N + 0.009\,SI - 8.17\,P - 1.53\,K - 1.44\,Ca - 0.552\,Cell - 0.4\,SPWT - 3.5. \quad (1)$$

While this discriminant function is relatively complex, a remarkable level of discrimination can be achieved using only two variables, nitrogen and calcium content:

$$\text{position} = 2.03\,N - 1.49\,Ca - 1.28 \quad (2)$$

This simple canonical discrimination function provides a correct assignment to 79% of the species analyzed here. As this function suggests, the mediterranean ecosystem species have a tendency to have lower levels of nitrogen and higher levels of calcium compared to the non-mediterranean sclerophyll species.

A second approach to assessing differences in leaf traits between mediterranean and non-mediterranean sclerophyll species is to carry out a mathematical ordination to group species with similar traits of leaf structure and chemistry. The approaches I have used are principal components analysis (PCA), and an eigenvector method of ordination termed reciprocal averaging (Hill 1973, Gauch 1982).

A good separation of mediterranean species can be achieved with a PCA ordination of all ten character traits for chaparral, matorral and maquis species against those for campina and sand pine scrub. This separation is shown in Fig. 4, with each letter plotting a centered species mean. Each character trait is also ordinated on the same axes using the species values as traits. A reasonably good separation can also be achieved using a standardized and centered PCA ordination with data for nitrogen, cellulose and lignin alone from the full set of data for all of the mediterranean species listed in Table 1 and for the non-mediterranean taxa from

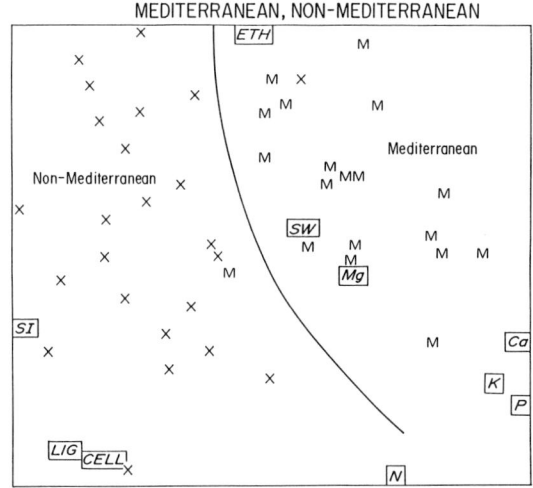

Fig. 4. PCA ordination of the character trait for leaves from three mediterranean communities (M) and two non-mediterranean communities (X). Each letter plots a single species value. See text for discussion.

Brazil and Florida. In this ordination the majority of South African species sort out distinctly from the remainder of the mediterranean species (Fig. 5).

A third approach to separating mediterranean and non-mediterranean leaf traits is to use a PCA ordination of five leaf nutrients (N, P, K, Na, Mg) for the same set of species presented in Fig. 5. Using these data, there is again a separation of higher-nutrient mediterranean species (California, Chile and France) from non-mediterranean sclerophylls in Florida and Brazil and from South African taxa (Fig. 6). In this ordination the Florida (F) and Brazilian (B) taxa separate relatively well. More interesting, however, is the separation of three groupings of South African species – proteoid (P), ericoid (E) and restioid (R). From the plots of mean character trait positions on the same axes (based on species values as traits for ordination), the mediterranean taxa are separated by relatively low levels of nitrogen and magnesium and relatively high levels of calcium and phosphorus. The South African taxa are separated by relatively low levels of phosphorus and high levels of magnesium.

From both the canonical discriminant analysis and the PCA ordinations, it seems clear that all sclerophyllous leaves are not structured alike and that general patterns of separation between medi-

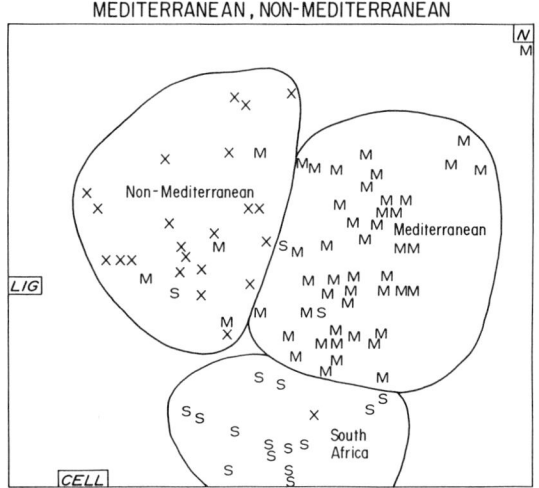

Fig. 5. PCA ordination of nitrogen, cellulose and lignin traits for leaves from three mediterranean communities (M), two non-mediterranean communities (X) and South African fynbos (S). Each letter plots a single species value. See text for discussion.

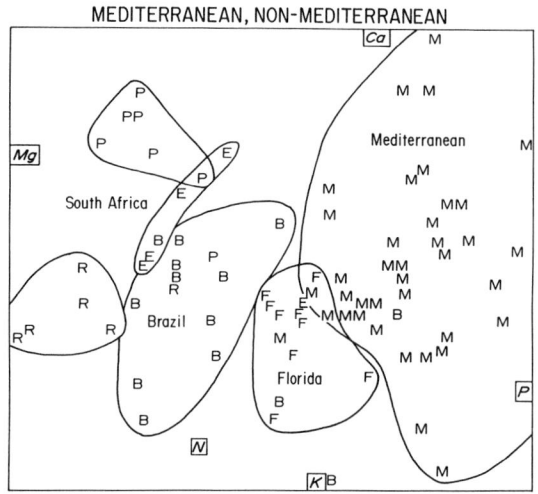

Fig. 6. PCA ordination of five leaf nutrient traits for leaves from three mediterranean communities (M), two non-mediterranean communities (Brazil = B, Florida = F), and South African fynbos. Fynbos species are separated into proteoid (P), restioid (R), and ericoid (E) groups.

terranean and non-mediterranean sclerophylls can be described. As more data sets on leaf structure and nutrient content are collected and analyzed, the specific nature of such differences can be better understood.

Relationships between leaf structure and nutrition

From the data collected from the mediterranean-type communities in Table 1, several patterns of relationships between individual nutrients and/or structural traits are apparent. One such relationship is the general balance between leaf nitrogen and phosphorus (Fig. 7). Such a relationship parallels that reported by Loveless (1961, 1962), a pattern where individual communities separate out well, and is consistent with current theories of physiological constraints on leaf structure and function (Mooney and Gulmon 1982). From a best-fit line in Fig. 7, California chaparral species are lower than expected in foliar nitrogen and ericoid taxa from South Africa are somewhat higher than expected.

A second pattern, more unexpected than the first, is the highly significant relationship between nitrogen and calcium (Fig. 8). Individual communities, however, do not separate out as well in this plot as in that shown in the previous figure. The worst fit for a group in Fig. 8 is for proteoid species (P) from South Africa.

A final pattern of relationship evident in the data set from mediterranean-type ecosystems is that between sclerophyll index (fibre to protein ratio × 100) and leaf phosphorus. The curvilinear relationship between these two parameters is a highly significant logarithmic fit, with some fair degree again of community separation (Fig. 9). This relationship of sclerophyll index to phosphorus has previously been reported for a mixture of sclerophyllous and mesophytic leaves by Loveless (1961, 1962). However, a major difference in the relationship reported here and that described by Loveless lies with the range of values included and the breaking point in the curve. Loveless reported mean levels of sclerophyll index (as fibre to protein ratio) at levels of about 50 until a threshold level of about 0.3% phosphorus was reached and SI values increased rapidly to as high as 350. In my data set, all levels of phosphorus measured were below the threshold level reported by Loveless, and values of sclerophyll index are much higher. In Fig. 9, mean levels of SI of about 200 reach a threshold of about 0.07% phosphorus before values climb sharply. The pres-

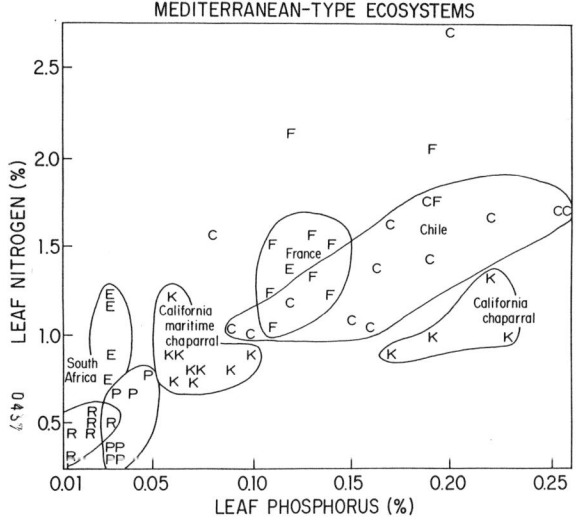

Fig. 7. Relationship between leaf nitrogen and leaf phosphorus concentrations in leaves of mediterranean sclerophyll communities from California (K), Chile (C), France (F), and South Africa (P, R, E).

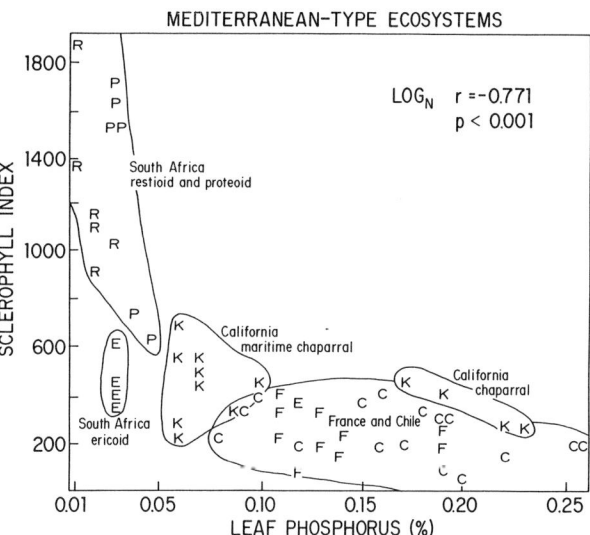

Fig. 9. Relationship between leaf sclerophyll index and leaf phosphorus concentration in mediterranean sclerophyll communities as in Figure 7.

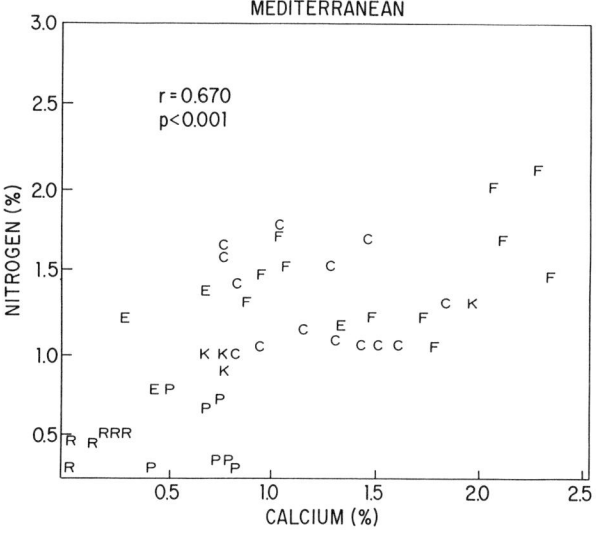

Fig. 8. Relationship between leaf nitrogen and leaf calcium concentrations in mediterranean sclerophyll communities as in Figure 7.

ence of the sharp increase in value of leaf sclerophyll index at relatively low levels of phosphorus in the two data sets suggest a possible metabolic control of high levels of sclerophylly in phosphorus deficient taxa. The sharp increase in sclerophyll index is much more a function of increasing fibre content than it is of lower protein levels.

Another possible explanation for the observed curvilinear relationship between sclerophyll index and phosphorus content is a hypothetical balance between nitrogen, phosphorus and fibre content across this nutrition gradient. If such a strong physiological linkage existed, then the curvilinear plot would be the functional result of a reciprocal plot. This suggestion should be testable.

Inter- and intra-regional patterns of leaf structure and nutrition

In previous figures it was seen that sclerophylls growing on highly-leached, oligotrophic soils in South Africa often separate well in ordinations from sclerophylls growing on better soils in Chile and California. This relationship was explored in more detail in further ordinations. Sclerophylls from the French maquis studied here were eliminated from these analyses because of the calcium-rich limestone substrates on which they grow. A PCA ordination of nine leaf traits (eliminating only specific leaf weight) from the remaining data lumps Chile and California sclerophylls, but widely sep-

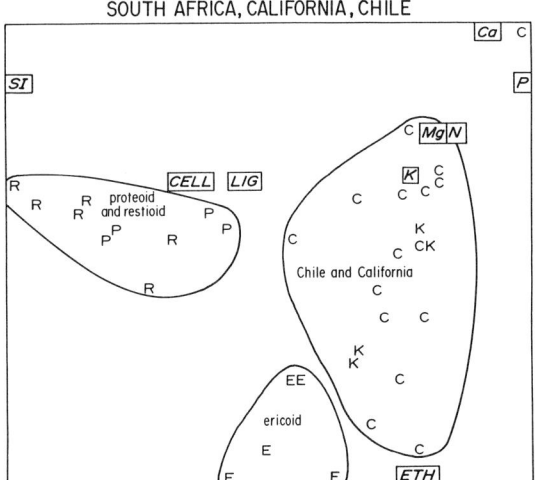

Fig. 10. PCA ordination of nine leaf traits for California chaparral (K), Chilean matorral (C) and South African fynbos (proteoid = P, restioid = R, ericoid = E). See text for discussion.

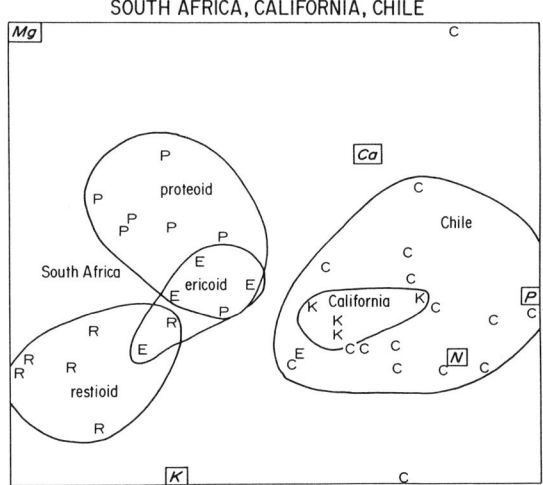

Fig. 11. Reciprocal averaging ordination of five leaf nutrient traits for California chaparral (K), Chilean matorral (C) and South African fynbos (proteoid = P, restioid = R, ericoid = E). See text for discussion.

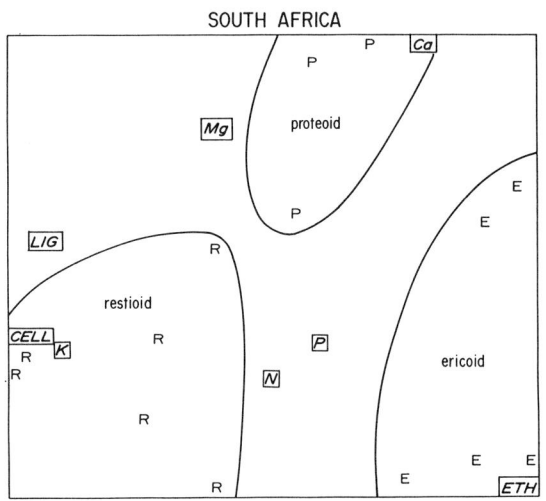

Fig. 12. Reciprocal averaging ordination of eight leaf traits (eliminating sclerophyll index and ether extractives) for proteoid (P), restioid (R), and ericoid (E) species from South African fynbos. See text for discussion.

arates two groupings of South African species (Fig. 10). Proteoid (P) and restioid (R) taxa separate from ericoid taxa on the basis of their high levels of lignin and cellulose and related high values of sclerophyll index, and on their low levels of ether extractive compounds in their foliage.

An ordination of this same data set using reciprocal averaging of five foliar nutrients similarly separates Chile-California sclerophylls well from those from South Africa (Fig. 11). With the data on lignin, cellulose, sclerophyll index and ether extractives not included, the ericoid (E) taxa fall out intermediate between the proteoid and restioid species.

If the South African sclerophylls alone are ordinated using reciprocal averaging on the basis of the five mineral nutrients plus lignin, cellulose and ether extractives, all three groups of taxa separate very well (Fig. 12). High ether extractives pull the ericoids to one corner, while higher calcium levels pull the proteoids away. The restioids are characterized by higher levels of lignin, cellulose and potassium. If only the five mineral nutrients are ordinated, good separation continues to be present, with relatively high nitrogen and phosphorus levels becoming significant correlates for the ericoids (Fig. 13). An interesting relationship present in the South African sclerophylls is the strong positive correlation between ether extractive content and foliar concentrations of nitrogen (Fig. 14). It is not clear at this time what selective factors might be responsible for this highly significant correlation. One possible hypothesis is that these ether extractives serve as anti-herbivore defenses in higher quality foliage. Another more conjectural hypoth-

165

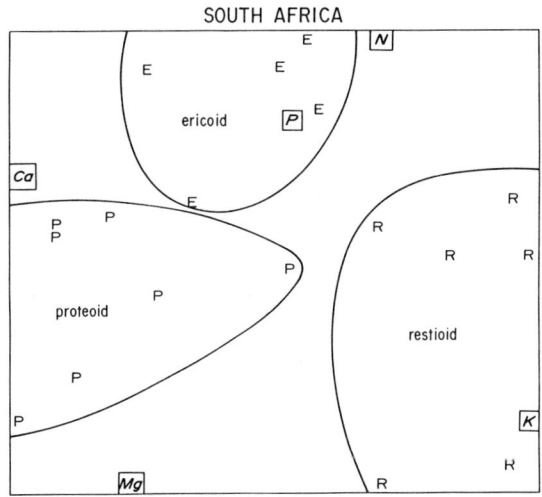

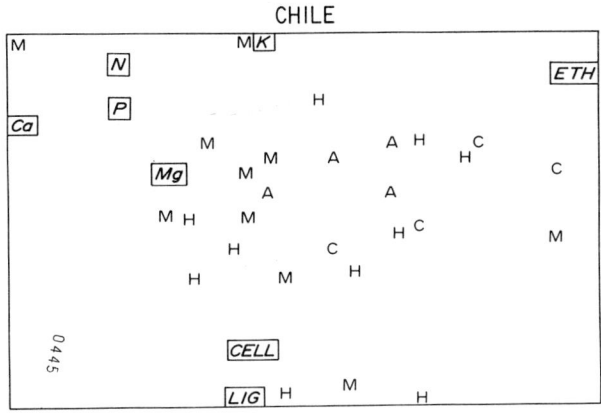

Fig. 13. Ordination of five leaf nutrient traits for proteoid (P), restioid (R), and ericoid (E) species from South African fynbos. See text for discussion.

Fig. 15. PCA ordination of eight character traits for Chilean woody perennials from matorral (M), coastal matorral (C), montane matorral (A), and hygrophilous forest (H) communities. See text for discussion.

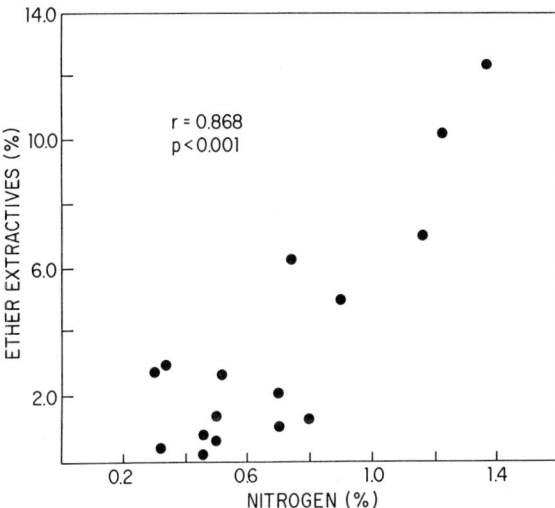

Fig. 14. Relationship of leaf ether extractive and nitrogen concentrations in 16 species of South African fynbos perennials.

monly quite distinct in the three groups. Belowground studies have also established a strong difference between the groups, with the ericoid taxa being mycorrhizal as opposed to the non-mycorrhizal proteoid and restioid root morphologies in the other two groups (Lamont 1982). More detailed studies of the nutritional ecology of these South African groups will be necessary to better resolve the open questions concerning physiological versus ecological tolerances of low nutrient availability.

Functional guilds of leaf structure and nutrition, such as evident in South Africa, do not seem to be present in the other data sets analyzed from mediterranean-climate ecosystems. This can be best illustrated by a PCA ordination of leaf data from four Chilean communities – matorral (M), coastal matorral (C), montane matorral (A), and hygrophilous forest (H). As shown in Fig. 15, there is no pattern of community separation. Species from all four communities are relatively well spread across the ordination.

esis would be a possible linkage of ether extractives with flammability in relatively short-lived ericoids which are specialists in post-fire habitats where nutrients are relatively more available.

At the very least, the observed patterns of ordination clusters between proteoid, restioid and ericoid species in South Africa suggest that three functional guilds of nutritional strategy may be present. Leaf form and plant growth-form are com-

References

Allen, S.E., Grimshaw, H.M., Parkinson, J.A. and Quarmby, C., 1974. Chemical Analysis of Ecological Materials. John Wiley, New York.

Anderson, A.B., 1981. White sand vegetation of Brazilian Amazonia. Biotropica 13: 199–210.

Axelrod, D.I., 1975. Evolution and biogeography of Madrean-

Tethyan sclerophyll vegetation. Annals of the Missouri Botanical Garden 62: 280–334.

Axelrod, D.I., 1977. Outline history of California vegetation. pp. 139–193, In: Barbour, M.F. and Major, J. (eds). Terrestrial Vegetation of California. Wiley, New York.

Beard, J.S., 1977. Tertiary evolution of the Australian flora in the light of latitudinal movements of the continent. J. Biogeogr. 4: 111–118.

Boucher, C., 1978. Cape Hangklip area. II. The Vegetation. Bothalia 12: 455–497.

Chabot, B.F. and Hicks, D.J., 1982. The ecology of leaf life spans. Ann. Rev. Ecol. Syst. 13: 229–259.

Chapin, F.S., 1980. The mineral nutrition of wild plants. Ann. Rev. Ecol. Syst. 11: 233–260.

Chapin, F.S. and Kendrowski, R.A., 1983. Seasonal changes in nitrogen and phosphorus fractions and autumn translocation in evergreen and deciduous taiga trees. Ecology 64: 376–391.

Dunn, E.L., Shropshire, F., Song, L. and Mooney, H.A., 1977. The water factor and convergent evolution in mediterranean-type vegetation. pp. 492–505. In: Lange, O.L., Kappen, L. and Schulze D.-D. (eds). Water and Plant Life: Problems and Modern Approaches. Springer-Verlag, Berlin.

Field, C. and Mooney, H.A., 1986. The photosynthesis-nitrogen relationship in wild plants, pp. 25–56. In: Givnish, T.J. (ed.). On the Economy of Plant Form and Function. Cambridge Univ. Press. Cambridge.

Field, C., Merino, J. and Mooney, H.A., 1983. Compromises between water-use efficiency and nitrogen-use efficiency in five species of California evergreens. Oecologia 60: 384–389.

Gauch, H.G., 1982. Multivariate Analyses in Community Ecology. Cambridge Univ. Press. 28 p.

Gauch, H.G., and Wentworth, T.R., 1976. Canonical correlation analysis as an ordination technique. Vegetatio 33: 17–22.

Griffin, J.R., 1978. Maritime chaparral and endemic shrubs of the Monterey Bay Region, California. Madroño 25: 65–81.

Harrison, A.T., Small, E. and Monney, H.A., 1971. Drought relationships and distribution of two mediterranean-climate California plant communities. Ecology 52: 869–875.

Hill, M.O., 1973. Reciprocal averaging: an eigenvector method of ordination. J. Ecol. 61: 237–249.

Kruger, F.J., 1979. South African heathlands. pp. 19–80. In: Specht, R.L. (ed.). Heathlands and Related Shrublands. Descriptive Studies. Elsevier, Amsterdam.

Kummerow, J., 1973. Comparative anatomy of sclerophylls of mediterranean climate areas. pp. 157–167. In: di Castri, F. and Mooney, H.A. (eds). Mediterranean Type Ecosystems. Springer-Verlag, New York.

Lamont, B., 1982. Mechanisms for enhancing nutrient uptake in plants, with special reference to Mediterranean South Africa and Western Australia. Bot. Rev. 48: 597–689.

Loveless, A.R., 1961. A nutritional interpretation of sclerophylly based on differences in the chemical composition of sclerophyllous and mesophytic leaves. Ann. Bot. 25: 168–184.

Loveless, A.R., 1962. Further evidence to support a nutritional interpretation of sclerophylly. Ann. Bot. 26: 551–561.

Miller, P.C. (ed.), 1981. Resource Use by Chaparral and Matorral. Springer-Verlag, New York. 455 p.

Miller, P.C., and Stoner, W.A., 1979. Canopy structural and environmental interactions. pp. 428–458. In: Solbrig, O.T., Jain, S. Johnson, G.B. and Raven, P.H. (eds). Topics in Plant Population Biology. Columbia Univ. Press, New York.

Mooney, H.A., and Dunn, E.L., 1970. Convergent evolution of mediterranean-climate evergreen sclerophyll shrubs. Evolution 24: 292–303.

Mooney, H.A., and Gulmon, S.L. 1982. Constraints on leaf structure and function. Bioscience 32: 198–206.

Mooney, H.A., and Rundel, P.W., 1979. Nutrient relations of the evergreen shrub, *Adenostoma fasciculatum*, in the California chaparral. Bot. Gaz. 140: 109–113.

Mooney, H.A., Field, C., Gulmon, S.L., Rundel, P.W. and Kruger, F.J., 1983. Photosynthetic characteristics of South African sclerophylls. Oecologica 58: 398–401.

Mooney, H.A., Kummerow, J., Johnson A.W., Parsons, D.J., Keeley, S., Hoffman, A., Hays, R.I., Giliberto, T. and Chu, C., 1977. The producers – their resources and adaptive responses. pp. 85–143. In: Mooney, H.A. (ed.). Convergent Evolution in Chile and California: Mediterranean-Climate Ecosystems. Dowden, Hutchinson and Ross, Stroudsburg, Penn.

Muller, C.H., 1939. Relations of the vegetation and climatic types in Nuevo León, Mexico. American Midland Naturalist 21: 687–729.

Muller, C.H., 1947. Vegetation and climate of Coahuila, Mexico. Madroño 9: 33–57.

Randall, E.L., 1974. Improved method for fat and oil analysis by a new process of extraction. J. HOAC 57: 1165–1168.

Rundel, P.W., 1981. The matorral zone of central Chile. pp. 175–201. In: di Castri, F., Goodall, D.W. and Specht, R.L. (eds). Mediterranean-type Shrublands. Elsevier, Amsterdam.

Rundel, P.W., 1982. Nitrogen utilization efficiencies in mediterranean-climate shrubs of California and Chile. Oecologia 55: 409–413.

Sabrado, M.A. and Medina, E., 1980. General morphology, anatomical structure and nutrient content of sclerophyllous leaves of the 'bana' vegetation of the Amazon. Oecologia 45: 341–345.

Schlesinger, W.H. and Chabot, B.F., 1977. The use of water and minerals by evergreen and deciduous shrubs in Okefenokee Swamp. Bot. Gaz. 138: 490–497.

Schlesinger, W.H. and Hasey, M.M., 1981. Decomposition of chaparral shrub foliage: losses of organic and inorganic constituents from deciduous and evergreen leaves. Ecology 62: 762–774.

Seddon, G., 1974. Xerophytes, xeromorphs and sclerophylls: the history of some concepts in ecology. Biol. J. Linn. Soc. 6: 65–87.

Small, E., 1972. Photosynthetic rates in relation to nitrogen recycling as an adaptation to nutrient deficiency in peat bog plants. Can. J. Bot. 50: 2227–2233.

Specht, R.L., 1979. The sclerophyllous (heath) vegetation of

Australia: The eastern and central states. pp. 125–210. In: Specht, R.L. (ed.). Heathlands and Related Shrublands. Descriptive Studies. Elsevier, Amsterdam.

Specht, R.L. and Moll, E.J., 1983. Mediterranean-type heathlands and sclerophyllous shrublands of the world: An overview. pp. 41–65. In: Kruger, F.J., Mitchell, D.T. and Jarvis, J.U.M. (eds). Mediterranean-Type Ecosystems: The Role of Nutrients. Springer-Verlag, Berlin.

Weisser, P. and Rundel, P.W., 1981. Estudio comparativo del matorral de Pichidangui con el de Los Molles. Ann. Mus. Hist. Nat. Valparaíso 13: 47–57.

CHAPTER 4

Vertebrates

Vertebrates

Co-ordinator: P.C. Catling

Contributors

P.C. Catling, B.J. Fox and D.C. Paton (Southeastern Australia)
D. Allen, A.E. Newsome and I.R. Noble (South Australia)
P. Christensen, S.J.J.F. Davies, R. How and G. Smith (Southwestern Australia)
R. Quinn, P.A. Stanton and G. Stewart (California)
E.R. Fuentes (Chile)
M.R. Warburg and D. Rankevich (Israel)
G.J. Breytenbach (South Africa)

Contents

Introduction	171
Southeastern Australia	172
South Australia	174
Southwestern Australia	175
California	176
Chile	179
Northern Israel	179
South Africa	180
References	181
Vertebrate sampling sites (Table 1)	183
Site specific vegetation and mammal data (Tables 2 & 2A)	185
Site specific vegetation and bird data (Tables 3, 3A, 3B & 3C)	187
Site specific vegetation and reptile / amphibia data (Tables 4 & 4A)	193

Introduction

It has been a very difficult task to assemble a chapter on vertebrates for the Data Source Book. The contributors wish to make it clear that the data are to be considered only as a *data guide* to a particular site and as a source of reference. These data should be used with great care for many reasons, some of which are outlined below, and particularly if comparisons are to be drawn across sites or between geographical regions.

The data for each study site are presented in two sections. Firstly, a succinct description of the environmental information which includes the specific site name; latitude and longitude; altitude (metres), area (hectares); aspect; general climate (UNESCO-FAO Bioclimatic map of the Mediterranean zone 1963), soils (Plumb, 1963; UNESCO-FAO Soil Map of the World, 1978); the major vegetation type (Plumb 1976); a listing of 3 or 4 of the most abundant plant species; the height and percent cover of each vegetation stratum; the period of time since the last disturbance (years) and the type of disturbance; the period of data collection, and a listing of the major references for the vertebrate data in the tables (see below).

Secondly, detailed data are provided in separate tables for mammals, birds, and reptiles and amphibia. Each table provides specific site name; general vegetation type; the period of time (years) since the last disturbance; the type of disturbance; understorey shrub cover (%); height of the dominant vegetation (metres) and a detailed description of the number and type of vertebrate species on each site. For mammals, this includes the type of mammal (e.g. small ground mammal) and the major dietary type for each species. For birds, resident and non-resident species are presented and a description of the major foraging zone and dietary type for each species, and for reptiles and amphibia the number of species and the number of species in each family. Some site pecularities are listed in the tables. The ordering of the sites in the tables is firstly by geographical location and then by decreasing structural complexity of habitat. This format is also followed in the index of environmental information.

Great care has been exercised in assembling the data; but for a collaborative contribution such as this it has been difficult to set rigid standards. For logistic reasons it has been necessary to make some rigid decisions and be definitive in our statements. For example, contributors were asked to place bird species in only one dietary or foraging category when most bird species exploit a number of foraging zones and have mixed diets. Another example was the problem for contributors to classify the major vegetation type of their study site. Obviously many study sites have a range of vegetation types. Such diversified habitats will of course provide a more diversified fauna than records from a single vegetation type. Also the number of visits to sites varied and some sites were not visited in all seasons. This was particularly important for records of the avifauna. Location of site also influenced the number of species recorded, as shown by the bird species list within Innes National Park (South Australia) which is located on a peninsula.

For the avifauna it was difficult to ascertain which are residents or seasonal migrants because: some species follow the availability of flowering associations (e.g. honeyeaters and lorikeets); the same species during good seasons can be residents; some species have members resident all year round with either an exodus or an invasion of the rest of the population in winter. Another problem was that small bird species, which are often in high densities and readily identified, can easily be classified as rare, uncommon or common On the other hand, raptorial species, which hunt over a large area, are usually classed as uncommon or rare although the density is most likely to be similar at all sites. Also, only land birds have been included in the tables, and insectivores include those birds that eat other invertebrates.

Despite the logistic restrictions, varying methods of collection and site peculiarities, we are confident about this contribution to the overall aim of the Data Source Book.

References

Plumb, T.W. (ed.), 1963. Soils. Atlas of Australian Resources (Second Series). Dept. National Development, Canberra.
Plumb, T.W. (ed.), 1976. Natural Vegetation. Atlas of Australian Resources (Second Series) Geographic section. Div. National Mapping, Dept. National Resources, Canberra.
UNESCO-FAO 1963. Bioclimatic Map of the Mediterranean Zone. Ecological Study of the Mediterranean Zone, Paris.
UNESCO-FAO 1978. Soil Map of the World. 1:5,000,000. Vol. I. Paris.

Southeastern Australia

Contributors: P.C. Catling, B.J. Fox and D.C. Paton.

Unpublished data from: A.E. Newsome.
Data for birds from Wyperfeld National Park and Nowa Nowa State Forest were kindly compiled by: I.J. Mason.
Data for reptiles and amphibians from Nadgee Nature Reserve, Myall Lakes National Park, Nowa Nowa State Forest and Wyperfeld National Park were kindly compiled by: J.C. Wombey.

Sites

1. *Nadgee Nature Reserve* is a coastal wilderness area of about 15,000 hectares. The major vegetation type is open forest. The topography is comprised mainly of upper Devonian Sandstones, mudstones, clay stones, shales and conglomerates which have been uplifted and deformed by an intrusion of granite (Fox 1978). The feature of the coast is the heathlands, which are perched behind sea cliffs about 20–50 m high and cover about 7.5% of the reserve. The vegetation is generally short with a high graminoid content due to frequent fires over the last 100 years. In recent years Nadgee has been severely burned by two severe wildfires (1972; 1980).

There are three major heathlands, each about

2 km² in size. Six to eight kilometres from the coast is a range 540 m high. A larger heathland of about 4 km² is perched on top of the range. Open forest predominates between the coastal and upland heathlands and very small patches of wet sclerophyll forest extend along watercourses and wet gullies. *Leptospermum* spp. and *Melaleuca* spp. grow in thickets along the cliffs and in the swales behind the beaches and there are also many swamps. Vertebrates have been recorded in all habitats.

Small mammal trapping grids were located in the major habitats. Elliott traps were set on grids 0.09 ha. in size in two lines of ten traps, each 7 m apart. Trapping was for three consecutive nights at least every three months and sometimes monthly. Large vertebrates were detected on raked plots of soil 1 m wide dug up across thoroughfares such as beaches and vehicle tracks. There were 112 plots each about 0.4 km apart. Plots were prepared one day and read the next for three to five consecutive days at least every three months. Vertebrate tracks on each plot were identified and the direction and number of every set of tracks recorded. Plots were not read after heavy rain, or high tides and wind on the beaches.

The recovery of populations of vertebrate species have been studied since 1972 (Newsome *et al.* 1975, Newsome & Catling 1979, Catling & Newsome 1981, Catling *et al.* 1982, Newsome & Catling 1983a, Catling 1986). Small patches of unburned vegetation are very important for survival and recolonization. Recolonization of burned areas by small mammals is not effective until vegetation has attained adequate structural complexity (Catling 1986). By the fifth year after the fires populations of most species had reached levels higher than before the fire. Recovery after the second fire is following the same trend as that for the first.

Analysis of data collected on dingoes (*Canis familiaris dingo*) have provided basic information about diet, predator/prey relationships and the effects of predation on prey populations after wildfire (Newsome *et al.* 1983b and c).

A list of heathland and forest birds and their status were compiled by Mr A.K. Morris between 1969 and 1971 and added to by others too numerous to mention. A computerised list held by the N.S.W. National Parks and Wildlife Service was also used. In addition a brief survey over a two week period which took into account observations and calls was carried out by I.J. Mason in October 1984.

2. *The Nowa Nowa study site* is a State Forest managed by the Forests Commission of Victoria. The site is about 30 km from the coast and comprised of 4,000 hectares of dry open forest on undulating land of low fertility, dissected by small streams. The dominant trees are *Eucalyptus obliqua, E. sieberi* and *E. consideniana,* with a heathy understory dominated by *Acacia longifolia, A. dealbata, Leptospermum phylicoides, Cassinia longifolia* and *Daviesia virgata*. The heathy understorey is considerably thicker along the creeks and drainage lines. The forest has not been logged or fired for at least 20 years.

Small and large mammals were surveyed as described for Nadgee Nature Reserve (see above). Birds were recorded by mist-netting (banded and released) birds from five sites over eight days, and from bird calls and observations during two one week trips in January 1983 and 1984. The assessment of migratory and resident species over such a short observation period was difficult. However birds which were common on both visits were considered resident and migratory species included those which are well documented as such in the literature.

3. *The Royal Botanic Garden's Annexe at Cranbourne* is regenerating open forest which was originally cleared for farming between 1964–1968. It is a flat to undulating site of about 200 hectares with soils of low fertility. At present it has elements of closed heathland (*Leptospermum myrsinoides* predominating) with emergent *Eucalyptus* and *Banksia* spp. The area receives over 700 mm of precipitation annually and has an average maximum and minimum mean temperature in January of 26°C and 15°C and 13°C and 7°C in July. A total of 68 species of bird occurred in and used the site although all species would not occur at one time, as numbers changed seasonally. Birds were recorded during a 2 hour census on 18 hectares 4 times a year.

4. *Wyperfeld National Park* is a gently undulating park of 57,000 hectares in mid western Victoria. It

is an area of low fertility and subject to regular droughts. The dominant habitat is open scrub composed of mallee eucalypts (*Eucalyptus incrassata; E. foecunda*) with some *Leptospermum coriaceum*. The area has been unburnt for more than 40 years. Birds were categorised by Mr I.J. Mason from lists compiled by Tarr (1967), Gell (1977) and Cheal *et al.* (1979). (*cf.* Environmental Information for references). Wyperfeld National Park has a large bird list because it is visited regularly by ornithologists; it contains varied habitats; has some coastal species reaching their northern limits; is frequented by southern and northern migrants alike and is periodically augmented by nomads in good seasons.

5. *Myall Lakes National Park* is a coastal reserve with diverse heathlands on low relief inner barrier dunes and forest on outer barrier high dunes. Fires are frequent in the area creating a mosaic with different regeneration ages and histories.

In the forest sixteen 1 ha plots were trapped covering five regrowth stages. Each 1 ha plot comprised 25 stations in a 5 × 5 grid with 20 m spacing. At each station a small mammal trap was placed, and baited with rolled oats and peanut butter and trapped for 4 nights. All animals captured were toe-clipped, sexed, weighed, reproductive condition noted and standard measurements taken. On the heathland and two trapping grids were established; one 7 ha the other 4 ha. Each grid had traps placed at 20 m spacing and trapped as described above. The 7 ha grid was trapped at approximately 2 monthly intervals and sometimes as frequently as fortnightly during the breeding season and the 4 ha grid 3 monthly. The vertebrate data provided are for the oldest stage at each site.

References

Catling, P.C. and Newsome, A.E., 1981. Responses of the Australian vertebrate fauna to fire: An evolutionary approach. In: Groves, R.H., Noble, I.R. and Gill, A.M. (eds). Fire and the Australian Biota. Australian Academy of Science, Canberra, pp. 273–310.

Catling, P.C., Newsome, A.E. and Dudzinski, M.L., 1982. Small mammals, habitat components and fire in south-eastern Australia. In: Proceedings of Symposium on Dynamics and Management of Mediterranean-type Ecosystems: An International Symposium. June 1981. San Diego, California, pp. 199–206.

Catling, P.C., 1986. *Rattus lutreolus*, the colonizer of heathland after fire in the absence of *Pseudomys* species? Aust. Wildl. Res. 13: 127–139.

Fox, A., 1978. The '72' fire of Nadgee Nature Reserve. Parks and Wildlife 2 (2): 5–24.

Newsome, A.E., McIlroy, J. and Catling, P., 1975. The effects of an extensive wildfire on populations of twenty ground vertebrates in south-east Australia. Proc. Ecol. Soc. Aust. 9: 107–123.

Newsome, A.E. and Catling, P.C., 1983a. Animal demography in relation to fire and shortage of food. Some indicative models. In: Kruger, F.J., Mitchell, D.T. and Jarvis, J.U.M. (eds). Mediterranean Type Ecosystems. The Role of Nutrients. Springer-Verlag (Ecological Studies), pp. 490–508.

Newsome, A.E., Catling, P.C. and Corbett, L.K., 1983b. The feeding ecology of the dingo. II. Dietary and numerical relationships with fluctuating prey populations in south-east Australia. Aust. J. Ecol. 8: 345–366.

Newsome, A.E., Catling, P.C., Corbett, L.K. and Burt, R.J., 1983c. The feeding ecology of the dingo. I. Stomach contents from trapping in south-east Australia and the non-target wildlife also caught in dingo traps. Aust. Wildl. Res. 10: 477–486.

South Australia

Contributors: D. Allen, A.E. Newsome and I.R. Noble.

References for study sites were supplied by: L. Best, S.A. Parker and A.C. Robinson.
Unpublished data from: T. Bradley.
Data from the references were kindly compiled for the tables by: I.J. Mason.

Data for reptiles and amphibians from Brookfield. Conservation Park and Innes National Park were kindly compiled by: J.C. Wombey.

Location of each site is important, in either restricting the number of species within an area or increasing them. For example, bird species within Innes National Park are low because the park is located on a peninsula and is just beyond the western distribution of a number of eastern species. Also the number of migrants or vagrants are respectively, either reaching their south-eastern limit or have no other surrounding habitat into which they can move. Similarly for Brookfield Conservation Park, although it is in the arid zone its bird fauna is influenced by the high number of seasonal migrants from the nearby Mt Lofty ranges. Sites with low numbers of regular nomads and migrants had the lowest number of recordings as all seasons were not covered. These include Middleback (18), Mount Rescue Conservation Park (17) and Billiatt Conservation Park (5).

Middleback Station on north-eastern Eyre Peninsula is a mosaic of open chenopod steppe and an acacia woodland with a chenopod understorey. Dominant shrub species are *Atriplex vesicaria*, *Maireana sedifolia* and *M. pyramidata*. In woodland the dominant species is *Acacia papyrocarpa* with *Alectryon oleifolius*, *Myoporum platycarpum* and monospecific stands of *Casuarina cristata*. All the chenopod shrublands on north-eastern Eyre Peninsula have been stocked with introduced herbivores (sheep, cattle) and the European Rabbit (*Oryctolagus cuniculus*) has also been present since settlement. Their impact plus that of introduced predators, such as European Fox (*Vulpes vulpes*) and Feral Cat (*Felis catus*), have modified the chenopod shrublands and their associated wildlife respectively.

In general, landform is gently undulating with pure stands of *A. vesicaria* in low lying areas and a mixture of *A. vesicaria* and *M. sedifolia* on higher ground in areas least affected by domestic herbivores. *A. vesicaria* is palatable to sheep and susceptible to trampling, disappearing in areas of heavy stock usage. *M. pyramidata* is an aggressive coloniser and occurs in areas of natural disturbance such as drainage lines, as well as along roads and around rabbit warrens.

At eighteen sites birds were censused on a marked transect 1000 m long and 100 m wide through homogenous habitat. All birds seen or heard within the transect were identified and recorded. Censusing was carried out from early September until early December 1983, when sedentary, migrant and nomadic species were present and breeding. Censuses were restricted to clear, windless days as it is known that wind has an effect on bird activity and the observer's ability to detect them. It was noticed that overcast conditions also reduced bird activity. Activity was highest in the first two hours following sunrise so censuses commenced half an hour after sunrise, and since each transect took 1 h to complete, usually only two counts were completed before activity decreased. Each transect was visited at least four times. On each re-sample of a transect the direction of traverse was reversed and the time of morning was varied to minimise the variability which may have been introduced by the time gradient along the transect.

Southwestern Australia

Contributors: P. Christensen, S.J.J.F. Davies, R. How and G. Smith.

This section does not intend to be encyclopaedic in its coverage of data relevant to the vertebrates of the mediterranean south-west, but rather to present information from selected data bases that encompasses different designs and methods. Information on the vertebrate fauna has been collated in a recent bibliography and research inventory covering most studies in the south-western mediterranean region (Daze 1984). Specific distributions of the mediterranean south-west mammals can also be obtained from Kitchener and Vicker (1981), birds from Serventy and Whittell (1976), amphibians from Tyler, Smith and Johnstone (1984) and some reptile families from Storr, Smith and Johnstone (1981, 1983, 1986).

The rationale, design and methods of two major surveys covering part of the mediterranean southwest have been published (Kitchener 1976; Biological Surveys Committee 1984). Differences in vertebrate sampling methods employed in these surveys suggest the need for caution when comparing their data. No direct comparison of small terrestrial or volant mammal assemblages is valid because of the contrasting collecting methods; a difference which also affects comparisons of the reptile assemblages. Some attempt has been made to assess the efficacy of small mammal collecting techniques within this region (How, Humphreys and Dell 1984).

Comparison of bird communities in mediterranean regions of other parts of the world has received considerable attention (Cody 1973, 1975; Blondel 1981) but few comparisons have been made with Australian mediterranean assemblages (Cody *et al.* 1983) or within Australian mediterranean regions. Data presented in Table II indicates that avifaunal assemblages in woodlands of the Western Australian wheatbelt are richer than assemblages in other vegetation types (*see also* Kitchener et al. 1982); a relationship not upheld when comparing selected sites in the mediterranean section near Kalgoorlie. The employment of different temporal and spatial census techniques in the latter study may account in part for this difference. Also the data from the forest sites were collected using slightly different methods and again clearly illustrates the need to evaluate methodologies before making comparisons.

References

Biological Surveys Committee, 1984. The biological survey of the Eastern Goldfields of Western Australia. Part I. Introduction and methods. Rec. West. Aust. Mus. Suppl. 18: 1–19.

Blondel, J., 1981. Structure and dynamics of bird communities in Mediterranean habitats. In: di Castri, F., Goodall, D.W. and Specht, R.L. (eds). Ecosystems of the World. 11. Mediterranean-type Shrublands. Elsevier Scientific Publishing Company, Amsterdam, pp. 361–385.

Cody, M.L., 1973. Parallel evolution and bird niches. In: di Castri, F. and Mooney, H.A. (eds). Mediterranean-type Ecosystems, Ecological Studies. Springer-Verlag, Berlin, pp. 307–338.

Cody, M.L., 1975. Towards a theory of continental species diversities: Bird distributions over Mediterranean habitat gradients. In: Cody, M.L. and Diamond, J. (eds). Ecology and Evolution of Communities. Cambridge University Press, London, pp. 214–257.

Cody, M.L., Breytenbach, G.J., Fox, B., Newsome, A.E., Quinn, R.D. and Siegfried, W.R., 1983. Animal communities: diversity, density and dynamics. In: Day, J.A. (ed.) Mineral Nutrients in Mediterranean Ecosystems. South African National Scientific Programmes Report No. 71, pp. 91–110.

Daze, D., 1984. A bibliography and research inventory of vertebrate fauna in Western Australia. Dept. of Conservation and Environment, Perth, Western Australia.

How, R.A., Humphreys, W.F. and Dell, J., 1984. Vertebrate surveys in semi-arid Western Australia. In: Myers, K., Musto, I. and Margules, C. (eds). Survey Methods for Nature Conservation.

Kitchener, D.J., 1976. Preface to the biological survey of the Western Australian Wheatbelt. Rec. W. Aust. Mus. Suppl. 2: 3–10.

Kitchener, D.J. and Vicker, E., 1981. Catalogue of Modern Mammals in the Western Australian Museum 1895–1981. Western Australian Museum, Perth.

Kitchener, D.J., Dell, J., Muir, B.G. and Palmer, M., 1982. Birds of Western Australian Wheatbelt reserves – implications for conservation. Biol. Conserv. 22: 127–163.

Serventy, D.L. and Whittell, H.M., 1976. Birds of Western Australia. University of Western Australia Press, Perth.

Storr, G.M., Smith, L.A. and Johnstone, R.E., 1981. Lizards of Western Australia, I. Skinks. University of Western Australia Press, Perth.

Storr, G.M., Smith, L.A. and Johnstone, R.E., 1983. Lizards of Western Australia. II. Dragons and Monitors. Western Australian Museum, Perth.

Storr, G.M., Smith, L.A. and Johnstone, R.E. 1986. Lizards of Western Australia. III. Geckos and Legless Lizards. Western Australian Museum, Perth.

Tyler, M.J., Smith, L.A. and Johnstone, R.E., 1984. Frogs of Western Australia. Western Australian Museum, Perth.

California

Contributors: R. Quinn, P.A. Stanton and G. Stewart.

Two Californian locations are described, one a 6,885 hectare mountainous site of elevations 458–1,678 m, and the other a 9 hectare foothill site of elevations 290–335 m. The higher of the two, the San Dimas Experimental Forest (SDEF), is dominated for the most part by a plant community of evergreen sclerophyll shrubs (chaparral). The smaller and lower site, the Voorhis Ecological Reserve, has a plant community dominated by drought deciduous shrubs (coastal sage scrub). The two sites are 11 km apart.

Sites

1. *San Dimas Experimental Forest.* The SDEF is located in the San Gabriel Mountains and possesses extremely rugged topography, with an elevational range of 1,220 m (Bentley 1961; Hill 1963). The parent material is mostly granitic, or pre-Cretaceous or pre-Cambrian crystalline metamorphics (Storey 1948). This unreferenced soil series is shallow, coarse-textured, azonal and well drained (Crawford 1962). The series has a weak, angular blocky structure, is chemically neutral near the surface, loose when dry, and the boundary between the soil and parent material is gradual and sometimes indistinct.

Climate has been studied in detail since 1935 (*see* Mooney and Parsons 1973 and Dunn *et al.* 1987 for reviews). Mean annual precipitation for a 32 year period was 670 mm, with 498 mm falling between 1 December and 30 March and only 13 mm occurring between June 1 and September 30. Mean maximum and minimum air temperatures for July are 30.5 and 13.8° C, and for January are 14.8 and 2.6° C. The bioclimate is xerothermic except at the very highest elevations (Thrower and Bradbury 1977). This means there are 150–200 dry days per year.

The vegetational characteristics were reviewed by Mooney and Parsons (1973), and a comprehensive flora published by Dunn *et al.* (1987). Most of the SDEF is covered with chaparral. The dominant vegetation on equator-facing slopes is a mixture of *Adenostoma fasciculatum* (Rosaceae) and *Ceanothus crassifolius* (Rhamnaceae) shrubs, while on pole-facing slopes the most common shrubs are *Quercus dumosa* (Fagaceae) and *Cercocarpus betuloides* (Rosaceae). Some of the lowest elevations on equator-facing slopes are dominated by coastal sage scrub. A few places near canyon bottoms and scattered sites, with unusually deep soils, support an oak savanna [*Quercus agrifolia* (Fagaceae)] with an understorey of annual grasses and forbs. The canyon bottoms of the larger drainages, most with seasonal streams support riparian woodland with one or more of the following species of trees: *Platanus racemosa* (Platanaceae), *Acer macrophyllum* (Aceraceae), *Salix* sp. (Salicaceae), *Alnus rhombifolia* (Fagaceae), and *Umbellularia californica* (Lauraceae). Some of the highest slopes have scattered groves of the coniferous tree *Pseudotsuga macrocarpa,* and several small areas have been planted with the native conifer *Pinus coulteri.* One hundred hectares of a major watershed were artificially converted from chaparral to perennial and annual grasses in the 1960's (Corbett and Rice 1966). This array of plant communities, provides a wide range of habitats for vertebrates and partially accounts for the large number of species present.

The SDEF has been the site of biological studies since 1935, and was designated as a Man and the Biosphere (MAB) Reserve in 1977. Data reported here for birds cover the periods 1936–1953 (Wright and Horton 1951, 1953), and 1969 to the present (Wirtz 1987; Dunn *et al.* 1987; Quinn, unpublished data). Data for mammals, reptiles, and amphibians are derived from Wright and Horton (1951, 1953), Dunn *et al.* (1987), Quinn (1987), and Stewart (unpublished data). Additional information about the mammals was taken from Vaughn (1954), and for reptiles and amphibians from Schoenherr (1976).

The total number of species of birds observed at least once over this half century time span (179) is considerably higher than the number of species (123) accounted for in Table 3, because this table excludes two classes of observations: species of birds that are strictly associated with water impoundments; and species once present that are now locally extinct. The observations of mammals, reptiles, and amphibians are quite thorough. They summarise systematic observations by a number of observers over five decades and include all of the various habitats present.

2. Voorhis Ecological Reserve. The Reserve was established in 1978, when it became part of the campus of California State Polytechnic University, Pomona and consists of nine ha of steep-sloped canyons located in the San Jose Hills, Los Angeles County, California, U.S.A. The canyons have broad bottoms with slopes of 0–5 percent, and steep sides with slopes ranging from 15 percent to nearly vertical.

The San Jose Hills are the result of orogenic activity which began in the mid-Pleistocene and continues to the present. Consequently the soils are poorly developed on the steep and rapidly eroding hillsides, and soils in the canyon bottoms result from recent alluvial deposits. The underlying parent material, which is exposed in some places, is a succession of marine clastic sediments of Mid-Miocene origin (Pomerening 1970). The soils are Soper loam on the gentler slopes (15–30 percent) and Gaviota loam on the slopes exceeding 30 percent (Pomerening 1970; United States Department of Agriculture 1960). These soils are shallow, well drained loams over weakly cemented conglomerate.

The bioclimate is classified as xerothermic (Thrower and Bradbury 1977) with 150–200 dry days per year. Mean annual precipitation is 457 mm which falls entirely as rain, with 343 mm between December 1 and March 30, and 9 mm between June 1 and September 30 (Pomerening 1970). The mean daily temperature in January is 11°C and in July is 25°C. The frost-free period averages 255 days, and winter temperatures seldom fall below −5°C.

The most common vegetation type is coastal sage scrub, a community of drought deciduous shrubs dominated here by *Artemisia californica* (Asteraceae) and *Eriogonum fasciculatum* (Polygonaceae). In the unburned area the projected vegetative cover of these shrubs is at or near 100 percent on the canyon slopes (Quinn, unpublished data). A few trees [*Platanus racemosa* (Platanaceae), *Juglans californica* (Juglandaceae) and *Quercus agrifolia* (Fagaceae)] are present at the canyon bottom, with scattered shrubs of *Heteromeles arbutifolia* (Rosaceae) on the upper canyon slopes.

Sheep and cattle have been grazed in the reserve since the arrival of the Spanish in 1771. Overgrazing combined with episodes of severe drought, caused a rapid replacement of most of the native perennial grasses by annual grasses and forbs from the Mediterranean Basin (reviewed in Heady 1977). The mature coastal sage scrub contains introduced grasses such as *Bromus* sp. and *Avena fatua*, and introduced forbs such as *Erodium* sp. (Geraniaceae) and *Brassica campestris* (Brassicaceae). In the winter and spring these introduced annuals occur as an understorey beneath and between the shrubs.

On 21 August 1981 a wildfire burned all of the vegetation in a nine hectare portion of the Reserve. The post-fire pattern of revegetation has generally followed that described by Westman (1981). The biota of a second canyon, located 400 m from and adjacent to the burned canyon, has been studied for comparative purposes. This second canyon is similar to the first in size, vegetation and topography, except that it opens to the southwest rather than the southeast, and it contains a small group of introduced *Eucalyptus globulus* (Myrtaceae) trees. It has not burned in at least 30 years.

Avian censuses were conducted at intervals of approximately one week in both canyons for 17 months following the fire (Moriarty *et al.* 1985, Stanton 1986). Mammals have been censused regularly since the fire, at intervals of approximately three months. Small mammals are captured along permanent trap lines approximately 200 m long, which extend across each of the canyons perpendicular to the direction of drainage. Trapping is done with pairs of Sherman traps spaced at 15 m intervals. The presence of larger mammals, reptiles, and amphibians has been determined by direct observation and by animal signs such as scats, tracks, and diggings.

A comparison of the census data for birds between the burned and unburned areas, as summarized in Table II, shows little apparent difference between the two. There were, however, substantial differences (Stanton 1986, Moriarty *et al.* 1985). For example only 56 percent of the 72 avian species were sighted in both areas, and there were differences in abundances between the species found in both. The mature coastal sage scrub area provided adequate habitat for more species and individual birds on a year-round basis. Only in spring was abundance and species richness in the burned area comparable to the unburned area (Stanton 1986).

Immediately after the fire both species richness and abundance of the small mammal community were reduced in the burned area. At the same time there was a temporary increase in the abundance of the opportunistic cricetid rodent *Peromyscus maniculatus*, and the canid *Canis latrans* was seen more often in the burn. These and other differences between the paired mammal communities have diminished with time, and are not apparent when comparing species numbers in Table 2.

There were no differences observed in the use of the two areas by reptiles and amphibians; however, these animals were not systematically censused. The relatively low species richness indicated in Table 4 is probably due to the limited range of habitats available.

The Reserve lies within a vast metropolitan area but is connected to non-urbanized portions of three ranges that extend in different directions and encompass thousands of hectares of wildlife habitat. These ranges are corridors for the movement of vertebrates of all kinds. Thus the Reserve is more analagous to a portion of an archipelago, rather than to an isolated ecological island. The only two species thought to be absent because of the lack of adequate contiguous habitat are the cats *Felis concolor* and *Lynx rufus*.

References

Bentley, J.R., 1961. Fitting brush conversion to San Gabriel watersheds. Misc. Paper No. 61. Pacific Southwest Forest and Range Experiment Station, USFS. Berkeley, California 8 pp.

Corbett, E.S. and Rice, R.M., 1966. Soil slippage increased by brush conversion. Res. Note PSW-128. Pacific Southwest Forest and Range Experiment Station, USFS. Berkeley, California, 8 pp.

Crawford, J.M., Jr., 1962. Soils of the San Dimas Experimental Forest. Misc. Paper No. 76. Pacific Southwest Forest and Range Experiment Station, USFS. Berkeley, California, 21 pp.

FAO-UNESCO, 1975. Scil Map of the World. Vol. II: Map II-2, North America. UNESCO, Paris.

Heady, H.F., 1977. Valley grassland. In: Barbour, M.G. and Major, J. (eds). Terrestrial Vegetation of California. John Wiley & Sons, New York, pp. 491–514.

Hill, L.W., 1963. The San Dimas Experimental Forest. Pacific Southwest Forest and Range Experiment Station, USFS. Berkeley, California, 25 pp.

Mooney, H.A. and Parsons, D.J., 1973. Structure and function of the California chaparral – an example from San Dimas. In: di Castri, F. and Mooney, H.A. (eds). Mediterranean-type Ecosystems; Origin and Structure. Springer-Verlag, New York, pp. 83–112.

Pomerening, J.A., 1970. Soil survey of California State Polytechnic College, Kellogg-Voorhis: Pomona. 188 pp.

Quinn, R.D., 1987. Habitat preferences and distribution of mammals in California chaparral. Pacific Southwest Forest and Range Experiment Station, USFS. Berkeley, California. (In press).

Storey, H.C., 1948. Geology of the San Gabriel Mountains, California, and its relation to water distribution. California Range and Experiment Station, USFS. Berkeley, California. 19 pp.

Thrower, N.J.W. and Bradbury, D.E., 1977. Chile-California Mediterranean Scrub Atlas. Dowden, Hutchinson & Ross, Inc. Stroudsburg, Pennsylvania.

United States Department of Agriculture, 1960. Soil Classification, a Comprehensive System. 265 pp. (Supplement issued in 1967 and ammended in 1968).

Westman, W.E., 1981. Diversity relations and succession in Californian coastal sage scrub. Ecology 62: 170–184.

For other references *see* Environmental Information.

Chile

Contributor: E.R. Fuentes.

Chilean small mammals are still relatively unknown regarding both average and variability in their characteristics. The data provided are best approximations to averages or educated guesses in many cases. In more than one case the categories constraint forced rigid decisions and the entries should be taken as hypotheses for further work and then, only within the framework provided. Diversities and densities of small mammals in Chilean matorral are very variable between sites, between months and between years and therefore they should be regarded with extreme caution.

Data are provided for a coastal site (Zapallar); two evergreen shrubland sites of different moistures (Santa Laura and Los Dominicos); one mountain site with a winter deciduous *Nothofagus* forest (El Roble) and one espinal site (Quebrada de La Plata) which is a successional location dominated by *Proustia cuneifolia* rather than *Acacia caven*. This was the closest site, geographically and physiognomically, to an espinal association. However, mammal species diversity is probably much higher than in a typical *Acacia caven* association.

Northern Israel

Contributors: M.R. Warburg and D. Rankevich.

Of the 30 species of reptiles found in the Mediterranean region of Israel, 14 are snakes and 14 lizards with one tortoise and chameleon species (see Table 2 in Warburg, 1978).

The study consisted of periodic sampling of reptiles in several plots 20×50 m each in various typical habitats of different regions within the Mediterranean regions of northern Israel. These habitats, ranging from grassland to shrubland and oak-woodland, were situated along a climatic gradient from 500 mm to 1000 mm rainfall. The region in the Gilboa Mountains was typical grassland and that in the Upper Galil was oak-woodland. In the Lower Galil and on Mt Carmel three habitats were chosen, in each grassland, shrubland and oak-woodland. The study lasted for four years and the plots were each visited at least ten times, some up to forty times, The abundance of each species collected is given in Table 4A.

Of the twenty four reptilian species collected, fourteen were lizards, eight snakes, one tortoise and one chameleon. Of these only the tortoise species, four lizard species (*Ophisaurus apodus, Lacerta laevis, Lacerta trilineata* and *Ablepharus kitaibelii*), and one snake species (*Coluber ravergieri*) can be considered typical of the Mediterranean region alone. All the remaining species can be found elsewhere as well. Further details can be seen in Warburg (1978).

Bird species richness and abundance was studied in 4 different regions of typical Mediterranean type habitats along a climatic gradient. These ranged from grassland habitats with 500 mm rainfall (on Mt Gilboa) to dense woodland with 1000 mm rainfall (in the Upper Galil). Three habitats of shrubland and oak-woodland in the Lower Galil (650 mm rainfall) and three similar habitats on Mt Carmel (750 mm rainfall) were also studied between the above locations. This study lasted over 2 years. Each of the 8 plots was 1 ha in area. Birds were identified and counted by strolling twice diagonally through the area in almost parallel lines whenever possible. Alternatively counts were made from observation positions for a period lasting about 3 hrs each time.

Altogether 1,369 individual birds were observed belonging to 52 species. Of these, 12 species were passing through the habitats during summer and 25 species were wintering (*see* Table 3B, C). Only 15 species can be found throughout the entire year. The habitats richest in both species and their numbers was in the Lower Galil. This habitat was an open parkland moderately grazed by cattle or goats. The bird lists and frequencies are given for summer and winter (see Table 3A). A full account can be found in Rankevich and Warburg (1983; p. 67).

Rodent species richness and abundance was studied in eight areas chosen along a climatic gradient in different and typical habitats within the Mediterranean region of northern Israel. The study sites ranged from grassland on Mt Gilboa with 500 mm rainfall, to oak-woodland in the Upper Galil with 1000 mm rainfall. In between these extremes three study sites were chosen in shrubland and woodland with 650 mm rainfall in the Lower Galil and similar sites on Mt Carmel with 750 mm rainfall. Each such area, 10 ha large, was surveyed periodically during the 2.5 year study, by laying traps 5 m apart in two 20 m transects. The trapping took place in each plot approximately during the same period. The rodents were identified, marked and released.

Altogether 175 individuals belonging to 10 species were captured in the study sites (*see* Table 2A). The largest number of species was found in the low rainfall grassland habitat in the Gilboa Mountains. This site was also richest in numbers. Among the species that had been trapped only two can be considered typical of the Mediterranean oak-woodland: *Apodemus sylvaticus* and *A. mystacinus*. *Meriones tristrami* is found in various grassland and sand dune habitats typical of the Mediterranean region. *Microtus guentheri* is mostly found in agricultural areas in the Mediterranean region. The other species can be found also in the arid region or elsewhere and are not confined to the Mediterranean habitats. Further details can be found in Warburg, Ben-Horin and Rankevich (1978).

References

Warburg, M.R., 1978. Diversity of the herpetofauna in the Mediterranean region of northern Israel. J. Arid Environments 1: 253–259.

Warburg, M.R., Ben-Horin, A. and Rankevich, D., 1978. Rodent species diversity in mesic and xeric habitats in the Mediterranean region of northern Israel. J. Arid Environments 1: 63–69.

South Africa

Contributor: G.J. Breytenbach.

The Swartberg mountains lie some 100 km inland from the southern Cape coast. The work reported comes from an ongoing project of the Directorate of Forestry of the Department of Environment Affairs. The Swartberg Fynbos site is situated 1,600 m above sea level on an equator facing slope and receives 800–900 mm rainfall. The Spekboomveld on the other hand is only 900 m above sea level also on an equator facing slope and only receives 250 mm rainfall.

The small mammal data represents samples at alpha diversity level but represents captures over several seasons. All the small mammals did therefore not necessarily co-occur. Small mammals were collected from a 0.18 ha grid. Large mammals and medium sized mammals were identified from tracks and direct observations. Only animals occurring within the near vicinity of the site were recorded.

References

Attwood, R., 1977. Birds recorded at Mt Rescue Conservation Park. S. Aust. Ornith. 27: 173–175.
Baxter, C.I., 1980. Birds of Belair Recreation Park. S. Aust. Ornith. 28: 90–98.
Chapman, A., Dell, J., Kitchener, D.J. and Muir, B.G., 1978. Biological Survey of the Western Australian Wheatbelt. Part 5. Dongolocking Nature Reserve. Rec. W. Aust. Mus. Suppl. 6, pp. 7–80.
Chapman, A., Dell, J., Kitchener, D.J. and Muir, B.G., 1980. Biological Survey of the Western Australian Wheatbelt. Part 11. Yorkrakine Rock, East Yorkrakine and North Bungulla Nature Reserves. Rec. W. Aust. Mus. Suppl. 12, pp. 5–76.
Chapman, A., Dell, J., Kitchener, D.J. and Muir, B.G., 1981. Biological Survey of the Western Australian Wheatbelt. Part 13. Billyacatting Hill Nature Reserve. Rec. W. Aust. Mus. Suppl. 13, pp. 5–54.
Cheal, P.D., Dary, J.C. and Meredith, C.W., 1979. Fire in the National Parks of North-West Victoria. N.P.W.S. of Victoria – Series of 5 Chapters.
Christensen, P., Annels, A., Liddelow, G. and Skinner, P., 1985. Vertebrate fauna in the southern forests of Western Australia – A survey. West. Aust. For. Dept. Bull. 94.
Clarke, G., 1967. Birds of Para Wirra National Park. S. Aust. Ornith. 24: 119–134.
Cody, M.L., 1974. Competition and the Structure of Bird Communities. Princeton: Princeton Univ.
Cogger, H.G., 1986. Reptiles and Amphibians of Australia. Reed, Sydney.
Dell, J., Chapman, A., Kitchener, D.J. and Muir, B.G., 1979a. Biological Survey of the Western Australian Wheatbelt. Part 8. Wilroy Nature Reserve. Rec. W. Aust. Mus. Suppl. 8, pp. 9–54.
Dell, J., Harold, G., Kitchener, D.J., Morris, K.D. and Muir, B.G., 1979b. Biological Survey of the Western Australian Wheatbelt. Part 7. Yornaning Nature Reserve. Rec. W. Aust. Mus. Suppl. 8, pp. 5–48.
Dell, J., Chapman, A., Kitchener, D.J., McGauran, D.J. and Muir, B.G., 1981. Biological Survey of the Western Australian Wheatbelt. Part 14. East Yuna and Bindoo Hill Nature Reserves. Rec. W. Aust. Mus. Suppl. 13, pp. 9–105.
Dell, J., How, R.A., Newbey, K.R. and Hnatiuk, R.J., 1985. The Biological Survey of the Eastern Goldfields of Western Australia. Part 3. Jackson-Kalgoorlie Study Area. Rec. W. Aust. Mus. Suppl. 23, pp. 1–168.
Dunn, P.H., Barro, S.C., Wells, W. II, Poth, M. and Wohlgemuth, P., 1987. The San Dimas Experimental Forest: a research compendium. USDA Forest Service General Technical Report. In review.
Ford, H. and Paton, D., 1976. Birds of Para Wirra Recreation Park: Changes in status over 10 years. S. Aust. Ornith. 27: 88–95.
Fox, A., 1970. Development, 1970–5 Nature Reserve No. 6 – Nadgee, N.S.W. National Parks and Wildlife Service Report.
Fox, B.J., 1982. Fire and mammalian secondary succession in an Australian coastal heath. Ecology 63: 1332–1341.
Fox, B.J. and McKay, G.M., 1981. Small mammal responses to pyric successional change in a eucalypt forest. Aust. J. Ecol. 6: 29–41.
Fox, B.J., Fox, M.D. and McKay, G.M., 1979. Litter accumulation after fire in a eucalypt forest. Aust. J. Bot. 27: 157–165.
Fox, M.D., 1981. Coexistence between two eucalypts in coastal open forest. Ph. D. Thesis, Macquarie University, Sydney.
Fox, M.D. and Fox, B.J., 1986. The effect of fire frequency on the structure and floristic composition of a woodland understorey. Aust. J. Ecol. 11: 77–85.
Fulk, G.W., 1975. Population ecology of rodents in the semi-arid shrublands of Chile. Occasional Papers Mus. Texas Tech. Univ. 33: 1–40.
Gell, P., 1977. Bird List: Wyperfeld National Park, Victoria. Aust. Bird Watcher 7: 52–55.
Glanz, W.E., 1977. In: Thrower, N.J.W. and Bradbury, D.E. (eds). Chile-Californian Mediterranean Scrub Atlas. A comparative Analysis. Dowden, Hutchinson and Ross Inc. Stroudsburg, Penn.
Gullan, P.K., 1978. Vegetation of the Royal Botanic Gardens Annexe at Cranbourne, Victoria. Proc. R. Soc. Vic. 90: 225–240.
Hatch, J.H., 1977. The birds of Comet Bore (Ninety Mile Plain). S. Aust. Ornith. 27: 163–172.
Hopkins, A.J.M. and Smith, G.T. (in prep.). The Natural Resources of Two Peoples Bay Nature Reserve. W.A. Dept. Conservation and Land Management Research Bull.
Hossfeld, P.S., 1964. Geology and physiography of the Para Wirra National Park. p. 29. In: Cotton, B.C. (ed.). South Australian National Parks and Wildlife Reserves. Govt. Printer, Adelaide.
How, R.A., Newbey, K.R., Dell, J., Muir, B.G. and Hnatiuk, R.J., 1987. The Biological Survey of the Eastern Goldfields

of Western Australia, Lake Johnston-Hyden Area. Rec. W. Aust. Mus. Suppl. (in prep.).

Kitchener, D.J., Chapman, A., Dell, J., Johnstone, R.E., Muir, B.G. and Smith, L.A., 1976. Biological Survey of the Western Australian Wheatbelt. Part 1. Tarin Rock and North Tarin Rock Reserves. Rec. W. Aust. Mus. Suppl. 2, pp. 11–87.

Kitchener, D.J., Chapman, A., Dell, J. and Muir, B.G., 1977. Biological Survey of the Western Australian Wheatbelt. Part 3. Vertebrate fauna of Bendering and West Bendering Nature Reserves. Rec. W. Aust. Mus. Suppl. 5, pp. 5–58.

Kitchener, D.J., Chapman, A., Dell, J. and Muir, B.G., 1979. Biological Survey of the Western Australian Wheatbelt. Part 10. Buntine, Nugadong and East Nugadong Nature Reserves, and Nugadong Forest Reserve. Rec. W. Aust. Mus. Suppl. 9, pp. 5–127.

Moriarty, D.J., Farris, R.E., Noda, D.K. and Stanton, P.A., 1985. Effects of fire on a coastal sage scrub bird community. Southwestern Naturalist 30: 452–453.

Muir, B.G., 1977. Biological Survey of the Western Australian Wheatbelt. Part 4. Vegetation of West Bendering Nature Reserve. Rec. W. Aust. Mus. Suppl. 5, pp. 5–31.

Muir, B.G., Chapman, A., Dell, J. and Kitchener, D.J., 1978. Biological Survey of the Western Australian Wheatbelt. Part 6. Durokoppin and Kodj Kodjin Nature Reserves. Rec. W. Aust. Mus. Suppl. 7, pp. 9–77.

Muir, B.G., Chapman, A., Dell, J. and Kitchener, D.J., 1980. Biological Survey of the Western Australian Wheatbelt. Part 12. Badjaling Nature Reserve, South Badjaling Nature Reserve, Yoting Town Reserve and Yoting Water Reserve. Rec. W. Aust. Mus. Suppl. 12, pp. 3–66.

Newbey, K.R., Dell, J., How, R.A. and Hnatiuk, R.J., 1984. The Biological Survey of the Eastern Goldfields of Western Australia. Part 2. Widgiemooltha-Zanthus Area. Rec. W. Aust. Mus. Suppl. 18, pp. 21–157.

Newsome, A.E. and Catling, P.C., 1979. Habitat preferences of mammals inhabiting heathlands of warm temperate coastal, montane and alpine regions of south eastern Australia. In: Specht R.L. (ed.). Heathlands and Related Shrublands of the World. A. Descriptive Studies. Elsevier Publ. Co., Amsterdam, pp. 301–316.

Playford, P.E., Horwitz, R.C., Peers, R. and Baxter, J.L., 1970. Geraldton, Western Australia 1:250,000 Geological Series Map and Explanatory Notes SH/50–1. Australian Government Printer, Canberra.

Rankevich, D., and Warburg, M.R., 1983. Diversity of bird species in mesic and xeric habitats within the Mediterranean region of northern Israel. J. Arid Environments 6: 161–171.

Rix, C.E., 1976. The birds of Sandy Creek Conservation Park. Aust. Bird Watcher 6: 209–222; 255–288; 330–354.

S.A. National Parks and Wildlife Service, 1982. Innes National Park Management Plan: Yorke Peninsula. National Parks and Wildlife Service, Department of Environment and Planning: South Aust.

S.A. National Parks and Wildlife Service, 1984. Conservation Parks of the Murraylands (Western Plains) Management Plans. Brookfield, Ridley and Swan Reach. N.P.W.S., Department of Environment and Planning: South Aust.

Schoenheer, A.A., 1976. The herpetofauna of the San Gabriel Mountains, Los Angeles County, California. Special publication of the Southwestern Herpetologist's Society. 95 pp.

Specht, R.L. and Cleland, J.B., 1963. Flora conservation in South Australia. II. The preservation of species recorded in South Australia. Trans. R. Soc. S. Aust. 87: 63–92.

Specht, R.L. and Perry, R.A., 1948. The plant ecology of part of the Mount Lofty Ranges (1). Trans. R. Soc. S. Aust. 72: 91–132.

Specht, R.L. and Rayson, P., 1957. Dark Island heath (Ninety-Mile Plain, South Australia. 1. Definition of the ecosystem. Aust. J. Bot. 5: 52–85.

Stanton, P.A., 1986. Comparison of avian community dynamics of burned and unburned coastal sage scrub. The Condor 88: 285–289.

Tarr, H.E., 1967. Bird List – Wyperfeld National Park. National Parks Authority, Victoria.

Thrower, N.J.W. and Bradbury, D.E., 1977. Chile-Californian Mediterranean Scrub Atlas. A Comparative Analysis. Dowden, Hutchinson and Ross Inc. Stroudsburg, Penn.

Vaughn, T.A., 1954. Mammals of the San Gabriel Mountains of California. Univ. Kansas Publ., Museum of Natural History 7: 513–582.

Warburg, M.R., 1978. Diversity of the herpetofauna in the Mediterranean region of northern Israel. J. Arid. Environments 1: 253–259.

Warburg, M.R., Ben-Horin, A. and Rankevich, D., 1978. Rodent species diversity in mesic and xeric habitats in the Mediterranean region of northern Israel. J. Arid Environments 1: 63–69.

Wirtz, W.O. II. 1987. Seasonal distribution, habitat utilization, and reproductive phenology of avifauna in southern California chaparral. Pacific Southwest Forest and Range Experiment Station, USFS. Berkeley, Cal. (In press.)

Wright, J.T. and Horton, J.S., 1951. Checklist of the vertebrate fauna of San Dimas Experimental Forest. Misc. paper No. 7. California Forest and Range Experiment Station, USFS. Berkeley, Cal. 15 pp.

Wright, J.T. and Horton, J.S. 1953. Supplement to checklist of the vertebrate fauna of San Dimas Experimental Forest. Misc. paper No. 13. California Forest and Range Experiment Station, USFS. Berkeley, Cal. 3 pp.

Table 1. Vertebrate sampling sites.

Location	Latitude	Longitude	Altitude (m)	Area (ha)	References
South-Eastern Australia					
Cranbourne, Vic. (Royal Bot. Gard. Annexe)	38° 07′ S	145° 18′ E	80	200	Gullan 1978, Paton unpubl.
Myall Lakes N.P., N.S.W.	32° 28′ S	152° 24′ E	30	16 × 1 ha	Cogger 1986, Fox 1981, 1982, Fox & Fox 1986, Fox & McKay 1981, Fox et al. 1979.
Nadgee N.R., N.S.W.	37° 26′ S	149° 56° E	0–500	15,000	Cogger 1986, Fox 1970, Mason & Wombey unpubl., Newsome & Catling 1979
Nowa Nowa S.F., Vic.	37° 38′ S	148° 06′ E	300	4,000	Catling & Newsome unpubl., Cogger 1986
Wyperfeld N.P., Vic.	35° 43′ S	142° 22′ E	100	57,000	Cheal et al. 1979, Cogger 1986, Gell 1977, Tarr 1967
South Australia					
Belair R.P., S.A.	35° 00′ S	138° 45′ E	255–490	835	Baxter 1980, S.A. NPWS unpubl. 1983, Specht & Perry 1948
Billiatt C.P., S.A.	35° 30′ S	140° 55′ E	98	59,131	Bradley unpubl., Specht & Cleland 1963
Brookfield C.P., S.A.	34° 20′ S	139° 23′ E	70–100	5,527	Bradley unpubl., Cogger 1986, S.A. NPWS 1984
Innes N.P., S.A.	35° 13′ S	136° 51′ E	0–60	9,141	Bradley unpubl., Cogger 1986, S.A. NWPS 1982
Middleback Stn, S.A.	32° 57′ S	137° 23′ E	0–50	13,500	Allen, Noble and Newsome, unpubl.
Mount Rescue C.P., S.A.	35° 50′ S	140° 21′ E	100	28,385	Attwood 1977, S.A. NPWS unpubl. 1984, Specht & Rayson 1957
Narcut (Comet Bore) C.P., S.A.	35° 30′ S	140° 55′ E	100	4	Hatch 1977
Para Wirra R.P., S.A.	34° 43′ S	138° 50′ E	c. 270	1,417	Clarke 1967, Ford & Paton 1976, Hossfeld 1964
Sandy Creek C.P., S.A.	34° 36′ S	138° 53′ E	180	104	Rix 1976
South-Western Australia					
Badjaling N.R., W.A.	31° 59′ S	117° 30′ E	240	272	Muir et al. 1980
Billyacatting Hill N.R., W.A.	31° 03′ E	118° 00′ S	400	2,075	Chapman et al. 1981
Bungalbin, W.A.	30° 18′ S	119° 43′ E	c. 500	9	Dell et al. 1985
Dongolocking N.R., W.A.	33° 04′ S	117° 41′ E	370–420	1,061	Chapman et al. 1978
Durokoppin N.R., W.A.	31° 24′ S	117° 44′ E	325–400	1,030	Muir et al. 1978
East Yuna N.R., W.A.	28° 20′ S	115° 01′ E	c. 296	1,740	Dell et al. 1981, Playford et al. 1970
Karri, W.A. (Karridale)	34° 09′ S	115° 03′ E	60–100	15,000	Christensen et al. 1985
McDermid Rock, W.A.	32° 02′ S	120° 42′ E	?	9	How et al. 1987
North Tarin Rock Res., W.A.	32° 58′ S	118° 16′ E	320–410	1,415	Kitchener et al. 1976
Nugadong N.R., W.A.	30° 12′ S	116° 39′ E	?	400	Kitchener et al. 1979
Perup, W.A.	34° 15′ S	116° 27′ E	260–300	40,000	Christensen et al. 1985
Two Peoples Bay N.R., W.A.	34° 58′ S	118° 10′ E	0–408	4,637	Hopkins & Smith in prep., Smith unpubl.
West Bendering N.R., W.A.	32° 24′ S	118° 22′ E	330–380	1,602	Kitchener et al. 1977, Muir 1977
Wilroy N.R., W.A.	28° 38′ S	115° 38′ E	c. 330	331	Dell et al. 1979a
Woodline Hills, W.A.	31° 54′ S	122° 24′ E	300–500	9	Newbey et al. 1984
Yorkrakine Rock N.R., W.A.	31° 26′ S	117° 31′ E	300	158	Chapman et al. 1980
Yornaning N.R., W.A.	32° 44′ S	117° 22′ E	400–430	248	Dell et al. 1979b
California					
San Dimas Expt. For., Cal.	34° 12′ N	117° 12′ W	458–1,678	6,945	Dunn et al. 1987, Schoenheer 1976, Vaughn 1954, Wirtz 1987, Wright & Horton 1951, 1953
Voorhis Ecol., Res., Cal.	34° 03′ N	117° 49′ W	290–335	9	Moriarty et al. 1985, Stanton 1986, Stewart unpubl.
Central Chile					
El Roble, Chile	33° 10′ S	71° 00′ W	2,000	0.33	Fuentes, Jaksić & Simonetti unpubl., Fulk 1975, Glanz 1977

Table 1. (Continued).

Location	Latitude	Longitude	Altitude (m)	Area (ha)	References
Los Dominicos, Chile	33°23' S	70°30' W	800–1,200	0.09	Fuentes, Jaksić & Simonetti unpubl.
Melipilla, Chile	33°40' S	71°15' W	520	6.7	Cody 1974, Jaksić unpubl.
Pichidangui, Chile	32°09' S	71°33' W	60	3.0	Cody 1974, Fuentes & Jaksić unpubl., Thrower & Bradbury 1977
Puchuncavi, Chile	32°45' S	71°23' W	1,050	1.86	Cody 1974, Jaksić unpubl.
Rinconada, Chile	33°31' S	70°50' W	500–600	1.0	Fuentes, Jaksić & Simonetti unpubl.
Santa Laura, Chile	33°04' S	71°00' W	900–1,000	0.66	Fuentes, Jaksić & Simonetti unpubl., Glanz 1977
Zapallar, Chile	32°35' S	71°28' W	50–60	0.33	Fuentes, Jaksić & Simonetti unpubl., Glanz 1977
Israel					
Gilboa Mtns, Israel	32°29' N	35°26' E	?	10 (mam.) & 1 (birds)	Rankevich & Warburg 1983, Warburg 1978, Warburg *et al.* 1978
Lower Galil, Israel – Allonim	32°44' N	35°10' E	?	10 (mam.) & 3 (birds)	Rankevich & Warburg 1983, Warburg 1978, Warburg *et al.* 1978
Mt Carmel, Israel – Bet Oren	32°44' N	35°00' E	?	10 (mam.) & 3 (birds)	Warburg & Rankevich unpubl., Warburg *et al.* 1978
– Muhraqa	32°40' N	35°05' E	?		
Upper Galil, Israel – Mt Meron	33°00' N	35°26' E	900	10 (mam.) & 1 (bird)	Warburg & Rankevich unpubl., Waburg *et al.* 1978
South Africa					
Swartberg For. Res., Cape Prov.	30°15' S	29°20' E	900 & 1,600	0.18 grid	Breytenbach unpubl.

Table 2. Site specific vegetation and mammal data from mediterranean regions.

Specific site name	General vegetation type	Site (type)	Age of site (years)	Understorey shrub cover (%)	Height of dominant vegetation (m)	Total species	Detailed data for mammal species
South-Eastern Australia							
Nadgee Nature Res.	Open Forest	fire	5	30<70	10<30	28	7S[4i, lih, lis(1), lo(1)], 6M[3i, lhi, lci(1), lcis(1)], 5L [3g, lhb, lc], 8A[3l, llf, 2n, 2ni], 2W [li, lci].
Myall Lakes Nat. Pk.	Open Forest	fire	8	30<70	10<30	21	7S [2i, 2s(1), 2o(1), lh], 5M [3i, lc(1), lo(1)], 4L [2g, lb, lc], 5A [3l, 2n], ?B.
Nowa Nowa State For.	Open Forest	fire logging	>20	10<30	10<30	23	3S [3i(1)], 5M [3i, lci(1), lcis(1)], 5L [3g, lhb, lc], 6A [3l, llf, ln, lni], 4B [4i].
Wyperfeld Nat. Pk.	Open Scrub	fire	>40	30<70	3<10	18	6S [4i, 2is(1)], 5M [2g (2), li, lci(1), lcis(1)], 5L [2g, lhb(1), lo(1), lc], 2A[2n].
Nadgee Nature Res.	Heathland	fire	5	70–100	1<3	19	6S [4i, lih, lis(1)], 7M [3i, lhi, lci(1), lcis(1), lg(1)], 3L(2g, lc], 3A [ll, ln, lni].
Myall Lakes Nat. Pk.	Heathland	fire	6	70–100	1<3	17	7S [2i, 2s(1), 2o, lh], 4M [2i, lc(1), lo(1)], 4L [2g, lb, lc], 1A [1l], 1W [lc], ?B.
Wyperfeld Nat. Pk.	Heathland	fire	>19	30<70	1<3	16	6S [4i, 2is(1)], 5M [2g(2), li, lci(1), lcis(1)], 5L [2g, lhb(1), lo(1), lc].
South Australia							
Brookfield Cons. Pk.	Open Scrub	grazing fire logging	12 15 15	–	3<10	16	2S [li, lis(1)], 4M [lg(1), li, 2ic(2)], 5L [3g(1), 2hb(1)], 1A [lf], 4B [4i].
Innes Nat. Pk.	Open Scrub	mining farming clearing	10	–	1<3	13	2S [lis(1), lo(1)], 4M [lg(1), li, lci(1), lcis(1)], 1L [lg], 1A [ln] 5B [5i].
South-Western Australia							
Karri	Tall Open Forest	mature	20	70–100	>30	17	6S [4o(2), 2i], 7M [3h(2), 2o, 2c(2)], 1L [lh], 3A [2f, li].
Perup	Open Forest	mature	12	10<30	10<30	26	10S [7i, 3o(1)], 12M [4h(1), 2o, 4C(3), 2i], 1L [h], 3A [2f, li].
Woodline Hills	Woodland	logged	30	10<30	10<30	10	1S [s(1)], 1M [h(1)], 4L [2h, 2c(2)], 4B [4i].
Dongolocking Nature Res.	Woodland	mature	>20	<10	10<30	12	5S [ln, 2i, lo, ls(1)], 2M [li, lh(1)], 2L [2h], 2B [2i], 1A [1].
Coastal	Low Woodland	mature	15	30<70	3<10	17	9S [6i, 3c(2)], 7M [2h(1), 2o, 3c(3)], 1L [1h].
N. Tarin Rock Res.	Mallee	mature	>12+	10–30<70	3<10	8	3S [ln, li, ls(1)], 3L [3h], 2B [2i].
Bungalbin	Mallee	mature	>50	<10	3<10	5	3S [2i, ls(1)], 1M [lh(1)], 1L [lc(1)].
Wilroy Nature Res.	Shrubland	mature	>20	10<30	1<3	12	4S [li, lo, 2s(1)], 3M [li, lh(1), lc(1)] 3L [2h, lc(1)], 2B [2i].
McDermid Rock	Shrubland	fire	–	10–30<70	1<3	7	4S [2i, 2s(1)], 1M [lh(1)], 2L [2h].
W. Bendering Nat. Res.	Heathland	fire	>25–30	10–30<70	<1–2	15	7S [ln, 3i, lo, 2s(1)], 3M [li, lh(1), lc(1)], 2L [2h], 1A [1], 2B [2i].
California							
San Dimas Exp. For.	Chaparral	mature fire fire	25 10 4	70–100	1<3	38	17S [li, 6o, 2s, 7h, lc], 9M [6c, 3o(1)], 3L [lc, lh, lo(1)], 1A [lh], 2F [li, lh], 6B [6i].
Voorhis Ecol. Res.	Coastal Sage Scrub	grazing	>30	70–100	1<3	26	12S [li, 7h, 3o, lc], 5M [5o(1)], 1L [lh], 2F [li, lh], 1A [lh], 5B [5i].
Voorhis Ecol. Res.	Coastal Sage Scrub/ Grassland	fire	4	70–100	<1	25	12S [li, 7h, 3o, lc], 5M [5o(1)], 1L [lh], 2F [li, lh], 5B [5i].
Central Chile							
El Roble	Deciduous Forest	*1	–	30<70	10<30	3–4	1S[li], 1SF [lh], 1–2M [1–2c].
Santa Laura	Matorral	*1	–	30<70	3<10	11	6S [2i, 4h], 5M [lh(1), 4c].
Rinconada	Espinal	*1	–	30<70	3<10	11	6S [2i, 3h, lo], 4M [2h(2), 2c],

Table 2. (Continued).

Specific site name	General vegetation type	Site (type)	Age of site (years)	Understorey shrub cover (%)	Height of dominant vegetation (m)	Total species	Detailed data for mammal species
Los Dominicos	Matorral	*1	–	30<70	3<10	14	IL [(le)]. 8S [2i, 4h, 2o], 5M [2h(2), 3c], IL [l(e)].
Zapallar	Coastal Scrub	*1	–	30<70	1<3	11	6S [2i, 3h, lol], lSF [lh], 3M[1h(1), 2c], IL [l(e)].

All data for M and L mammals are estimates based on experience in similar areas [e = locally extinct]. No trapping data available.
*1 All sites disturbed to an unknown degree by (either/or all) agriculture, grazing and woodcutting, present and past.

Northern Israel

Specific site name	General vegetation type	Site (type)	Age of site (years)	Understorey shrub cover (%)	Height of dominant vegetation (m)	Total species	Detailed data for mammal species
Mt Carmel	1. Dense Oakland	–	–	70–100	–	6	
	2. Less dense Oakland	–	–	70–100	–	5	6S*2
	3. Tall Old Oakland	–	–	70–100	–	4	
Upper Galil	Oak Woodland	–	–	70–100	–	4	4S*2
Lower Galil	1. Oak Woodland	sheep &	–	30<70	–	1	
	2. Shrubland	cattle	–	10<30	>0.5	0	2S*2
	3. Open Parkland	grazed	–	10<30	–	1	
Gilboa Mts.	Grasslands	–	–	<10	>0.5	7	7S*2

*2 See also Table 2A for details.

South Africa

Specific site name	General vegetation type	Site (type)	Age of site (years)	Understorey shrub cover (%)	Height of dominant vegetation (m)	Total species	Detailed data for mammal species
Swartberg For. Res.	Spekboomveld	mature	50+	<10	1<3	9	3S [1i, 2o], 2M [1c, 1h], 4L [1c, 3h].
Swartberg	Open-scrub	fire	15	70–100	1<3	15	8S [2i, 1h, 2o, 3s], 3M [2c, 1h], 4L [1c, 3h].
Swartberg For. Res.	Open-scrub	fire	>1	30<70	<1	16	9S [3i, 3s, 3o], 3M [2c, 1h], 4L [1c, 3h].

Key - Major type of mammal in upper case. Dietary type in lower case in brackets [×]. Number of introduced species in brackets (×).

Type of Mammal
S = Small ground mammal – small field trap size
M = Medium ground mammal
L = Large ground mammal
A = Arboreal
B = Bats
F = Fossorial
W = Aquatic

Dietary type
h = herbivore
s = seeds
i = insectivore
o = omnivore
c = carnivore
n = nectar
f = fruit
l = leaves
b = browser
g = grazer
x = fish

Table 2A. Abundance and species richness of rodents in four different regions within the Mediterranean zone in Northern Israel.

Species Annual Rainfall (mm)	Mt. Carmel (750)	U. Galil (1000)	L. Galil (650)	Gilboa Mts. (500)	Total
Acomys cahirinus	4	–	3	49	56
Apodemus mystacinus	12	27	–	1	40
Apodemus sylvaticus	4	23	–	2	29
Rattus rattus	3	1	1	–	5
Rattus norvegicus	1	–	–	–	1
Mus musculus	4	–	–	6	10
Meriones tristrami	–	–	–	3	3
Gerbilus dasyurus	–	–	–	27	27
Microtus guentheri	–	1	–	–	1
Cricetulus migratorius	–	–	–	3	3
Total individuals	28	52	4	91	175
Total species	6	4	2	7	10

Table 3. Site specific vegetation and bird data from mediterranean regions.

Specific site name	General vegetation type	Site (type)	Age of site (years)	Under-storey shrub cover (%)	Height of dominant vegetation (m)	Total resident species	Detailed data for resident species	Total non-resident species	Detailed data for non-resident species
South-Eastern Australia									
Nadgee Nature Res.	Open forest	fire	5	30<70	10<30	59	(G7, GUS8, GUT1, GUC2, GS2, GT4, GTC1, GTB2, SU2, SUC2, UTC7, UTB1, TC15, TCB1, TBA1, AG1, C2).	21	(GUS2, GUT1, GT1, SUT3, SUC1, UTC5, UTB1, TC4, BA1, A2).
							(i24, in7, if3, inf2, ic4, s5, si2, sf4, f1, n1, c4, o2).		(i11, in2, if4, inf1, ic2, o1).
Nowa Nowa State For.	Open forest	fire logging	>20	10<30	10<30	43	(G4, GUS5, GUT1, GUC2, GS2, GT4, GTC1, GTB1, SU1, SUC1, UTC6, UTB1, TC10, TCB1, TBA1, C2).	13	(GUS1, GT1, SU1, SUT2, SUC1, UTC1, UTB1, TC3, BA2).
							(i19, in6, if3, ic4, s3, sf3, n1, c1, o3).		(i6, in4, if2, ic1).
Royal Bot. Garden's Annexe (Cranbourne)	*1	farming	17	70–100	3<10*2	52	(G6, GUS7, GUT1, GUC2, GS2, GT4, GTC1, GTB1, SU2, SUC4, UTC6, UTB1, TB1, TC10, TCB2, TBA1, C1)	23	(G2, GUS2, GUC1, GSC1, GT2, GTC1, SUT2, SUC2, UTC1, UTB1, TC3, A5).
							(i21, in10, if4, inf2, ic3, s3, si2, sf1, f1, c1, o4).		(i15, in1, if1, inf1, ic2, isf1, s1, si1).

*1 Regenerating open forest after being cleared for farming (1964–68). Currently has elements of closed heathland, *Leptospermum myrsinoides* dominating, with emergent *Eucalyptus* spp. (*see* Gullan 1978).
*2 Dominant vegetation is the understorey (shrub layer) of *Leptospermum myrsinoides* (50 cm–3 m high), *Melaleuca squarrosa* 2–7 m) and *Leptospermum juniperinum* (2–7 m).

Specific site name	General vegetation type	Site (type)	Age of site (years)	Under-storey shrub cover (%)	Height of dominant vegetation (m)	Total resident species	Detailed data for resident species	Total non-resident species	Detailed data for non-resident species
Wyperfeld Nat. Pk.*3	Open scrub	fire	>40	30<70	3<10	60	(G7, GUS8, GUT3, GUC2, GS1, GT5, GTC1, GTB1, SU2, SUT2, SUC4, UTC5, UTB1, TB2, TC12, TBA1, AG3).	23	(G4, GUS1, GS2, GSC1, GTC1, SUT3, SUC2, UTC1, AG1, BA1, C2, CAB2, A2).
							(i27, in8, if2, inf2, ic5, s4, si3, sf1, sfn1, o5, c2).		(i6, in3, if1, inf1, ic2, s6, si1, sf1, n2).

*3 This location's bird fauna is influenced by seasonal movements of coastal and arid species alike, and varied habitats that abut.

Specific site name	General vegetation type	Site (type)	Age of site (years)	Under-storey shrub cover (%)	Height of dominant vegetation (m)	Total resident species	Detailed data for resident species	Total non-resident species	Detailed data for non-resident species
Nadgee Nature Res.	Heathland	fire	5	70–100	1<3	27	(G7, GUS3, GS3, GT1, SU3, SUC3, UTC2, UTB1, TB1, TC1, TBA1, AG1).	21	(GUS4, GUC2, GS1, GT1, GTC1, SUT1, SUC1, UTC3, TB1, TC1, AG2, A3).
							(il2, in5, inf1, ic1, s2, si3, c2, o1).		(i10, in2, if2, inf1, ic2, s1, si1, sf1, c1).
Wyperfeld Nat. Pk. *3	Heathland	fire	>19	30<70	1<3	37	(G2, GUS7, GUT2, GUC2, GS1, SU2, SUT3, SUC6, UTC3, UTB1, TB1, TC5, AG1, A1).	23	(G2, GUS2. GS1, GSC1, GT3, GTC1, SUT2, SUC2, UTC1, TB1, TC4, TBA1, AG2, A1).
							(i21, in8, if2, inf2, ic2, s2).		(i11, in1, if1, ic2, s2, si2, sf1, sfn1, o2).
South Australia									
Para Wirra Rec. Pk.	Woodland	fire grazing	1 12	–	10<30	56	(G7, GUS6, GUT2, GUC1, GS1, GT6, GTC2, GTB1, SU2, SUC4, UTC6, TB3, TC11, TBA1, AG1, A1, C1).	20	(G2, GUS2, GS1, GT1, SUT2, SUC2, UTC1, UTB1, AG3, BA1, C2, A2).
							(i22, in7, if3, ic3, isf1, s7, si3, sf1, f1, n1, c3, o4).		(i7, in2, inf1, ic4, s2, si1, n1, c1, o1).

Table 3. (Continued).

Specific site name	General vegetation type	Site (type)	Age of site (years)	Under-storey shrub cover (%)	Height of dominant vegetation (m)	Total resident species	Detailed data for resident species	Total non-resident species	Detailed data for non-resident species
Belair Rec. Pk.	Woodland	fire	–	–	10<30	57	(G8, GUS5, GUT2, GUC2, GS1, GT5, GTC3, GTB1, SU2, SUC4, UTC6, UTB1, TB2, TC10, BA1, C3, A1). (i22, in9, if2, inf1, ic4, isf1, s7, sf2, f1, n2, c1, o5).	22	(G2, GUS1, GT1, SUT2, SUC1, UTC4, TB1, TC4, TBA1, AG1, C1, A3). (i10, in1, if1, inf1, ic1, si4, sf1, n1, c2).
Sandy Cr. Cons. Pk.	Woodland	fire grazing	>12 >17	–	10<30	58	(G6, GUS7, GUT2, GUC3, GS1, GT6, GTC2, GTB1, SU2, SUC2, UTC7, UTB1, TB2, TC11, TBA1, AG2, C1, A1). (i22, in6, if2, inf1, ic4, isf1, s8, si4, sf1, f1, n1, c2, o5).	30	(G3, GUS3, GS1, GSC1, GT1, SUT2, SUC5, UTC3, TB1, TC1, AG4, C3, A2). (i10, in7, if2, ic3, s2, si2, n2, c2).
Billiatt Cons. Pk.	Open scrub	grazing	>40	–	3<10	49	(G6, GUS7, GUT1, GUC2, GS3, GT2, GTC1, SU1, SUT2, SUC5, UTC6, TB1, TC9, TBA1, AG2). (i25, in8, if2, inf1, ic1, s3, si1, sf1, c3, o4).	6*⁴	(G1, GSC1, SUT1, SUC1, TC1, AG1). (i1, in1, if1, inf1, s1, c1).

*⁴ The numbers of non-resident species (nomads and seasonal migrants) will vary greatly in the arid zone, dependent on time of year and rainfall. As this survey only encompasses a minimal number of days, the low number of non-resident species depicted here is to be expected.

Specific site name	General vegetation type	Site (type)	Age of site (years)	Under-storey shrub cover (%)	Height of dominant vegetation (m)	Total resident species	Detailed data for resident species	Total non-resident species	Detailed data for non-resident species
Mount Rescue Cons. Pk.	Open scrub	fire	–	30<70	3<10	60	(G9, GUS7, GUT2, GUC3, GS2, GT4, GTC2, SU1, SUT1, SUC6, UTC7, UTB1, TB2, TC8, TBA1, AG3, A1). (i28, in8, if2, inf1, ic4, isf1, s4, si1, sf1, f1, o6, c3).	17	(G1, GUS5, GTC1, SUC2, TC1, AG1, BA2, C1, CAB2, A1). (i5, in3, if1, inf1, si2, sf2, n1, c2).
Narcut Cons. Pk.	Open scrub*⁵	–	–	–	3<10	58	(G8, GUS7, GUT2, GUC3, GS5, GT3, GTC1, SU2, SUT1, SUC4, UTC7, TB2, TC9, TBA1, AG3). (i28, in8, if2, inf1, ic4, s4, si1, sf1, f1, c3, o5).	23	(G2, GUS4, GS2, GSC1, GT2, SUT1, SUC2, TC2, AG2, BA1, C1, CAB2, A1). (i6, in4, if1, inf2, ic2, s3, si2, sf1, n1, c1).

*⁵ Note that this location has a mixture of two habitats (Mallee and Heathland) hence a mixture of both mallee and heathland bird species are included in this list.

Specific site name	General vegetation type	Site (type)	Age of site (years)	Under-storey shrub cover (%)	Height of dominant vegetation (m)	Total resident species	Detailed data for resident species	Total non-resident species	Detailed data for non-resident species
Brookfield Cons. Pk.	Open scrub	grazing fire logging	12 15 15	–	3<10	69	(G10, GUS9, GUT3, GUC3, GS2, GT6, GTC3, GTB1, SU2, SUT1, SUC4, UTC5, UTB1, TB2, TC10, TBA2, AG3, C1, A1). (i31, in8, if2, inf2, ic4, isf1, s5, si5, sf2, f1, c2, o6).	23	(G3, GUS2, GS2, GSC1, GT1, SUT1, SUC2, UTC1, TB1, TC1, GA2, BA1, C1, CAB2, A2). (i6, in5, if1, inf1, ic2, s2, si3, n1, c2).
Middleback Station	Open scrub	grazing	–	–	3<10	42	(G5, GUS8, GUT2, GUC3, GS2, GT3, GTC2, SU1, SUT1, SUC3, UTC3, TB2, TC3, TBA1, AG2, A1). i20, in4, if1, inf1, ic2, s3, si2, sf2, f1, c2, o4).	18	(G3, GUS5, GS1, SUT1, SUC1, AG2, AB2, A3). (i9, in5, s1, si1, c2).

Table 3. (Continued).

Specific site name	General vegetation type	Site (type)	Age of site (years)	Under-storey shrub cover (%)	Height of dominant vegetation (m)	Total resident species	Detailed data for resident species	Total non-resident species	Detailed data for non-resident species
Innes Nat. Pk.	Open scrub	mining farming clearing	10	–	1< 3	48	(G8, GUS6, GUT2, GUC3, GS3, GT3, GTC1, GTB1, SU2, SUC4, UTC6, UTB1, TB1, TC3, AG3, A1). (i16, in6, if2, inf2, ic5, s4, si3, sfn1, c3, o6).	23	(G3, GUS3, GUCî, GS1, GT1, GTC1, SU1, SUC2, UTC1, TC3, TBA1, AG2, BA1, C1, A1). (i11, in3, s3, si2, sf1, n1, c2).
South-Western Australia									
Karri	Tall open forest	mature	20	70–100	> 30	52	22G (2s, 15i, 1o, 3c, 1g), 3S (1s, 2i), 4U (1i, 2n, 1f), 4T (3i, 1c), 11C (6i, 2n, 3b), 3B (2i, 1c), 4A (2i, 2c); 1W (1h) (Total only).		
Perup	Open forest	mature	12	10< 30	10< 30	87	36G (3h, 4s, 19i, 1o, 7c, 2g), 2S (2i), 7U (3i, 3n, 1f), 6T (4i, 1c, 1b), 15C (8i, 4n, 3b), 6B (3i, 3c), 7A (2i, 5c), 8W (2h, 1s, 2i, 3c) (Total only).		
Billyacatting Hill Nature Res.	Woodland	mature	> 50	< 10	10< 30	23	10G (5s, 4i, 1c), 3S (2i, 1s), 2U (2i), 2T (1o, 1c), 1C (1i), 4B (2i, 2c) 1A (1i).	12	2G (1i, 1c), 2U (2i), 1T (1n), 3C (3i), 2B (1i, 1c), 2A (2c).
Woodline Hills	Woodland	logged	30	10< 30	10< 30	17	4G (2i, 1s, 1o), 3S (3i), 2U (2i), 3C (3i), 3B (2c, 1i), 1A (1c), 1T (1i).	10	2S (1i, 1n) 2U (1s, 1n), 4C (2i, 1n, 1f), 1B (1c), 1A (1i).
Dongolocking Nature Res.	Woodland	mature	> 20	< 10	10< 30	17	9G (3i, 4s, 2c), 1S (1i), 4T (3i, 1c), 2C (2i), 1B (1c).	21	4G (3i, 1s), 1S (1n), 3U (3i), 1T (1i), 3C (3i), 4B (2c, 2i), 5A (3i, 2c).
Coastal	Low woodland	mature	15	30< 70	3< 10	71	27G (1h, 2s, 16i, 1o, 5c, 2g), 5S (1s, 4i), 8U (3i, 4n, 1f), 4T (2i, 1c, 1b), 12C (7i, 2n, 3b), 3B (2i, 1c), 3A (1i, 2c), 9W (3h, 3i, 3c) (Total only).		
Bungalbin Hills	Mallee	mature	> 50	< 10	3< 10	20	7G (5i, 1s, 1c), 5S (4i, 1n), 3U (2i, 1o), 1T (1i), 2C (2i), 2B (2c).	7	1S (1n), 4C (3i, 1n), 1B (1i), 1A (1c).
Nugadong Nature Res.	Shrubland	grazed	> 50	< 10	3< 10	25	10G (5i, 4s, 1c), 6S (5i, 1s), 2U (2i), 2T (1o, 1c), 1C (1i), 4B (2i, 2c).	16	3G (1h, 1l, 1i), 1S (1s), 2U (2i), 3C (3i), 3B (2c, 1i), 4A (2c, 2i).
Badjaling Nature Res.	Shrubland	mature	> 25	< 10	1< 3	8	3G (2s, 1c), 3S (3i), 1B (1c), 1A (1i).	5	1S (1n), 2U (2i), 1C (1i), 1A (1c).
Yorkrakine Rock Nat. Res.	Shrubland	mature	> 25	< 10	1< 3	7	2G (1s, 1i), 2S (2i), 1U (1i), 1B (1i), 1A (1i).	1	1S (1n).
McDermid Rock	Shrubland	fire	–	10– 30< 70	1< 3	21	7G(4i, 2s, 1o), 10S (9i, 1o), 3C (3i), 1B (1c).	6	1G (1o), 2S (2n), 1B (1c), 2A(2i).
Yornaning Nature Res.	Heathland	mature	> 25	< 10	1< 3	6	3G (1s, 1i, 1c), 2S (2i), 1A (1i).	8	3S (3n), 2U (2i), 1C (1i), 2A (1i, 1c).
Two Peoples Bay Nature Res.	Heathland	fire	> 10	30< 70	1< 3	39	(4G, 7GUS, 1GUT, 1GUC, 5GS, 3GT, 3SO, 3SUC, 4UTC, TC2, 1C, 1TB, 1TAB, 2GA, 1A). (18i, 4in, 2if, 1inf, 4ic, 3s, 2si, 1n, 2o, 2c).	23	(1G, 1GUS, 3GT, 1GTB, 1SUT, 1SUC, 3UTC, 1TC, 2C, 1TB, 5GA,1A, 2BA). (7i, 3in, 4ic, 1si, 2sf, 1n, 5c).

Table 3. (Continued).

Specific site name	General vegetation type	Site (type)	Age of site (years)	Under-storey shrub cover (%)	Height of dominant vegetation (m)	Total resident species	Detailed data for resident species	Total non-resident species	Detailed data for non-resident species
California									
San Dimas Ex. For.	Chaparral	mature	25	70–100	1<3	58	16A (5c, 2ci, 9i), 2B (2c), 13G (2g1, 2s, 2ci, 2o, 2i, 3si), ITG (1i), 4GS (1i, 2si, 1isf), 3US (3n), 5T (2i, 1ih, 2fi), 4ST (1is, 2i, 1fi), 10S (4i, 2s, 1if, 3si).	65	14S (10i, 2s,1fi, 1si), 12A (6c, 6i), 19G (4if, 4s,2o, 8si, 1sf), 3US (3n), 10T (6i, 1isf, 2is, 1ih), 1ST (1is), 2GA (2if), ITG (1if), 2GS (1si, 1ifs), 1SA (1fi).
		fire	10						
Voorhis Ecol. Res.	Coastal sage scrub	fire grazing	4 >30	70–100	1<3	24	10G (6g, 2o, 1i, 1c), 2C (1n, 1i), 7S (5i, 1o, 1b), 2T (2i), 3A (3c).	35	7G (4g, 1b, 1i, 1o), 7A (2c, 1o, 4i), 16C (7f, 1n, 7i, 1b), 3B (3i), 2S (2i).
Voorhis Ecol. Res.	Coastal sage scrub/ Grassland	fire	4	70–100	<1	20	5S (3i, 1b, 1f), 2A (2C), 9G (6g, 2o, 1i), 2C (1n, 1f), 2T (2i).	36	10A (7i, 3c), 6S (5i, 1b), 9C (4f, 2i, 2n, 1b), 7G (4g, 1b, 1i, 1o), 4B (4i).
Central Chile									
Puchuncavi	Coastal Matorral	–	–	30<70?	3<10	(10)*[7]18	7G (4s, 2i, 1ic), 4S (3i, 1fl). 1U (1s), 4C (3i, 1n), 2A (2i).		
Melipilla	Espinal	–	–	10<30?	3<10	(6)*[7]16	8G (3i, 1ic, 4s), 1GT (1i), 3S (3i), 1U (1s), 3c (3i).		
Los Dominicos	Matorral	*[6]	–	30<70	3<10	(9)*[7]29	15G (9s, 4i, 1sfi, 1ic), 1GT (1i), 5s (4i, 1fl), 2U (1s, 1sc), 5C (3i, 2n), 1A (1i).		
Pichidangui	Coastal scrub	–	–	30<70?	<1	(7)*[7]5	5G (2s, 2i, 1ic).		

*[6] This Site is disturbed to an unknown degree.
*[7] Raptors in brackets.

Northern Israel									
Mt. Carmel		1. Dense oakland	–	70–100	–				
		2. Less dense oakland	–	70–100	–	9	(12 summer visitors, 10 winter visitors)*[8]		
		3. Tall old oakland	–	70–100	–				
Upper Galil	Oak woodland		–	70–100	–	1	(2 summer visitors, 3 winter visitors)*[8]		
Lower Galil		1. Oak woodland	sheep & cattle grazed	30<70	–				
		2. Shrubland		10<30	>0.5	13	(11 summer visitors, 21 winter visitors)*[8]		
		3. Open parkland		10<30	–				
Gilboa Mts	Grasslands	–	–	<10	>0.5	8	(7 summer visitors, 15 winter visitors)*[8]		

*[8] *See also* Tables 3 (A, B and C).
Key – Major foraging zone in upper case. Dietary type in lower case in brackets (x).

Foraging zone
G = Ground
S = Shrubs <2 metres
U = Understorey
T = Trunks and main branches
C = Outer canopy and branches
B = Below canopy air space
A = Above canopy air space
W = Water

Dietary type
h = herbivore
s = seeds
i = insects
o = omnivore
c = carnivore
n = nectar
f = fruit
l = leaves
b = seed on bush and trees
g = seed on ground

Resident species – those which are found in an area every visit, whether common or uncommon.
Non resident species – only includes species that regularly occur in an area but are not residents e.g. seasonal migrants and transition species. Vagrants are excluded.

Table 3A. Bird species richness and abundance in four areas within the Mediterranean region of Northern Israel.

No. of specimens Annual Rainfall (mm)	Gilboa Mts (500)	Lower Galil (650)	Mt. Carmel (750)	Upper Galil (1000)	Total
summer	57	313	201	8	579
winter	136	486	148	19	789
Total	193	799	349	27	1368
No. of species					
summer	12	24	21	3	27
winter	23	34	19	4	40
Total species	27	45	32	6	52

Table 3B. Avifauna of Northern Israel during summer.

Species	Gilboa Mts 1 habitat	Lower Galil 3 habitats	Mt. Carmel 3 habitats	Upper Galil 1 habitat	Total
Alaudidae					
Galerida cristata	13	4	–	–	17
Pycnonotidae					
Pycnonotus barbatus	8	31	32	–	71
Laniidae					
Lanius excubitor	7	–	–	–	7
Lanius nubicus	–	18	5	–	23
Lanius senator	3	12	–	–	15
Sylviidae					
Prinia gracilis	2	9	6	–	17
Hippolais sp.	–	1	1	–	2
Sylvia communis	–	1	–	–	1
Sylvia hortensii	–	1	1	–	2
Sylvia melanocephala	1	11	5	1	18
Muscicapidae					
Muscicapa striata	–	1	2	–	3
Turdidae					
Monticola solitarius	–	–	2	–	2
Cercotrichas galactotes	–	1	1	–	2
Turdus merula	–	45	51	5	101
Nectariniidae					
Nectarinia osea	8	5	2	–	15
Paridae					
Parus major	–	13	7	–	20
Emberizidae					
Emberiza calandra	2	1	–	–	3
Fringillidae					
Carduelis carduelis	–	21	12	–	33
Carduelis chloris	–	3	–	–	3
Ploceidae					
Passer domesticus	3	8	2	–	13
Corvidae					
Garrulus glandarius	–	5	16	2	23
Phasianidae					
Alectoris chukar	6	4	2	–	12
Columbidae					
Streptopelia decaocto	3	26	14	–	43
Streptopelia turtur	1	76	31	–	108
Streptopelia senegalensis	–	11	2	–	13
Upupidae					
Upupa epops	–	6	3	–	9
Picidae					
Dendrocopos syriacus	–	–	4	–	4
No. of specimens	57	314	201	8	580
No. of species	12	24	21	3	27

Table 3C. Avifauna of Northern Israel during winter.

Species	Gilboa Mts 1 habitat	Lower Galil 3 habitats	Mt. Carmel 3 habitats	Upper Galil 1 habitat	Total
Alaudidae					
Alauda arvensis	8	2	–	–	10
Galerida cristata	14	20	–	–	34
Motacillidae					
Anthus pratensis	–	1	–	–	1
Motacilla alba	–	5	–	–	5
Pycnonotidae					
Pycnonotus barbatus	11	49	18	–	78
Laniidae					
Lanius excubitor	5	–	–	–	5
Sylviidae					
Prinia gracilis	1	10	2	–	13
Sylvia curruca	–	1	5	–	6
Sylvia atricapilla	4	15	5	–	24
Sylvia melanocephala	4	3	1	–	8
Sylvia rupelli	–	1	–	–	1
Phylloscopus collybita	3	11	5	1	20
Phylloscopus bonelli	–	1	–	–	1
Muscicapidae					
Ficedula hypoleuca	–	4	–	–	4
Turdidae					
Saxicola torquata	5	5	–	–	10
Monticola solitarius	1	–	–	–	1
Phoenicurus ochruros	1	–	1	–	2
Phoenicurus phoenicurus	3	8	1	–	12
Erithacus rubecula	2	50	19	8	79
Luscinia svecica	–	1	–	–	1
Turdus merula	2	87	33	9	131
Turdus philomelos	3	5	–	–	8
Nectariniidae					
Nectarinia osea	7	4	–	–	11
Paridae					
Parus major	2	36	13	–	51
Emberizidae					
Emberiza caesia	–	–	1	–	1
Fringillidae					
Fringilla coelebs	1	11	4	–	16
Carduelis carduelis	14	10	–	–	24
Carduelis spinus	–	6	1	–	7
Carduelis chloris	26	31	7	1	65
Serinus serinus	–	34	1	–	35
Ploceidae					
Passer hispaniolensis	6	–	–	–	6
Passer domesticus	1	–	–	–	1
Corvidae					
Garrulus glandarius	–	39	9	–	48
Ardeidae					
Ardeola ibis	–	20	–	–	20
Phasianidae					
Alectoris chukar	12	2	21	–	35
Coturnix coturnix	–	1	–	–	1
Scolopacidae					
Scolopax rusticola	–	2	–	–	2
Columbidae					
Streptopelia decaocto	–	4	–	–	4
Upupidae					
Upupa epops	–	4	–	–	4
Picidae					
Dendrocopos syriacus	–	3	1	–	4
No. of specimens	136	486	148	19	789
No. of species	23	34	19	4	40

Table 4. Site specific vegetation and reptiles and/or amphibia data from mediterranean regions.

Specific site name	General vegetation type	Site (type)	Age of site (years)	Understorey shrub cover (%)	Height of dominant vegetation (m)	Total species	Detailed data on numbers in each family
South-Eastern Australia							
Nadgee Nature Res.	Open forest	fire	5	30<70	10<30	30	6A (1K, 4L, 1M), 9B, 2D, 11G, 1I, 1J.
Myall Lakes Nat. Pk.	Open forest	fire	8	30<70	10<30	68	13A (10L, 1M, 2N), 28B, 3D, 18G, 3H, 2I, 1J.
Nowa Nowa State For.	Open forest	fire logging	>20	10<30	10<30	44	8A (1K, 7L), 17B, 2D, 14G, 2I, 1J.
Wyperfeld Nat. Pk.	Open scrub	fire	>40	30<70	3<10	70	18A (8K, 10L), 8B, 5D, 20G, 9H, 8I, 2J.
Nadgee Nature Res.	Heathland	fire	5	70–100	1<3	29	5A (5L), 12B, 1D, 10G, 1I.
Myall Lakes Nat. Pk.	Heathland	fire	6	70–100	1<3	39	8A (8L), 16B, 2D, 10G, 1H, 1I, 1J.
Wyperfeld Nat. Pk.	Heathland	fire	>19	30<70	1<3	49	15A (7K, 8L), 3B, 5D, 14G, 6H, 5I, 1J.
South Australia							
Innes Nat. Pk.	Open scrub	mining farming clearing	10	–	1<30	42	8A (3K, 4L, 1M), 2B, 4D, 18G, 5H, 4I, 1J.
Brookfield Cons. Pk.	Open scrub	grazing fire logging	12 15 15	–	3<10	53	8A (3K, 5L), 7B, 6D, 21G, 6H, 4I, 1J.
South-Western Australia							
Karri	Tall open forest	mature	20	70–100	>30	19	2A (2L), 9B, 1C, 6G, 1J.
Perup	Open forest	mature	12	10<30	10<30	25	5A (1K, 4L), 8B, 1C, 7G, 1H, 2I, 1J.
Woodline Hills	Woodland	logged	30	10<30	10<30	19	1A (1L), 2B, 2D, 7G, 7H.
Dongolocking Nature Res.	Woodland	mature	>20	<10	10<30	30	3A (1K, 2L), 7B, 2D, 7G, 6H, 4I, 1J.
Coastal	Low woodland	mature	15	30<70	3<10	24	3A (3L), 6B, 1C, 10G, 1H, 2I, 1J.
Bungalbin	Mallee	mature	>50	<10	3<10	17	1A (1L), 1B, 2D, 5G, 7H, 1I.
East Yuna Nature Res.	Shrubland	fire	>35	10<30	1<3	36	2A (2L), 1B, 6D, 12G, 10H, 4I, 1J.
Durokoppin Nature Res.	Shrubland	fire	>50	30<70	1<3	31	3A (3L), 2B, 5D, 8G, 9H, 4I.
McDermid Rock	Shrubland	fire	–	10–30<70	1<3	17	1A (1L), 1B, 4D, 4G, 5H, 2I.
California							
San Dimas Exp. For.	Chaparral	mature fire fire	25 10 4	70–100	1<3	29	14A, 7B, 5E, 1G, 1F, 1O.
Voorhis Ecol. Res.	Coastal sage scrub	grazing	>30	70–100	1<3	10	5A, 2B, 1E, 1G, 1O.
Voorhis Ecol. Res.	Coastal sage scrub/ grassland	fire	4	70–100	<1	10	5A, 2B, 1E, 1G, 1O.
Central Chile							
El Roble	Deciduous forest	*1	–	30<70	10<30	5	1A, 4E.
Santa Laura	Matorral	*1	–	30<70	3<10	7	2A, 4E, 1F.
Rinconada	Espinal	*1	–	30<70	3<10	6	1A, 4E, 1F.
Los Dominicos	Matorral	*1	–	30<70	3<10	6	1A, 4E, 1F.
Zapallar	Coastal scrub	*1	–	30<70	1<3	4	1A, 3E, 1H.

*1 All sites disturbed to an unknown degree by (either/or all) agriculture, grazing and wood cutting, present and past.

Specific site name	General vegetation type	Site (type)	Age of site (years)	Understorey shrub cover (%)	Height of dominant vegetation (m)	Total species	Detailed data on numbers in each family
Northern Israel							
Mt. Carmel	1. Dense oakland	–		70–100	–		
	2. Less dense oakland	–		70–100	–	3*2	
	3. Tall old oakland	–		70–100	–		
Upper Galil	Oak woodland	–		70–100	–	0*2	
Lower Galil	1. Oak woodland	sheep & cattle		30<70	–		
	2. Shrubland			10<30	>0.5	6*2	
	3. Open parkland	grazed		10<30	–		
Gilboa Mts	Grasslands			<10	>0.5	3*2	

*2 *See also* Table 4A for details.

Key
Families
A = Snakes E = Iguanidae I = Pygopodidae M = Boidae
B = Amphibians F = Teiidae J = Varanidae N = Colubridae
C = Tortoise G = Scincidae K = Typhlopidae O = Anugidae
D = Agamidae H = Geckonidae L = Elapidae

Table 4A. Abundance and species richness of reptiles in four different regions within the Mediterranean zone in Northern Israel. Annual rainfall (mm) in brackets.

Species	Gilboa (500)	L. Galil (650)	Mt. Carmel (750)	U. Galil (1000)	Total
Testudo graeca	3	19	1	–	23
Chamaeleo chamaeleon	–	3	2	–	5
Ptyodactylus hasselquistii	7	1	2	–	10
Hemidactylus turcicus	2	1	–	–	3
Cyrtodactylus kotschyi	1	–	5	–	6
Agama stellio	2	22	22	–	46
Ophisops elegans	–	2	15	12	29
Lacerta laevis	–	6	18	19	43
Lacerta trilineata	–	1	3	–	4
Ophisaurus apodus	1	5	–	–	6
Ablepharus kitaibelii	3	21	35	4	63
Mabuya vittata	8	1	3	–	12
Eumeces schneideri	3	6	1	–	10
Chalcides ocellatus	4	6	3	–	13
Chalcides guentheri	6	3	1	1	11
Ophiomorus latastii	1	–	–	–	1
Typhlops simoni	3	1	–	–	4
Malpolon monspessulanus	1	1	–	–	2
Rhynchocalamus melanocephalus	–	1	–	–	1
Eirenis lineomaculata	–	1	1	–	2
Eirenis rothi	–	2	–	–	2
Coluber najadum	–	–	2	–	2
Coluber ravergieri	–	1	–	–	1
Coluber jugularis	2	–	1	–	3
Total no. of specimens	47	104	115	36	302
Total no. of species	15	20	16	4	24

CHAPTER 5

Soil and litter invertebrates

Soil and litter invertebrates

Co-ordinators: Jonathan D. Majer and Penelope Greenslade

Contributors

W. Breytenbach and G.J. Breytenbach (South Africa)
D. Donnelly and J.H. Giliomee (South Africa)
Penelope Greenslade (South Australia)
J.D. Majer (Western Australia)
M. Narog (California)
A.C. Postle (Western Australia)
T.J. Ridsdill-Smith (Western Australia)
F. Sáiz (Chile)

Contents

1	**Introduction**	197
2	**General taxa treatments**	198
2.1	Density and seasonality of soil and litter invertebrates in Western Australian jarrah forest	198
2.2	Density and seasonality of epigaeic fauna in South Australia	199
2.3	The composition of fynbos epigaeic invertebrates in the south western cape of South Africa	200
2.4	The epigaeic invertebrate fauna in South African Fynbos of different ages after fire	201
2.5	The composition of soil, litter and epigaeic fauna in three different vegetation associations in the semi-arid mediterranean region of Chile	204
2.5.1	Soil and litter fauna	204
2.5.2	Epigaeic fauna	207
2.6	The composition of epigaeic fauna in California	211
2.7	The composition of epigaeic fauna in mediterranean southern France	211
3	**Individual taxa treatments**	212
3.1	The Collembola	212
3.1.1	Abundance	212
3.1.2	General seasonality	214
3.1.3	Taxonomic composition	214
3.1.3.1	Size of the fauna	214
3.1.3.2	Seasonality of individual groups	214
3.1.3.3	Adaptations	215
3.1.3.4	Resilience	215
3.2	The scarabaeine dung beetles	216
3.2.1	Dung beetle faunas of mediterranean climate areas	216
3.2.2	Seasonal occurrence and abundance of dung beetles in different mediterranean climate areas	217
3.2.3	Discussion	218
3.3	The Formicidae	219
3.3.1	Ant abundance	219
3.3.2	Ant species richness	219
3.3.3	Ant generic composition	220
3.3.4	Seasonality of the ant fauna	222
3.3.5	Influence of fire on the ant fauna	222
4	**Synthesis**	224
5	**References**	224

1 Introduction

Soil and litter invertebrates are a large, but generally inconspicuous, component of the biota of mediterranean regions whose existence is of crucial importance to the functioning of these ecosystems. Some groups are involved in the regulation of nutrient cycles, the maintenance of soil structure or are linked through their herbivorous activities with the productivity and dynamics of plant communities. Other taxa play an important role in seed dispersal and survival, in pollination or they may provide an important food source for insectivores. It is therefore fitting that this volume includes a chapter which compares the soil and litter invertebrates of the various mediterranean regions.

From the outset we were struck by the lack of comparable data from the regions in which we were interested. Consequently, some of the data were collected especially for this chapter. In view of the shortage of time and resources to undertake this task we had to choose a cost-effective sampling method to obtain the data. Pitfall traps were generally used to obtain a rapid census of the invertebrate fauna. This technique, which has a number of limitations, samples surface-active (epigaeic) animals and, to a lesser extent, those from the soil and litter. Catches are influenced by the relative activity and trapability of the animals as well as their abundance. The nature of the ground cover also influences the efficiency of these traps. We have therefore expressed the results as 'catches' of animals rather than 'densities' as the data may not always provide a true measure of the latter.

We solicited published and unpublished information on taxonomic composition, abundance, seasonality and the influence of fire on this component of the fauna and obtained contributions from South Africa, France, California, Chile, South Australia and Western Australia. Although the contributions are not all uniform in collection and treatment of data, they provide the opportunity to compare broad patterns of faunistic composition in these regions. This information may be supplemented by the existing reviews of Cody et al. (1977), Di Castri and Di Castri (1981) and Force (1981). Furthermore, a series of short papers on soil and litter invertebrates from Australian mediterranean-type ecosystems is available in Greenslade and Majer (1985).

The first section, by A.C. Postle, covers the density and seasonality of soil and litter animals in Western Australia's jarrah (*Eucalyptus marginata*) forest. This is followed by a contribution by Penelope Greenslade on South Australian epigaeic fauna. The first South African contribution, by W. and G.J. Breytenbach, examines the variation in composition of pitfall catches on sites of different altitude and rainfall. A second contribution from South Africa, by D. Donnelly and J.H. Giliomee, deals with the abundance and composition of epigaeic fauna in fynbos of different ages after fire. A section on the Chilean fauna by F. Sáiz describes the composition of soil and litter animals in three different vegetation associations. Finally, some data from pitfall traps from California and southern France are provided by M. Narog and Penelope Greenslade respectively.

The second part of this chapter covers specific taxa, namely Collembola, scarabaeine dung beetles and ants, compiled respectively by Penelope Greenslade, T.J. Ridsdill-Smith and J.D. Majer.

2 General taxa treatments

2.1 Density and seasonality of soil and litter invertebrates in Western Australian jarrah forest

A.C. Postle

The density and seasonality of soil and litter invertebrates has been investigated in 11 jarrah (*E. marginata*) forest plots near Dwellingup, Western Australia (32° 42′ S, 116° 02′ E) in relation to differences in soil type.

Invertebrates were sampled at monthly intervals from June 1980 to December 1981. At 20 randomly selected points in each plot all litter was removed from a 19 × 19 cm quadrat, bulked and sealed in a polyethylene bag. Invertebrates were extracted in a funnel from the bulked samples. The temperature in the funnel was raised from ambient to 40° C over a one week period. Soil cores (54 mm diameter, 97 mm deep) were collected from below all litter samples and extruded into plastic sleeves. Animals were extracted from individual soil cores using a multiple canister heat extractor in which the temperature regime was the same as that for the litter extraction.

Density of all taxa varied between plots so data from an average plot are presented here as a typical example of the results. For brevity, results from only five months, representing different seasons, are given. Numbers of both soil and litter invertebrates in samples varied with season (Table 1). Peak numbers occurred in early winter with a lower peak in early spring.

Invertebrates were over 40 times more abundant in the soil than litter. Samples in both the soil and litter consisted predominantly of mites (up to 92.6% and 77.5% of total soil and litter fauna respectively) and the second most abundant group was the Collembola (up to 16.9% and 39.0% of total soil and litter fauna respectively).

The only comparable figures from southwest Western Australia are those of Majer (1984) for Karragullen (32°07′S, 116′04′E) jarrah forest, and Springett (1976) for Swan Coastal Plain (31°47′S, 115°52′E) *Eucalyptus/Banksia* woodland. Majer's samples were collected using the same methods as the Dwellingup ones although Springett sampled fauna in the top 25 cm of soil with the litter layer unseparated. Majer's combined soil and litter density values ranged from 5,519 m^{-2} in June to 28,495 m^{-2} in November while Springett found 77–100,000 animals m^{-2}. Densities in the present study (Postle 1986) ranged from 40,676 m^{-2} in December to 136,860 m^{-2} in April.

2.2 Density and seasonality of epigaeic fauna in South Australia

Penelope Greenslade

The epigaeic fauna was sampled at Kuitpo/Kyeema forest, about 100 km south of Adelaide in the Mt. Lofty Ranges (33°15′S, 138°43′E). The vegetation was an open woodland of *Eucalyptus obliqua* and *E. cosmophylla* with scattered *Exocarpos cupressiformis*. The understorey was dense in places and included species of *Hakea*, *Xanthorrhoea*, *Banksia*, *Leptospermum*, *Lepidosperma* and *Hibbertia*. Three sites were chosen in an area where a prescribed burn was to be carried out, and were

Table 1. Mean number of invertebrates extracted per m^2 by funnels from soil (top 97 mm) and litter at different seasons in jarrah forest, Western Australia (data from Postle 1986).

Taxon	Spring Oct. 1980		Summer Dec. 1980		Summer Feb. 1981		Autumn Apr. 1981		Winter Jul. 1981	
	Soil	Litter	Soil	Litter	Soil	Litter	Soil	Litter	Soil	Litter
Nematoda	–	2	–	–	–	–	87	6	–	–
Arachnida Pseudoscorpionida	218	39	–	–	87	20	87	59	44	17
Arachnida Opilionida	22	2	–	–	–	–	–	–	22	–
Arachnida Acarina	53,886	750	36,491	977	78,788	2,061	119,950	1,691	55,698	1,395
Arachnida Araneae	–	43	–	28	–	59	–	42	–	31
Crustacea Isopoda	22	17	–	–	–	20	44	25	22	–
Diplopoda	22	47	87	9	–	45	87	62	44	58
Chilopoda	130	9	44	–	44	6	44	9	175	4
Symphyla	437	31	306	–	742	9	1,309	9	1,615	2
Pauropoda	873	–	44	–	1,179	168	1,309	76	1,571	21
Collembola	4,409	790	1,615	–	5,238	269	8,905	605	12,222	566
Protura	196	–	87	–	306	–	567	–	153	–
Diplura	131	–	44	–	44	3	44	3	–	–
Insecta Thysanura	–	–	–	–	–	6	–	3	–	–
Insecta Blattodea	–	2	–	–	–	–	–	–	–	–
Insecta Isoptera	371	–	–	–	–	3	–	11	–	–
Insecta Dermaptera	22	–	–	–	–	–	–	–	–	–
Insecta Orthoptera Tettigonioidea	–	2	–	–	–	–	–	–	–	–
Insecta Embioptera	–	–	–	–	–	–	–	–	–	2
Insecta Psocoptera	196	9	131	90	175	143	–	11	–	–
Insecta Homoptera	–	2	–	3	87	9	393	3	44	–
Insecta Heteroptera	22	2	–	–	–	34	87	56	44	31
Insecta Thysanoptera	–	18	87	48	–	56	–	25	–	4
Insecta Coleoptera adults	240	97	87	3	349	65	349	98	415	34
Insecta Coleoptera larvae	153	107	262	56	218	112	87	70	44	9
Insecta Diptera adults	–	9	87	–	655	9	–	6	22	6
Insecta Diptera larvae	66	20	–	6	–	6	262	70	153	52
Insecta Lepidoptera adults	–	4	–	3	–	–	–	–	–	–
Insecta Lepidoptera larvae	–	16	44	28	–	95	–	22	22	6
Insecta Hymenoptera Formicidae	–	3	–	9	44	193	87	177	–	4
Insecta Hymenoptera others	–	6	–	–	44	6	–	23	22	–
Totals	61,416	2,027	39,416	1,260	88,000	3,397	133,698	3,162	72,332	2,242
	63,443		40,676		91,397		136,860		74,574	

paired with sites outside the area which acted as controls. At each site five pitfall traps (1.8 cm diameter, containing an alcohol preservative) were placed in a line about a metre apart and run for seven days during most months between January, 1981 and July, 1984.

Total invertebrates trapped in two of the unburnt plots (controls) are shown in Table 2. Data for five months have been selected to represent the seasons used in section 2.1.

Collembola were by far the most abundant group trapped over all seasons followed by Formicidae, Acarina, Dermaptera and Coleoptera. The Collembola were most abundant in traps in spring and autumn with lowest numbers in mid summer. Formicidae and Dermaptera were trapped far more frequently in summer than any other season and the Acarina and Coleoptera varied irregularly throughout the year.

2.3 The composition of fynbos epigaeic invertebrates in the south western cape of South Africa

W. Breytenbach and G.T. Breytenbach

The information presented here was gathered during an ongoing project of the South African Forestry Research Institute, with the objective of examining the distribution patterns of arthropods along an altitudinal gradient.

The study sites are situated in the Groot Swartberg forest near Oudtshoorn in the southwestern Cape Province, South Africa (33° 20' S, 22° 02' E). The vegetation at all the sites represents typical fynbos with an overstorey of *Protea* species, and an understorey of Restionaceae and Ericaceae. All the sites (1–4) were in mature fynbos (15–20 years post fire), and the soils have a lithocutanic B horizon, overlayed by a shallow sandy A horizon. The sites in the Groot Swartberg forest follow an altitudinal gradient with site 1 at ca. 900 m, site 2 at ca. 1,150 m, site 3 at ca. 1,220 m and site 4 at ca. 1,600 m above mean sea level (amsl) on south facing slopes. The reserve has rain throughout the year, but there is a definite decrease in rainfall during June and July (Breytenbach 1982).

Table 2. Pooled catch of arthropods sampled by pitfall traps (70 trap-days per date) in two plots within Kuitpo/Kyeema forest, South Australia, during several seasons between 1981–1982.

Taxon	Sampling date					
	Oct. 1981	Dec. 1981	Feb. 1982	Apr. 1982	Jul. 1982	Total
Arachnida Acarina	126	56	118	21	16	337
Arachnida Araneae	1	6	9	3	2	21
Crustacea Isopoda	1	–	–	–	–	1
Crustacea Amphipoda	1	–	–	–	–	1
Diplopoda	1	–	–	1	1	3
Chilopoda	–	–	1	–	–	1
Collembola	519	331	82	467	277	1676
Insecta Thysanura	4	–	4	–	6	14
Insecta Blattodea	1	–	–	1	–	2
Insecta Dermaptera	23	87	81	9	3	203
Insecta Orthoptera	–	11	–	–	–	11
Insecta Psocoptera	–	–	1	–	–	1
Insecta Hemiptera	–	2	2	–	1	5
Insecta Thysanoptera	–	–	3	–	–	3
Insecta Coleoptera adults	46	18	25	33	7	129
Insecta Coleoptera larvae	1	8	1	2	1	13
Insecta Diptera adults	29	20	6	6	10	71
Insecta Hymenoptera Formicidae	57	71	229	8	5	370
Insecta Hymenoptera others	2	–	2	–	–	4
Total	812	610	564	551	329	2866

Arthropods were sampled by a range of pitfall trap designs during summer, autumn and winter. Plastic buckets of 18 cm diameter and 21 cm depth were partly filled with 250 ml of preservative (ethylene glycol). Three additional pitfall traps were set 10 m apart at each site. These were baited with rotten meat, human faeces and a banana, rum plus sugar mixture, respectively.

The five most abundant groups represented in the samples were Formicidae, Coleoptera, Diptera, Collembola and Araneae. Arthropod numbers in traps were highest during summer and lowest during winter (2,482 vs 189 total individuals, Table 3). There were also seasonal differences in the number of orders represented in the samples with 17 in summer, 15 in autumn and 13 in winter.

Along the altitudinal gradient the most individuals (716) were caught at site 4 (1,600 m amsl) during summer. This changed during autumn and winter when the most individuals (526 and 97 respectively) were caught at site 2 (1,150 m amsl).

2.4 The epigaeic invertebrate fauna in South African fynbos of different ages after fire

D. Donnelly and J.H. Giliomee

The epigaeic invertebrate fauna of the Jonkershoek Valley, South Africa (33° 57′ S, 18° 55′ E) was sampled by means of pitfall traps at six sites (A–F), each covered with fynbos vegetation of a different age after fire (Table 4). Traps were arranged in grids of 20 (5 × 4) with 20 m between rows and 7 m between columns. They had a diameter of 10 cm and contained water plus detergent and were opened for 24 hours approximately once a month from June 1979 to July 1980. The sites and the traps are described in more detail elsewhere (Donnelly and Giliomee 1985a).

Collembola were easily the most common invertebrates (Table 4), but much of this was the result of trapping very large numbers of juveniles in a few traps following rain. Group totals suggest that ants are the second most common group in terms of catch, followed by Phoridae (Diptera) and

Table 3. Numbers of arthropods sampled in South African fynbos during summer, autumn and winter using pitfall traps.

	Summer					Autumn					Winter				
Site number	1	2	3	4	Total	1	2	3	4	Total	1	2	3	4	Total
Arachnida Acarina	6	12	2	8	28	18	7	17	4	46	–	1	–	–	1
Arachnida Araneae	28	24	37	33	122	–	9	5	10	24	1	1	1	1	4
Crustacea Isopoda	1	4	–	–	5	–	3	1	1	5	2	1	1	–	4
Diplopoda	1	3	1	4	9	–	1	–	–	1	–	–	–	1	1
Chilopoda	1	–	1	1	3	–	–	1	–	1	–	–	–	–	0
Collembola	98	31	28	33	190	17	8	13	2	40	9	16	6	3	34
Insecta Ephemeroptera	2	7	–	7	16	–	–	–	–	0	–	–	–	–	0
Insecta Blattodea	7	9	15	10	41	–	–	4	4	8	1	2	1	–	4
Insecta Isoptera	10	6	–	1	17	1	7	–	–	8	–	–	–	–	0
Insecta Mantodea	–	–	–	–	0	–	–	1	–	1	–	–	–	–	0
Insecta Dermaptera	–	1	6	3	10	–	–	–	–	0	–	1	–	–	1
Insecta Orthoptera	20	12	8	4	44	3	4	4	1	12	–	–	–	–	0
Insecta Hemiptera	7	2	6	–	15	–	2	6	–	8	1	1	–	1	3
Insecta Thysanoptera	–	2	–	2	4	–	–	–	–	0	–	–	–	–	0
Insecta Coleoptera	42	87	240	317	686	128	88	150	48	414	6	44	13	2	65
Insecta Siphonaptera	–	–	–	1	1	13	18	39	18	88	–	1	–	–	1
Insecta Diptera	78	216	87	73	454	–	–	–	–	0	6	24	1	1	32
Insecta Trichoptera	–	–	–	–	–	–	–	–	–	0	–	1	–	–	1
Insecta Lepidoptera	–	–	–	–	–	1	5	1	8	15	–	–	–	–	0
Insecta Hymenoptera	–	–	–	–	–										
Insecta Formicidae	339	132	147	219	837	61	374	89	95	619	14	4	1	19	38
Total	640	548	578	716	2482	242	526	331	191	1290	40	97	24	28	189

Site co-ordinates
Site 1 = 33° 22′ 05″ S, 22° 05′ 40″ E.
Site 2 = 33° 21′ 55″ S, 22° 09′ 10″ E.
Site 3 = 33° 21′ 55″ S, 22° 03′ 55″ E.
Site 4 = 33° 21′ 10″ S, 22° 02′ 40″ E.

then Coleoptera. Large numbers of flying insects, such as Diptera and Staphylinidae (comprising 43.8% of the Coleoptera caught in this survey), suggest that these groups were possibly attracted to the liquid in the traps.

Some groups showed characteristic habitat preferences. Collembola were far more common in the more recently burnt areas, becoming progressively less abundant with increasing age of vegetation. Smaller numbers of mites (Acarina) were also caught in the older sites. Catches of other taxa (e.g. Diplopoda and especially amphipod Crustacea) increased with increasing age of vegetation and corresponding density of the litter layer.

The ants collected comprised 45 species (Donnelly and Giliomee 1985a), which is similar to the number of species found by Cody et al. (1977) in a comparable site of chaparral vegetation in California.

The adult beetles caught belonged to 30 families (Table 5). The greatest number of individuals belonged to the Staphylinidae, followed by Scarabaeidae, Chrysomelidae – mainly Halticinae, and Carabidae. The Staphylinidae also contained the greatest number of species (29), followed by Curculionidae (27), Chrysomelidae (24), and Scarabaeidae (15). Staphylinidae and Scarabaeidae were attracted to decaying material so catches of these animals could have been inflated by their attraction to the fluid in the traps (Greenslade and Greenslade 1971).

Total arthropod biomass was greatest on the

Table 4. Total number of arthropods caught in pitfall traps in South African fynbos which were opened for 10 separate 24-hour periods during the course of 1979 and 1980.

Taxon			Number of individuals caught at each site						Total
			South-west facing sites			North-east facing sites			
			A (37 year old vegetation)	B (4 year old vegetation)	C (2 year old vegetation)	D (2 year old vegetation)	E (4 year old vegetation)	F (21 year old vegetation)	
Arachnida	Pseudoscorpionida		6	10	2	3	1	12	34
Arachnida	Opilionida		7	4	1	2	6	–	20
Arachnida	Acarina		74	348	153	178	162	55	970
Arachnida	Solpugida		9	9	15	13	12	7	65
Arachnida	Araneae		88	163	113	98	125	89	676
Crustacea	Amphipoda		59	10	4	–	–	77	150
Crustacea	Isopoda		9	4	2	1	9	4	29
Diplopoda			10	3	2	1	1	26	43
Chilopoda	Scutigeromorpha		1	–	–	4	2	1	8
Chilopoda	Lithobiomorpha		–	4	–	–	1	1	6
Chilopoda	Scolopendromorpha		4	–	1	2	7	3	17
Symphyla			1	–	–	–	–	–	1
Collembola	Arthropleona		312	749	7687	3327	998	395	13468
Collembola	Symphypleona		5	91	89	71	38	4	298
Insecta	Archaeognatha		3	23	14	26	23	12	101
Insecta	Thysanura		–	3	–	–	–	1	4
Insecta	Orthoptera		43	48	41	30	20	65	247
Insecta	Phasmatodea		–	–	–	–	–	1	1
Insecta	Dermaptera		3	25	4	2	28	7	69
Insecta	Blattodea		10	6	4	7	27	12	66
Insecta	Isoptera		–	–	–	9	10	2	21
Insecta	Psocoptera		6	4	–	1	3	5	19
Insecta	Hemiptera		72	62	74	56	35	52	351
Insecta	Thysanoptera		1	4	–	7	7	4	23
Insecta	Neuroptera	larvae	–	1	–	–	1	–	2
		adults	–	–	–	1	–	–	1
Insecta	Coleoptera	larvae	35	35	24	24	21	41	180
		adults	146	151	199	106	210	283	1095
Insecta	Strepsiptera		2	–	–	–	–	–	2
Insecta	Siphonaptera		1	–	–	–	–	–	1
Insecta	Diptera	Phoridae	226	375	322	487	282	827	2519
		others	174	113	87	159	82	160	775
Insecta	Lepidoptera	larvae	2	–	–	1	–	1	4
		adults	–	5	1	–	1	4	11
Insecta	Hymenoptera	Formicidae	963	1453	1698	2075	1415	1107	8711
		others	32	33	18	55	31	34	203
Total			2304	3736	10555	6746	3558	3292	30191

younger firebreak sites, notably sites C, D and E (Fig. 1), although the biomass for any site was influenced by the chance trapping of only one or two large carabids or lepidopteran larvae. Comparison of the biomass of epigaeic invertebrates between sites should be approached with caution in view of the presence of a few large species and also because of the limitations of the pitfall trapping technique.

Changes in the catch of the more common taxa with time are given in Fig. 2, and that of the seven most abundant ant species in Fig. 3. The change in total biomass of the invertebrates caught in the pitfall traps with time is shown in Fig. 4.

There was some suggestion of an early-summer and mid-summer peak in total invertebrate biomass. Some of the taxa, e.g. Phoridae, Symphypleona, Coleoptera, etc., showed definite winter peaks in catch, while Acarina and Arthropleona showed less marked winter peaks. Coleoptera, notably carabids, are generally inactive as adults in winter in temperate regions (Thiele 1977) but the mediterranean winters, as is found in Jonkershoek, are very mild.

Conversely, other taxa showed summer peaks, e.g. Archaeognatha, Orthoptera, Blattodea, Formicidae, Solpugida, Araneae, etc. Catches of Formicidae were highest around mid to late summer for several species. Briese and Macauley (1980) and Whitford *et al.* (1980) also recorded increased activity of ants in summer. Although *Anoplolepis custodiens* showed a marked midsummer peak

Table 5. The numbers of individuals in the various beetle families sampled using pitfall traps in six South African fynbos sites.

Beetle family	Number of representative species	Number of individuals caught at each site						Total
		South-west facing sites			North-east facing sites			
		A (37 year old vegetation)	B (4 year old vegetation)	C (2 year old vegetation)	D (2 year old vegetation)	E (4 year old vegetation)	F (21 year old vegetation)	
Carabidae	8	2	2	17	9	18	13	61
Cicindelinae	3	–	–	2	1	1	–	4
Paussinae	1	–	–	–	–	3	–	3
Anisotomidae	1	1	–	–	–	–	–	1
Scydmaenidae	8	–	1	3	3	3	2	12
Pselaphidae	2	–	2	3	2	–	–	7
Staphylinidae	29	41	109	94	29	117	129	519
Trogidae	1	2	1	–	–	–	1	4
Scarabaeidae	15	61	8	21	19	8	101	218
Byrrhidae	2	–	–	–	2	–	2	4
Elateridae	2	13	1	1	–	4	19	38
Cantharidae	1	–	–	–	–	1	–	1
Malachiidae	1	1	–	–	1	3	–	5
Nitidulidae	2	1	–	–	1	–	–	2
Cucujidae	1	–	–	–	1	–	–	1
Phalacridae	1	–	–	–	–	1	–	1
Coccinellidae	2	1	–	1	–	–	–	2
Tenebrionidae	12	4	14	7	4	2	–	31
Lagriidae	1	–	–	–	1	–	–	1
Pythidae	1	–	–	–	–	–	1	1
Melandryidae	1	–	–	–	–	1	–	1
Mordellidae	1	1	4	–	–	1	–	6
Meloidae	1	–	–	–	1	–	–	1
Anthicidae	7	8	1	–	4	11	–	24
Cerambycidae	2	2	–	–	1	–	–	3
Chrysomelidae	7	–	–	3	4	–	2	9
Halticinae	17	2	2	26	12	24	2	68
Curculionidae	27	6	5	21	9	7	9	57
Scolytinae	1	–	–	–	–	–	2	2
Undetermined families	7	–	1	–	2	5	–	8
Total		146	151	199	106	210	283	1095

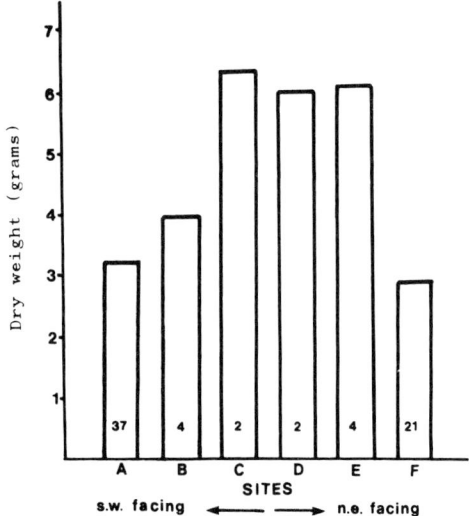

Fig. 1. Comparison of the total biomass (g) of the arthropods caught in pitfall traps in the Jonkershoek Valley, South Africa, for the 10 sampling periods combined. The age after fire (in years) of the vegetation is indicated in each bar (from Donnelly 1983).

here, Louw (1968) showed the same species to have spring and autumn peaks of abundance in the central Orange Free State. Other species, e.g. *Camponotus maculatus* and *Monomorium schultzei*, did not exhibit recognisable differences in seasonal abundance. *Iridomyrmex humilis*, the Argentine ant, was unusual in having a winter peak. The winter activity of *I. humilis* in Swartboschkloof was possibly related to the winter flowering of *Protea repens* which was very common throughout the area, and which this ant was seen to frequent (Mostert *et al.* 1980). Skaife (1961) reported the Argentine ant to be more or less inactive in winter, though Du Merle (1982) showed that seasonal variation in the foraging activity of ants varied according to the plant species present.

2.5 The composition of soil, litter and epigaeic fauna in three different vegetation associations in the semi-arid mediterranean region of Chile

F. Sáiz

Fray Jorge National Park is located in the mediterranean semi-arid region of Chile (30° 38′ S, 71° 40′ W). It ranges between 0–600 m altitude, and is characterised by the presence of a relic forest surrounded by different types of vegetation. The Park had not recently been affected by fire or significantly altered by man. Three types of vegetation were studied:

(a) Temperate rainforest (Cloud forest) with *Aextoxicon punctatum* as the dominant tree plus *Drimys winteri, Myrceugenia correaefolia,* etc. The altitude is 600 m (Bosque higrofilo);

(b) Xerophytic scrub with *Baccharis concava, Kageneckia oblonga, Haplopappus* sp., *Fuchsia lycioides* and *Proustia pungens.* The altitude is 500 m (Matorral xerofilo); and

(c) Thorny scrub with *Porlieria chilensis* and *Adesmia bedwelli.* The altitude ranges from 200–300 m (Matorral espinoso).

Two studies were performed simultaneously in the Park, one concerned with the epigaeic fauna, which was sampled with pitfall traps, and the other with the soil and litter fauna, which was extracted by funnels. The data have already been published elsewhere so they are presented here in summary form.

2.5.1 Soil and litter fauna

Invertebrates were sampled at four stations within each site every 45 days from August, 1967 to December, 1968 (Sáiz 1975a). At each station three 50 l samples were taken from the following strata: A = litter or moss; B = 0–5 cm soil depth; and C = 5–10 cm soil depth. Arthropods were extracted from the samples in funnels with a 25 W bulb over five days.

The number of invertebrates caught are presented in Table 6 and the principal conclusions are as follows (Sáiz 1975a):

– Soil and litter fauna consisted principally of Acarina and Collembola.
– The number of taxa decreased from the temperate rain-forest to the thorny scrub, and from the litter surface downwards.
– In the thorny scrub Prostigmata and Tarsonemini were the most abundant mites, while in the temperate rainforest the Oribatida and Uropodina were the dominant groups.
– Catches of Prostigmata and Oribatid mites exhibited different responses to humidity. A

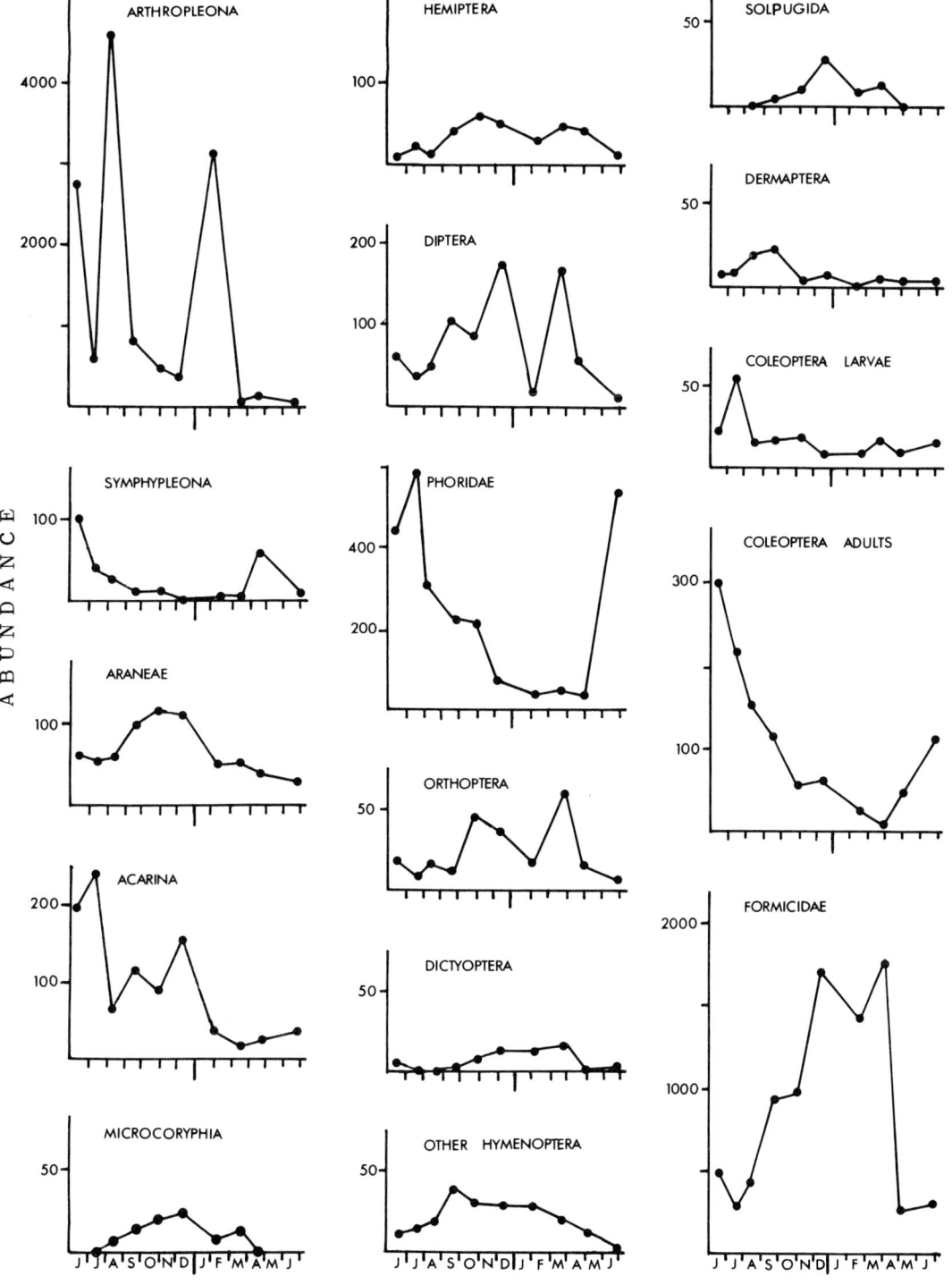

Fig. 2. Temporal variation in catch of the more common taxa obtained from pitfall traps in the Jonkershoek Valley, South Africa (from Donnelly 1983).

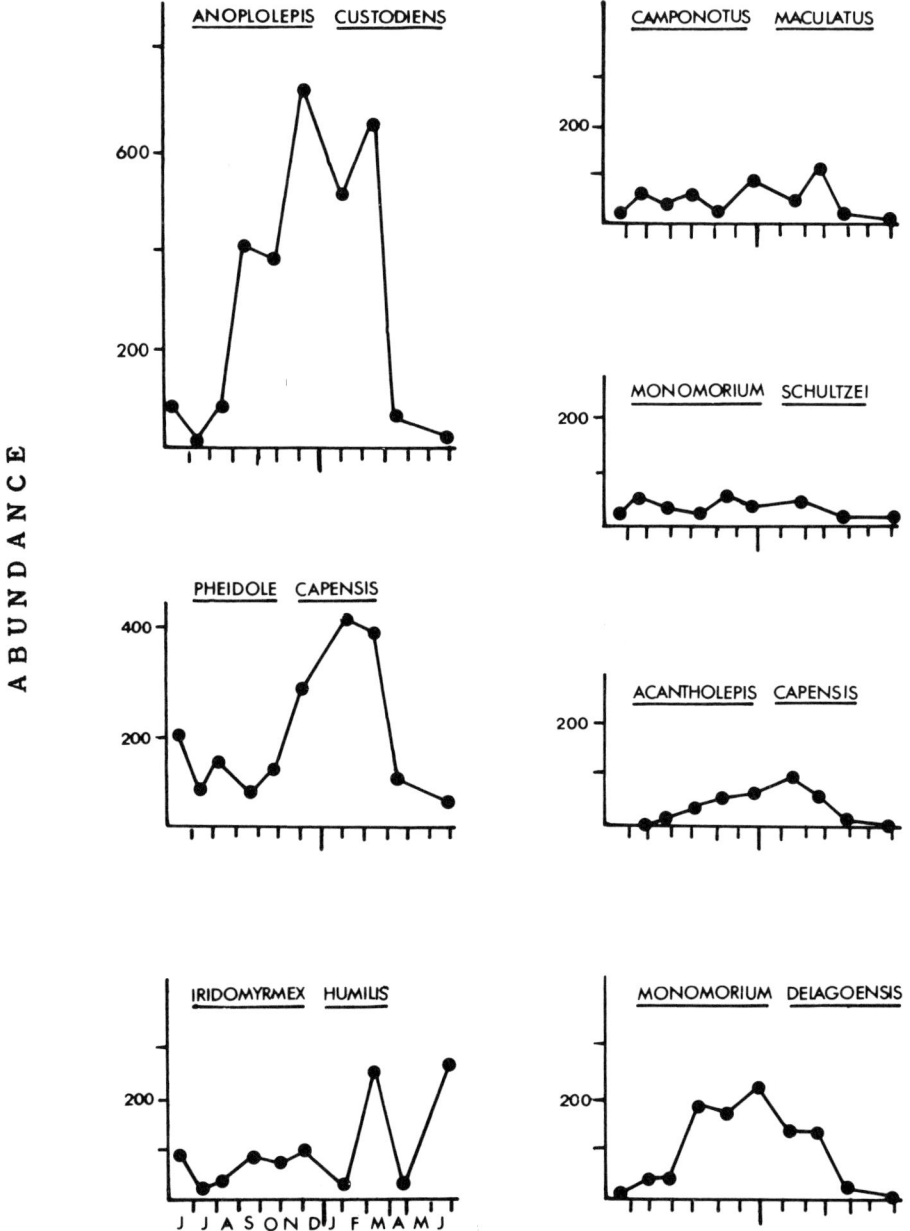

Fig. 3. Temporal variation in catch of the seven most common ant species obtained from pitfall traps in the Jonkershoek Valley, South Africa (from Donnelly 1983).

greater abundance of Prostigmata than Oribatida was found in the thorny scrub and the reverse was the case in temperate rainforest.
- Oribatida and Uropodina mites were typical of the temperate rainforest, especially in the litter or moss layer.
- Changes in abundance of Gamasina mites appeared to follow changes in prey abundance such as springtail, proturans, oribatids and larvae of insects.
- Acaridida and Tarsonemini mites were most abundant in the xerophytic environments.

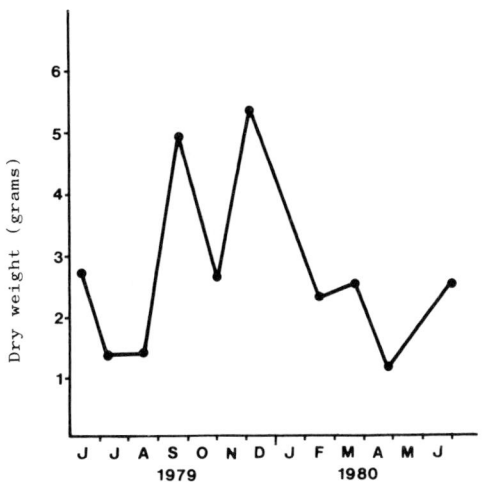

Fig. 4. Biomass of the arthropods caught in pitfall traps in the Jonkershoek Valley, South Africa, during the years 1979 and 1980 (from Donnelly 1983).

- The total number of Collembola increased with humidity (thorny scrub to temperate rainforest). Poduromorpha were dominant in xerophytic environments and Entomobryomorpha in the rainforest.
- Protura were concentrated in the rainforest, particularly at lower levels.
- Psocoptera were most abundant in the xerophyllous environments.
- The community diversity increased trom thorny scrub to temperate rainforest and this was principally due to increases of Oribatida, Uropodina and Gamasina.
- Staphylinidae, Pselaphidae and Ptiliidae were the most abundant families of Coleoptera.
- Seasonal changes in arthropod abundance were less affected by variations in climate in the humid environments than in the drier ones (Fig. 5). This is due to the buffering of microclimate by the forest canopy.

2.5.2 Epigaeic fauna
Epigaeic arthropods were collected by pitfall traps run every 45 days from August, 1967 to December, 1968. A total of 13 traps was used: three in the thorny scrub, four in the xerophytic scrub and six in the rainforest. The traps consisted of metal cylinders of 10 cm diameter and 15 cm depth, containing five percent formaldehyde.

The mean number of individuals caught per trap over the annual period of the research is given in Table 7. Acarina were not recorded and the numbers of Collembola do not represent the real density of this group on the surface of the soil. The principal conclusions are as follows:
- Diptera, Coleoptera and Araneae were the most abundant groups in traps run in the three environments. Diplopoda were particularly abundant in traps from the rainforest and Isopoda in traps from the xerophytic scrub (Sáiz 1975b).
- Diplopoda characterised the rainforest and often comprised more than 40% of the catch of epigaeic fauna. The Sphaerotrichopidae, containing the species *Chilorus ovallenus* and *Oligodesmus michelbacheri,* comprised 97.5% of the Diplopoda catch in the rainforest (Silva and Sáiz 1975).
- The taxonomic composition of the Araneae of the rainforest was different from that of the two scrub ecosystems (Sáiz and Calderon 1976).
- The percentage composition of the predominant Coleoptera caught in each community was:

Thorny scrub:
Ptinidae	74.4%
Tenebrionidae	15.7%
Others	9.9%

Xerophytic scrub:
Staphylinidae	28.5%
Ptinidae	23.9%
Curculionidae	14.4%
Lathridiidae	12.8%
Others	20.4%

Temperate rainforest:
Scaphidiidae	41.4%
Staphylinidae	31.6%
Lathridiidae	8.0%
Others	19.0%

A detailed study of staphylinids is reported in Sáiz (1971).
- The rainforest catch produced the greatest number of species and individuals of beetles.
- In the thorny scrub the Ptinidae largely comprised one species, while in the Tenebrionidae, *Apocrypha baloghi, Nycterinus rugiceps* and *Praocis spinolai* were the dominant species.
- In the xerophytic scrub the most important spe-

cies of Coleoptera were *Atheta obscuripennis* and *Bolitobius seriaticollis* (Staphylinidae), *Melanophthalma australis* (Lathridiidae) and *Puranius* sp. (Curculionidae).
– In the rainforest, one species of Scaphidiidae, *Eudera sculptilis* and *Loncovilius discoideus* (Staphylinidae) and *Aridius subfasciatus* (Lathridiidae) were the most abundant.
Another seasonal study of epigaeic Coleoptera at different seasons in three environments in central

Table 6. Total numbers of soil and litter arthropods sampled from August, 1967 to December, 1968 by funnels in Fray Jorge National Park, Chile. A = litter or moss; B = 0–5 cm soil depth; C = 5–10 cm soil depth.

Taxon	Thorny scrub Matorral			Xerophytic scrub Matorral			Temperate rainforest- Cloud forest			Total
	A	B	C	A	B	C	A	B	C	
Arachnida										
Pseudoscorpionida	–	2	3	1	6	5	33	76	30	156
Acarina										
Prostigmata	1038	817	483	389	792	737	531	415	243	5445
Gamasina	21	19	9	70	166	123	360	434	239	
Uropodina	–	–	–	11	16	35	211	422	148	843
Tarsonemini	415	165	12	121	226	72	279	414	177	1881
Acaridida	187	88	33	23	75	32	96	194	68	796
Oribatida	164	622	256	510	1598	710	3941	2888	1623	12312
Araneae	4	5	–	1	2	–	12	3	1	28
Crustacea										
Isopoda	1	–	–	9	17	1	7	12	1	48
Diplopoda	–	–	–	5	7	2	29	27	12	82
Chilopoda	–	1	–	2	2	1	1	2	3	12
Symphyla	–	–	–	–	1	–	–	–	1	2
Pauropoda	1	14	3	7	19	8	–	1	3	56
Collembola										
Entomobryomorpha	16	47	6	89	245	143	472	1326	721	3065
Poduromorpha	168	118	23	19	38	6	14	135	230	751
Symphypleona	4	1	–	1	20	2	37	23	7	95
Protura	–	1	–	1	–	9	1	83	138	233
Diplura	–	–	–	–	–	1	–	–	–	1
Insecta										
Thysanura	3	–	–	4	4	–	–	–	–	11
Psocoptera	61	53	27	76	54	21	2	2	2	298
Homoptera	5	5	2	74	36	53	46	114	58	393
Heteroptera	–	–	–	4	–	–	3	–	–	7
Thysanoptera	72	7	–	35	8	4	2	–	–	128
Coleoptera adults										
Staphylinidae	1	–	1	2	–	1	10	9	6	30
Pselaphidae	–	–	–	1	1	–	–	6	2	10
Dermestidae	–	–	–	1	–	–	–	–	–	1
Lathridiidae	–	1	–	–	1	–	–	–	–	2
Byrrhidae	–	–	–	–	2	–	–	–	–	2
Ptiliidae	2	–	–	–	–	1	4	6	3	16
Curculionidae	1	–	–	–	–	1	1	–	1	4
Scaphidiidae	1	–	–	–	–	–	2	–	–	3
Cryptocephalidae	1	–	–	–	–	–	–	–	–	1
Pedilinae	–	1	–	–	–	–	–	–	–	1
Cerambycidae	–	–	1	–	–	–	–	–	–	1
Anobiidae	–	–	1	–	–	–	–	–	–	1
Scydmaenidae	–	–	–	–	–	–	–	1	1	2
Alticidae	–	–	–	–	–	–	–	–	1	1
Coleoptera larvae	16	4	7	1	1	5	3	12	12	61
Diptera adults	–	2	1	1	1	–	2	3	4	14
Diptera larvae	5	18	43	17	27	24	26	21	45	226
Lepidoptera larvae	1	–	–	1	–	–	1	1	–	4
Hymenoptera Formicidae	–	–	–	2	3	1	–	1	1	8
Hymenoptera others	–	–	3	–	–	–	1	1	1	6
Total	2188	1991	914	1478	3368	1998	6127	6632	3782	28478

Data from Sáiz (1975a).

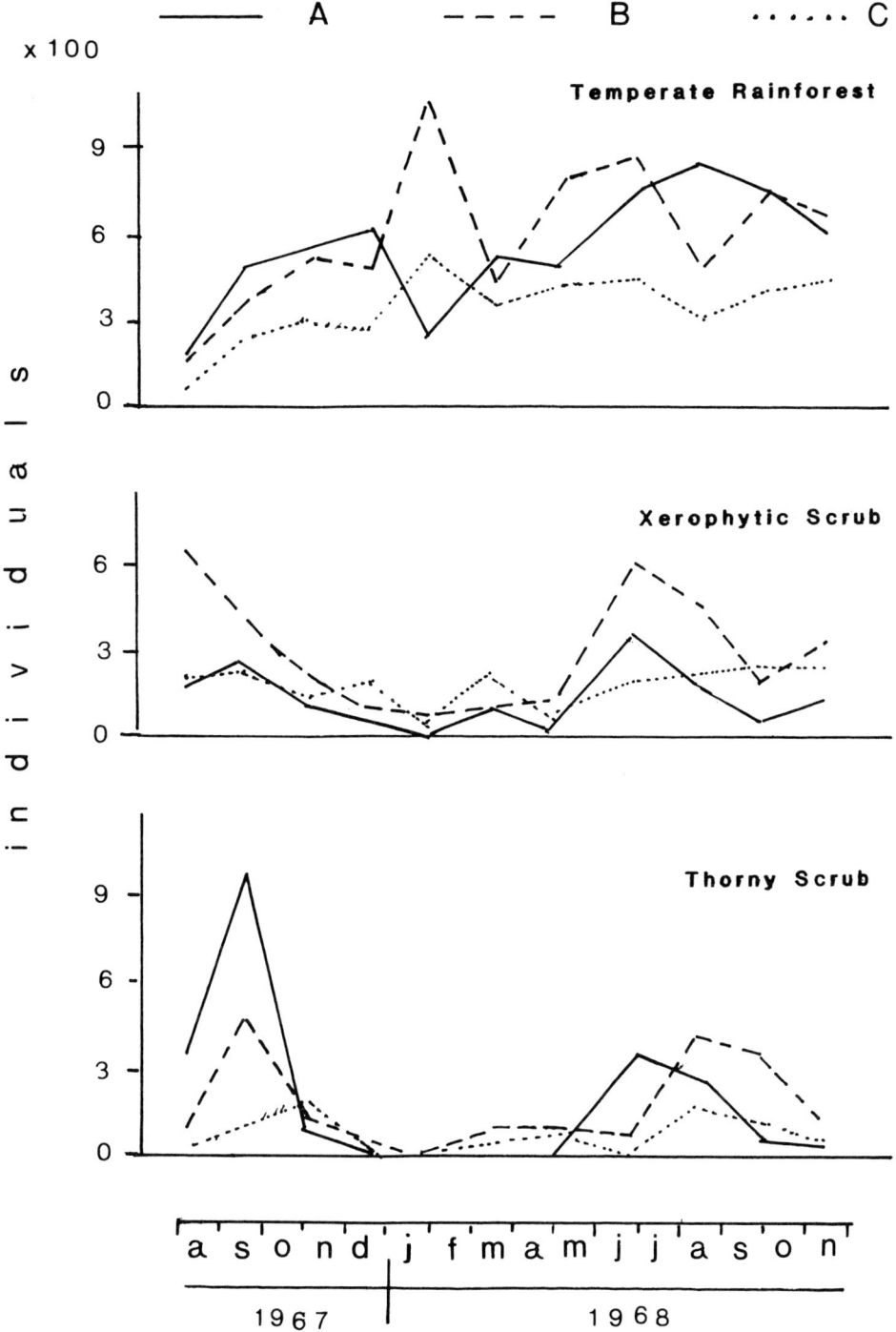

Fig. 5. Seasonal changes in density of soil and litter fauna in Fray Jorge National Park, Chile. A = moss or litter; B = 0–5 cm soil depth; C = 5–10 cm soil depth (from Sáiz 1975a).

Table 7. Mean number of arthropods per pitfall trap run from August, 1967 to December, 1968 in Fray Jorge National Park, Chile.

Taxon		Temperate rainforest, Cloud forest	Xerophytic scrub Matorral	Thorny scrub Matorral	Total
Arachnida	Pseudoscorpionida	16.3	1.3	12.3	29.9
Arachnida	Opilionida	1.5	3.5	–	5.0
Arachnida	Solpugida	–	7.8	8.6	16.4
Arachnida	Araneae	338.3	92.8	110.0	541.1
Arachnida	Scorpionida	–	0.8	0.6	1.4
Crustacea	Isopoda	52.8	110.3	3.0	166.1
Diplopoda	Sphaerotrichopidae	357.0	24.3	–	381.3
Diplopoda	Polydesmidae	47.0	–	–	47.0
Diplopoda	Siphonotidae	–	3.0	9.3	12.3
Diplopoda	Polyxenidae	4.0	1.3	–	5.3
Chilopoda		2.0	2.0	1.6	5.6
Collembola		23.3	12.0	80.0	115.3
Insecta	Thysanura	–	3.0	58.0	61.0
Insecta	Orthoptera	0.8	24.3	36.0	61.1
Insecta	Psocoptera	2.2	7.3	4.6	14.1
Insecta	Homoptera	1.8	27.0	53.0	81.8
Insecta	Heteroptera	41.0	10.3	5.3	56.6
Insecta	Thysanoptera	–	30.8	2.6	33.4
Insecta	Siphonaptera	–	0.3	0.3	0.6
Insecta	Diptera adults	1477.2	169.5	301.6	1948.3
Insecta	Diptera larvae	46.0	5.3	4.6	55.9
Insecta	Lepidoptera adults	3.3	5.8	12.0	21.1
Insecta	Lepidoptera larvae	32.5	11.8	9.6	53.9
Insecta	Hymenoptera Formicidae	36.3	81.8	100.6	218.7
Insecta	Hymenoptera others	71.3	18.8	22.6	112.7
Insecta	Coleoptera larvae	25.0	15.8	1.6	42.4
Insecta	Coleoptera Ptiliidae	35.3	1.3	–	36.6
Insecta	Coleoptera Scaphidiidae	439.3	9.8	0.3	449.4
Insecta	Coleoptera Staphylinidae	336.3	31.5	3.0	370.8
Insecta	Coleoptera Pselaphidae	8.8	1.3	–	10.1
Insecta	Coleoptera Lathridiidae	84.6	36.8	2.3	123.7
Insecta	Coleoptera Colydiidae	35.8	–	–	35.8
Insecta	Coleoptera Tenebrionidae	9.0	12.0	48.0	69.0
Insecta	Coleoptera Ptinidae	0.5	68.5	218.3	287.3
Insecta	Coleoptera Alticidae	26.8	0.8	–	27.6
Insecta	Coleoptera Curculionidae	39.0	41.0	7.6	87.6
Insecta	Coleoptera Carabidae	0.2	3.0	1.0	4.2
Insecta	Coleoptera Catopidae	1.9	–	–	1.9
Insecta	Coleoptera Nilionidae	12.5	–	–	12.5
Insecta	Coleoptera Scydmaenidae	2.5	2.0	0.3	4.8
Insecta	Coleoptera Cantharidae	9.0	–	–	9.0
Insecta	Coleoptera Phengodidae	0.2	–	–	0.2
Insecta	Coleoptera Ostomidae	0.2	0.3	2.0	2.5
Insecta	Coleoptera Cleridae	0.2	–	5.6	5.8
Insecta	Coleoptera Elateridae	1.0	–	0.6	1.6
Insecta	Coleoptera Byrrhidae	8.0	3.3	3.0	14.3
Insecta	Coleoptera Cryptophagidae	1.5	0.3	3.0	4.8
Insecta	Coleoptera Erotylidae	1.1	13.2	–	14.3
Insecta	Coleoptera Murmidiidae	–	–	1.6	1.6
Insecta	Coleoptera Discolomidae	0.2	–	0.3	0.5
Insecta	Coleoptera Anthicidae	–	–	0.6	0.6
Insecta	Coleoptera Melandryidae	4.0	7.3	–	11.3
Insecta	Coleoptera Salpingidae	0.6	0.3	–	0.9
Insecta	Coleoptera Alleculidae	0.3	–	–	0.3
Insecta	Coleoptera Anobiidae	0.2	–	1.0	1.2
Insecta	Coleoptera Trogidae	0.6	1.1	1.6	3.3
Insecta	Coleoptera Aphodiinae	–	1.0	–	1.0
Insecta	Coleoptera Cerambycidae	–	1.5	0.3	1.8
Insecta	Coleoptera Chrysomelidae	1.5	–	4.3	5.8
Insecta	Coleoptera Criocerinae	0.2	–	–	0.2
Insecta	Coleoptera Cryptocephalinae	–	–	0.3	0.3
Insecta	Coleoptera Scolytinae	0.5	–	0.3	0.8
Total		3641.4	907.2	1143.1	5691.7

Data from Sáiz (1975b), Sáiz (unpubl. data), Silva and Sáiz (1975).

Chile (evergreen forest, deciduous forest and matorral), over two years of research is reported in Sáiz (1977). It refers mainly to the families Staphylinidae, Carabidae, Tenebrionidae and Curculionidae. The information is compared with similar environments in California, which were sampled simultaneously.

2.6 The composition of epigaeic fauna in California

M. Narog

Epigaeic invertebrates were collected by pitfall traps in the Tanbark Flat and Wolfskill areas of the U.S. Forest Service Experimental forest, San Dimas, California (34°06′ N, 117°48′ W). These areas have not been burnt for at least 25 years. Three types of vegetation were studied:
(a) *Quercus chrysolepis,* Canyon live oak;
(b) *Ceanothus crassifolius,* California lilac; and
(c) *Adenostoma fasciculatum,* Chamise.

Two transects of five 1.8 cm diameter pitfall traps containing an alcohol/glycerol preservative were placed out in each vegetation association. Traps were opened for five days in each plot during February 1984.

The total number of individuals caught for each of the three vegetation associations is shown in Table 8. The most abundant group in the catch was the Collembola. The Diptera were probably attracted to the traps. There were fewer ants than Coleoptera trapped which is a result of trapping being carried out in winter.

2.7 The composition of epigaeic fauna in mediterranean southern France

Penelope Greenslade

The epigaeic fauna was sampled at four sites. Two were at Valmy, 20 km east of Banyuls (42°29′ N, 3°08′ W), and consisted of a burnt and an unburnt plot, both vegetated with cork oak (*Quercus suber*). The other two were near the Mas d'Abeilles, 8 km north-west of Banyuls. One of these was beside a creek which carried evergreen oak (*Quercus ilex*) and the other was about 50 m away on a hillside covered by dense maquis-type vegetation. The last area had been burnt in 1976 by a wildfire.

Each site was sampled for six days by 1.8 cm diameter pitfall traps containing an alcohol/glycerol preservative.

In this summer sample, Collembola were the most abundant group followed by Acarina, Formicidae and adult Diptera (Table 9). Few Coleoptera were trapped. The Diptera were probably attracted to the preservative in the traps. There was little indication of any differences between the burnt and unburnt sites at Valmy at the level of order. More differences were shown between the two sites at Mas d'Abeilles, which carried quite different vegetation. There were more groups trapped which

Table 8. Mean number of arthropods sampled by pitfall traps in three Californian vegetation associations sampled during February 1984.

Taxon		Canyon live oak	California lilac	Chamise	Total
Arachnida	Opilionida	–	–	1	1
Arachnida	Acarina	3	13	11	27
Arachnida	Araneae	4	6	3	13
Diplopoda		–	1	–	1
Collembola		32	23	10	65
Insecta	Dermaptera	–	2	–	2
Insecta	Embioptera	2	–	–	2
Insecta	Psocoptera	–	2	1	3
Insecta	Hemiptera	–	1	3	4
Insecta	Coleoptera adults	18	11	5	34
Insecta	Coleoptera larvae	6	1	–	7
Insecta	Diptera adults	25	39	26	90
Insecta	Diptera larvae	2	–	1	3
Insecta	Lepidoptera adults	–	–	1	1
Insecta	Hymenoptera Formicidae	2	3	3	8
Insecta	Hymenoptera others	–	2	–	2
Total		94	104	65	263

were typical of leaf litter (Collembola, Diplopoda, Pseudoscorpionida, Protura and Diplura), in the oak site and more groups typical of open areas (Formicidae, Orthoptera) at the maquis site.

3 Individual taxa treatments

3.1 The Collembola

Penelope Greenslade

Information on Collembola from mediterranean ecosystems of the world is extremely patchy. Data on composition and density of faunas from Western and South Australia, and also from southern France have recently become available. Comparisons can be made here since similar sampling methods were used. However, the only data from South Africa and California that are suitable for making comparisons are those on the relative abundance of this taxon in pitfall traps (Table 10). Di Castri and Di Castri (1981) have reviewed the literature on microarthropods of mediterranean ecosystems in detail. Consequently in this discussion I shall concentrate on French/mediterranean comparisons based on more recently published data.

3.1.1 Abundance

Soil and litter fauna have been studied in southern France by Poinsot-Balaguer and Tabone (1985, 1986) and in the south of Western Australia by Postle (1986 and this chapter) using funnel extraction. Both sites which were examined were open woodland or forest; in France these communities carried evergreen and white oak and in Australia they carried *Eucalyptus* species. At both sites soil animals, including the Collembola, were more abundant in the soil than in the leaf litter; these being, on average, four times more so in France and forty times more so in Australia. The differences may reflect slightly different ways in which the samples were taken and extracted. The total density of Collembola (to a depth of 5 cm) ranged from 2,000–50,000 m^{-2} in France and 1,600–13,000 m^{-2} in Australia, depending on season. Collembola were the second most abundant group after the

Table 9. Total number of arthropods sampled by pitfall traps in four plots in southern France. Samples were taken during June 1985.

Class	Order	Valmy		Mas d'Abeilles		Total
		Burnt	Unburnt	Oak	Maquis	
Annelida		–	1	–	–	1
Arachnida	Pseudoscorpionida	2	–	2	–	4
Arachnida	Opilionida	–	–	6	6	12
Arachnida	Acarina	80	43	95	129	347
Arachnida	Araneae	6	3	3	6	18
Crustacea	Isopoda	1	–	2	–	3
Diplopoda		1	2	13	–	16
Chilopoda		–	1	–	–	1
Collembola		146	132	177	107	562
Protura		–	–	1	–	1
Diplura		–	–	3	–	3
Insecta	Dermaptera	10	1	–	12	23
Insecta	Orthoptera	1	–	–	–	1
Insecta	Embioptera	1	–	–	–	1
Insecta	Psocoptera	1	4	1	2	8
Insecta	Hemiptera	–	1	4	1	6
Insecta	Thysanoptera	1	2	1	–	4
Insecta	Coleoptera adults	2	1	5	7	15
Insecta	Coleoptera larvae	–	–	–	1	1
Insecta	Diptera adults	21	36	47	45	149
Insecta	Diptera larvae	–	3	2	–	5
Insecta	Lepidoptera adults	–	–	1	–	1
Insecta	Lepidoptera larvae	–	–	–	1	1
Insecta	Hymenoptera Formicidae	33	7	10	95	145
Insecta	Hymenoptera others	–	2	3	1	6
Total		306	239	376	413	1334

mites; in France about a third of microarthropod individuals were Collembola with mites comprising one half. In Australia, the mites were numerically dominant comprising at least five sixths of all microarthropods while Collembola consisted at most of a fifth, and usually less than a tenth, of the total fauna. Poinsot-Balaguer and Tabone (1985) compared relative abundances in detail and include some data from Chile. Hutson and Kirkby (1985) found a similar dominance of mites over Collembola in soil and leaf litter on sites in South Australia, with Collembola being only a fifth as numerous as the mites and varying in density from about 5,000–10,000 m^{-2}. At all three vegetation types studied in Chile (Table 6), Collembola were the second most abundant order in soil and litter, while the mites

Table 10. Composition of pitfall trap catches in different mediterranean regions.

Taxon	Western Australia forest[a]	South Australia forest[b]	South Africa fynbos[c]	South France maquis[d]	California lilac[e]	Chile matorral[f]
Nematoda	–	–	–	–	–	–
Annelida	–	–	–	–	–	–
Onychophora	–	–	–	–	–	–
Arachnida Scorpionida	–	–	–	–	–	11
Arachnida Pseudoscorpionida	–	–	34	–	–	41
Arachnida Opilionida	–	–	20	6	–	4
Arachnida Acarina	358	118	970	129	13	*
Arachnida Solpugida	–	–	65	–	–	8
Arachnida Araneae	90	9	676	6	6	286
Crustacea Isopoda	7	–	29	–	–	10
Crustacea Amphipoda	–	–	150	–	1	–
Diplopoda	3	–	43	–	–	29
Chilopoda	–	1	31	–	–	7
Symphyla	–	–	1	–	–	–
Collembola	1868	82	13766	107	23	*
Insecta Archaeognatha	–	–	101	–	–	–
Insecta Thysanura	1	4	4	–	–	53
Insecta Odonata	–	–	–	–	–	–
Insecta Blattodea	3	–	66	–	–	15
Insecta Mantodea	–	–	–	–	–	1
Insecta Isoptera	50	–	21	–	–	–
Insecta Dermaptera	296	81	69	12	2	–
Insecta Orthoptera	125	–	247	–	–	91
Insecta Phasmatodea	–	–	1	–	–	1
Insecta Psocoptera	61	1	19	2	2	8
Insecta Hemiptera	52	2	351	1	1	1411
Insecta Thysanoptera	27	3	23	–	–	6
Insecta Neuroptera adults	–	–	1	–	–	–
Insecta Neuroptera larvae	2	–	2	–	–	3
Insecta Coleoptera adults	268	25	1095	7	11	1213
Insecta Coleoptera larvae	2	1	180	1	1	106
Insecta Strepsiptera	–	–	2	–	–	–
Insecta Siphonaptera	–	–	1	–	–	–
Insecta Mecoptera	1	–	–	–	–	–
Insecta Diptera adults	52	6	3294	45	39	991
Insecta Diptera larvae	11	–	–	–	–	161
Insecta Lepidoptera adults	–	–	11	–	–	–
Insecta Lepidoptera larvae	–	–	4	1	–	76
Insecta Hymenoptera Formicidae	540	229	8711	95	3	1577
Insecta Hymenoptera others	18	–	203	1	2	125
Total	3835	562	30191	413	104	6234

[a] From Postle (1986).
[b] From Greenslade (this chapter).
[c] From Donnelly and Giliomee (this chapter).
[d] From Greenslade (this chapter).
[e] From Narog (this chapter).
[f] From Sáiz (unpubl. data).
* These data were not recorded in the Chilean samples.

were on average ten times as numerous. Densities for Chile cannot be compared with other regions because the samples were taken to a different depth. Hutson and Veitch (1983) provide density data from sites in South Australia but it also is in a form which cannot be compared easily with other published data.

Winter pitfall catches in California showed Collembola to be the second most abundant order after Diptera while catches in South Africa and Western Australia indicated that Collembola were the most abundant group followed by ants and then mites (Table 10). In South Australia, they were the third most abundant after ants and mites; in Chile they were not counted.

In conclusion, it appears that in all regions and seasons Collembola generally are the second most abundant order of microarthropods in the soil and leaf litter, whatever sampling method is used.

3.1.2 General seasonality
Two peaks in abundance of Collembola in leaf litter and soil have been observed in Australia and France (Specht 1985, Poinsot-Balaguer and Tabone 1986, Postle this chapter). Usually the highest densities were either in the spring or autumn, while the winter density was lower. The lowest densities were found in summer. Highest numbers in France seemed to be in spring and winter and in Australia in autumn and winter (Poinsot-Balanguer and Tabone 1986, Postle this chapter). Considerable variation between years is illustrated by Hutson and Veitch's (1983) data.

Epigaeic fauna collected by pitfall traps in South and Western Australia followed the same trends (Greenslade and Majer 1980, Greenslade and Rosser 1984) with highest numbers in spring and autumn and lowest in winter and summer; again there was much variation between years.

3.1.3 Taxonomic composition

3.1.3.1 Size of fauna
Poinsot-Balaguer and Tabone (1986) found 37 species of Collembola in soil and leaf litter on their sites in southern France. These were all widely distributed species of which fifteen were cosmopolitan, six holarctic, 10 european and three mediterranean in occurrence. On sites in Western Australia (Postle 1986), over 50 species were found in the litter (four cosmopolitan) and 36 in soil (one cosmopolitan) given a total of 55 altogether (Penelope Greenslade unpublished data). In forest eucalypt at Kuitpo/Kyeema, South Australia, over 60 species of epigaeic Collembola have been collected in pitfall traps from a single site (Penelope Greenslade unpublished data). On similar sites in Western Australia, 57 species were recognised from pitfalls and only six had cosmopolitan distributions. The rest were either native or endemic to Australia (Greenslade and Majer 1980). Epigaeic Collembola were sampled for five years from low vegetation at Belair in South Australia. Twenty two species were found, of which three have cosmopolitan distributions (Penelope Greenslade 1985).

To conclude, it appears that at all levels the Collembola fauna in Australia is much more diverse than that in France, and has a much higher proportion of native and endemic species.

3.1.3.2 Seasonality of individual groups
Information collected from pitfall traps and funnel extraction of leaf litter in South and Western Australia has shown that there are three distinct faunas present during the year (Greenslade 1974, 1977, 1985, 1986, Greenslade and Majer 1980). The spring and autumn faunas are similar in composition while different groups of species are present in winter and summer.

There are no data on the change in composition of the fauna with season in France. The vegetation-associated fauna in Australia has no distinct winter species, there being only summer and autumn/winter/spring groups of species (Penelope Greenslade 1985). The summer species have taxonomic affinities with species found in the arid and semi-arid parts of Australia while the autumn/winter/spring fauna is related to species from cool temperate, southern regions. The spring and autumn species are typical of *Eucalyptus* woodland in Australia.

Data collected using the same sampling method (pitfall traps, described by Greenslade and Greenslade 1973) during summer indicate that Australian genera have ecological and taxonomic analogues in

France (Table 11). Eighteen species, represented by 385 individuals, were collected in Australia (1.0 indivs trap^{-1} day^{-1}). Since numbers trapped are dependent on activity, they vary with the weather. Composition by family was similar in France and Australia, with most individuals being from the family Entomobryidae. More poduromoph forms were collected in France.

3.1.3.3 Adaptations
Information on the life history and behaviour of Collembola of mediterranean areas is sparse; in Australia most species still have to be described. The few known examples are given below. Genera of epigaeic smithurid Collembola found in France and Australia in summer belong to the tribe Bourletiellini. These species possess well developed tracheae which may assist in respiration under dehydrating conditions by permitting development of a more impermeable cuticle. The genera possessing these characteristics are *Deuterosminthurus* in France and *Corynephoria* in Australia (Greenslade 1977 and unpublished data).

The lucerne flea, *Sminthurus viridis*, is a much studied species and is typical of mediterranean areas. It is native to Europe and has been accidentally introduced into South Africa, South America and Australia (Wallace 1967, 1968, Wallace and Mahon 1971) where it has spread to all regions experiencing a mediterranean climate. It is a winter-active species and survives the severe dry conditions of summer as a desiccation resistant egg. *Entomobrya unostrigata* is native to Europe and has also been introduced to Australia and California. It appears to have behavioural adaptions in which it survives the summer by aggregating in micro-sites of high humidity, under stones or even in houses (Penelope Greenslade unpublished data).

Certain species of Hypogastruridae and Isotomidae in France and Australia exhibit ecomorphosis. This entails an ecdysis into a morphologically distinct form which is inactive and passes the summer deep in the soil. Cassagnau (1986) has recently reviewed the literature on this topic. Other species of Isotomidae (e.g. *Folsomides* spp.), although characteristic of arid areas, are also found at low abundances in localities experiencing a mediterranean climate. They may pass dry periods in an anhydric inactive condition (Greenslade 1974, Poinsot 1968).

3.1.4 Resilience
A few weeks after fire there is often a peak in abundance of Collembola and it has been suggested that this is the result of increased microbial biomass produced by the flush of available nutrients (Trabaud 1984). In Australia this peak in abundance is illustrated by pitfall catches from South Australia (Greenslade and Rosser 1984) although part of the peak here is probably due to increased captures caused by the reduction in the litter and ground layer vegetation. The effect of this fire persisted for at least two years. Hutson and Kirkby (1985) observed little difference in the abundance of soil and litter Collembola by the fourth year after a fire but there are no data on any change in species composition from this work.

When comparing the Collembola trapped in pitfall traps on burnt and unburnt sites in southern France (Table 9), the total numbers of individuals and species trapped were similar while the species composition was different and could be related to the differing amounts of leaf litter or ground cover present (Penelope Greenslade unpublished data). This effect was evident three years after the fire. For instance, on a site carrying cork oak (Valmy, Table 9), litter species of *Sphaeridia* and *Sminthuri-*

Table 11. Genera of Collembola collected in pitfall traps from sites in southern France and South Australia in summer.

Valmy, France	Kuitpo/Kyeema, South Australia
SMINTHURIDAE	
Sphaeridia	*Corynephoria*
Prorastriopes	
Deuterosminthurus	
Sminthurinus	
ENTOMOBRYIDAE	
Entomobrya	*Drepanura*
Lepidocyrtus	*Lepidobrya*
Tomocerus	*Lepidocyrtoides*
Orchesella	*Acanthocyrtus*
	Willowsia
ISOTOMIDAE	*Acanthomurus*
	Cryptopygus
HYPOGASTRURIDAE	
Xenylla	*Xenylla*

nus, immature Entomobryidae and *Tomocerus* were only present on the unburnt site. Similarly, when comparing evergreen oak and neighbouring maquis vegetation (Mas d'Abeilles, Table 9), *Lepidocyrtus*, *Orchesella* and *Tomocerus* species were more abundant in the former vegetation. In contrast, *Deuterosminthurus*, *Prorastriopes* and *Entomobrya* species were only found in the recently burnt, more open, maquis vegetation.

3.2 The scarabaeine dung beetles

T.J. Ridsdill-Smith

In the Scarabaeidae, the coprophagous habit is most common in the Scarabaeinae and Aphodiinae. The taxonomy of the Aphodiinae is uncertain and so only the Scarabaeinae, which is the most specialised subfamily, is considered here. Most of the data presented were collected as part of a programme to introduce dung beetles into Australia to reduce accumulation of cattle dung on pastures. Scarabaeine beetles are beneficial in this respect since most species fly to fresh dung and may bury it for food or as provisions for their larvae. Only species known to occur in areas with mediterranean climates, defined here as winter rainfall and with a dry season of up to 200 days as outlined in the UNESCO-FAO maps (Anon 1963), are included. Regional faunas are described from areas known to contain undisturbed and partially disturbed natural habitats, as well as pastures, but it was not possible to separate species by their habitat preferences. Species recently introduced to Australia and to California from Africa and Europe are not discussed. The individual site in Australia was in an undisturbed habitat, while sites in other parts of the world were in partially disturbed habitats. Data are summarised by grouping species by tribes, since species within tribes show similarity in methods of provisioning of their larvae.

3.2.1 Dung beetle faunas of mediterranean climate areas

The composition of the dung beetle fauna varies between the different areas with mediterranean-type climates. The dung beetle fauna of the southwest Cape of South Africa consists of 46 species of which 43% are Scarabaeini (predominantly ball-rolling species) (Table 12). Four other tribes are represented. A small number of these species have distributions restricted to the winter rainfall area, but most have distributions which include areas with summer rainfall (A. Davis, pers. comm.).

There are 17 species in the south of Western Australia, of which 59% are Scarabaeini and 41% are Onthophagini (Table 12). These species are all endemic to the south of Western Australia (Matthews 1972, 1974). In contrast the species in the

Table 12. Numbers of dung beetle species listed by tribes from mediterranean climate areas of the world.

Tribe	Number of species							
	Africa	Australia		Europe			America	
	SW Cape[a]	SW[b]	SE[c]	France[d]	Corsica[e]	Spain[f]	California[g]	Chile[h]
Onthophagini	6	7	19	17	7	29	1	–
Onitini	7	–	–	1	2	7	–	–
Oniticellini	4	–	–	2	1	3	1	–
Coprini	9	–	–	2	2	2	–	1
Scarabaeini	20	10	–	3	3	11	1	1
Total	46	17	19	25	15	52	3	2

[a] From A. Davis (pers. comm.).
[b] From Matthews (1972, 1974), Ridsdill-Smith *et al.* (1983).
[c] From Matthews (1972).
[d] From Lumaret (1977).
[e] From Lumaret (1980).
[f] From Baguena (1967), Martin (1982).
[g] From A. Hardy (pers. comm.).
[h] From H. Howden (pers. comm.).

mediterranean areas of South Australia are all Onthophagini (Table 12). A number of these species have distributions extending outside the winter rainfall areas (Matthews 1972).

In the Mediterranean basin there are 25 dung beetle species in the garrigue area of southern France, 15 species in the maquis area of Corsica and 52 species in Spain (Table 12). The Onthophagini make up 68% of the fauna in the garrigue in France, 47% in the maquis in Corsica and 56% in Spain (Table 12). The Scarabaeini were the next most abundant tribe, and species of Onitini, Oniticellini and Coprini were present in all three areas. These faunas include species whose distributions spread outside winter rainfall areas of Europe, especially into temperate climatic areas (Lumaret 1977, Martin 1982). California has three species from three different tribes (A. Hardy, pers. comm.) and Chile has two species from two different tribes (H. Howden, pers. comm.) (Table 12).

3.2.2 Seasonal occurrence and abundance of dung beetles in different mediterranean climate areas
Data on seasonal occurrence and abundance have been collected from one site in undisturbed or partially disturbed open woodland or heath in each of four of the areas (Table 13). Dung was available mainly from marsupials at the site in Western Australia, but at the other sites dung was available from cattle, sheep or horses which were allowed to graze during the year. Pitfall traps baited with fresh dung were set at each site but, although general methods were similar, there were variations in numbers and size of traps, time of exposure, and dung type.

Mean numbers of beetles per trap are shown for three-monthly periods to indicate seasonal activity (Table 14). Some data which indicate the seasonal activity patterns are also given for beetles hand-collected throughout the year at the site in Chile at Concon. No further data are available about the presence of other beetles at this site.

At Geelbeck in South Africa, Scarabaeini was the most abundant tribe in terms of individuals and numbers of species (Table 14). Onthophagini was abundant in terms of individuals, but not species. Within each tribe one species dominated catches.

Scarabaeus rugosus accounted for 86% of the Scarabaeini, and 98% of the Onthophagini were *Onthophagus minutus* (A. Davis, pers. comm.). Scarabaeini were the most abundant in spring, and Onthophagini in winter and spring (Table 14).

At Watheroo in Western Australia the Onthophagini were dominant in terms of individuals and numbers of species. *Onthophagus rupicapra* formed 74% of the Onthophagini, and *Mentophilus hollandiae* was the only species in the tribe Scarabaeini (Ridsdill-Smith and Hall, unpubl. data). Beetles of both tribes were trapped mainly in autumn and winter (Ridsdill-Smith and Hall 1984b) (Table 14).

At Cazevieille in France the Onthophagini were most abundant in terms of numbers and species (Table 14). *Onthophagus lemur* accounted for 57% of the Onthophagini, and 97% of the Scarabaeini were *Scarabaeus laticollis* (A. Kirk, pers. comm.). Onthophagini were most abundant in spring and summer, the Oniticellini in summer, and Scarabaeini and Coprini in autumn and spring (Table

Table 13. Site information for dung beetle study sites.

	Geelbeck (S. Africa) 33°06′ S 18°03′ E	Watheroo (W. Australia) 30°12′ S 115°50′ E	Cazevieille (France) 43°46′ N 03°47′ E	Concon (Chile) 32°56′ S 71°33′ W
Vegetation[a]	Fynbos	Scrub-heath	Garrigue	Scrub
Soil[b]	Albic arenosol Qa7-1a	Ferralic arenosol Qf43-1b	Eutric cambisol Bel41-2/3b	Eutric regosol Rel-1b
Climate[c]	Thermomed	Xerothermomed	Mesomed	Xerothermomed
No. dry days	125–150	150–200	40–75	150–200
Mean annual temp. (°C)	17	18	13	14
Annual rainfall (mm)	290	463	978	490
Sampling period	Dec. 78–Nov. 79	June 82–May 84	Jan. 79–Dec. 79	Oct. 76–Apr. 78
Dung bait used	Cattle	Human	Cattle	Cattle

[a] From R. Specht (in prep.).
[b] From Anon. (1978), [c] from Anon. (1963).

14). Most beetles were trapped in spring, and few in winter. Lumaret (1977) collected more species of Scarabaeinae in the spring than in other seasons in the garrigue region of France.

The most widespread scarabaeine dung beetle in California is *Canthon simplex* of the tribe Scarabaeini (A. Hardy, pers. comm.). No Scarabaeinae were caught in dung baited traps set in oak savanna at Browns Valley (39° 12′ N, 121° 24° W) between 1971 and 1973 (Merritt and Anderson 1977). The species of Scarabaeini collected at Concon in Chile is *Megathopa villosa* (Ovalle and Solervicens 1980). It was abundant in spring (Table 14).

3.2.3 Discussion

There is little evidence that climate is the prime determinant of species composition of local dung beetle faunas since there were many species in the Ethiopian region, and very few in the Neotropical and Nearctic regions, all with similar mediterranean climates. Species composition varied between the two different mediterranean climate areas of the Australian region, and between three different areas in the Palaearctic region. Among the subfamily Scarabaeinae, species in the tribes Scarabaeini and Onthophagini were most common at all sites with a different species being most numerous in each tribe at each site. Scarabaeini were trapped more frequently in the Ethiopian, Nearctic and Neotropical regions and Onthophagini in the Australian and Palaearctic regions. The greatest proportion of beetles were trapped in spring, autumn or winter at all sites. Recently several scarabaeine dung beetles, which are trapped mostly in the summer in Africa and Europe, have been successfully introduced to mediterranean climate areas of Australia (Ridsdill-Smith and Hall 1984a) and California. Therefore the patterns of seasonal occurrence of beetles reported here at each site are probably species limited and not climate limited.

Species composition in an area results from a number of factors, including the movement of species from adjacent areas with contrasting climates (or deliberate introductions), the vegetation and habitat, the dung type and availability and the soil type. The importance of habitat is difficult to separate from the other factors, but there is evidence that dung beetles endemic to the south of Western Australia do not move readily from undisturbed habitats into cleared pasture land (Ridsdill-Smith and Hall 1984a). The species composition reported here from natural habitats would therefore be expected to differ to some extent from that which occurs in pasture.

Table 14. Seasonal activity of dung beetles trapped over 1 or 2 years at single sites.

Country (site)	Tribe	Mean number beetles/trap				Abundance %	No. of species
		Winter	Spring	Summer	Autumn		
Africa (Geelbeck)	Onthophagini	5	4	2	2	31	3
	Onitini	1	1	1	1	9	4
	Oniticellini	1	1	1	1	9	3
	Coprini	2	1	–	1	9	4
	Scarabaeini	4	10	1	3	42	14
	Total	30%	40%	12%	18%	100%	28
Australia (Watheroo)	Onthophagini	16	2	–	21	83	3
	Scarabaeini	4	1	–	3	17	1
	Total	43%	6%	–	51%	100%	4
France (Cazevieille)	Onthophagini	1	122	35	13	77	6
	Oniticellini	–	–	3	1	2	1
	Coprini	1	2	–	3	3	1
	Scarabaeini	–	17	2	21	18	3
	Total	1%	64%	18%	17%	100%	11
Chile (Concon)	Scarabaeini	1	65	1	4	–	1

3.3 The Formicidae

J.D. Majer

3.3.1 Ant abundance

Pitfall traps run for intervals of time in mediterranean areas give some idea of the abundance of ants relative to that of other epigaeic fauna. Table 10 provides representative data from the various mediterranean regions. The actual numbers trapped are not directly comparable due to the differing trapping intensities and types of trap used. Nevertheless, the data give a good indication of the numerical prevalence of ants on the ground stratum in each region.

With the exception of the Californian site, ants were among the three most abundant ground-surface taxa. They were exceeded in abundance at some of the sites by Acarina and Collembola although these groups were not counted in the Chilean samples. It is not clear whether the low catch of ants in California is related to a genuine low abundance or to low activity levels during the winter sampling period.

3.3.2 Ant species richness

Comparison of ant species richness between regions is difficult in view of the differing sampling intensities used in the various studies which have been carried out. Hunt (1977) compared the Chilean and Californian ant faunas in four vegetation associations in each country. In Israel, Ofer et al. (1978) sampled ant faunas over 14 different habitats from the Mediterranean coast to the Jordan Valley. Tohmé (1969) studied the ant fauna in four different geographic zones of Lebanon and, although he does not present data for the individual study sites, a total checklist of ants for Lebanon is provided. Tohmé's (1969) data complement those of Ofer et al. (1978). In South Africa, Donnelly and Giliomee (1985b) sampled the ant fauna in the Jonkershoek Valley, in six fynbos vegetation sites representing differing ages after fire. P.J.M.

Table 15. Comparison of ant species richness in study sites within each mediterranean region.

Site no.	Western Australia		South Australia	South Africa	Israel	Lebanon	California	Chile
	open forest[a]	heath[b]	open forest[c]	fynbos[d]	transect[e]	various[f]	chaparral & forest[g]	matorral & forest[h]
1	25	31	23	19	14	–	9	6
2	31	30	21	26	11	–	15	7
3	23	–	–	26	20	–	45	16
4	35	–	–	32	16	–	21	10
5	24	–	–	29	22	–	–	–
6	30	–	–	14	14	–	–	–
7	30	–	–	–	15	–	–	–
8	23	–	–	–	16	–	–	–
9	31	–	–	–	20	–	–	–
10	37	–	–	–	13	–	–	–
11	–	–	–	–	12	–	–	–
12	–	–	–	–	19	–	–	–
13	–	–	–	–	13	–	–	–
14	–	–	–	–	13	–	–	–
Total species for all sites	73	57	>44	45	49	–	50	23
Mean species per site	28.9	30.5	>22.0	24.3	15.6	–	22.5	9.8
Estimate of total ants in mediterranean region of the State or Country	>500		>1000	100	?	100	206	62

[a] From Majer (1980a).
[b] From Majer et al. (1982 and unpublished data).
[c] From P.J.M. Greenslade (1985 and unpublished data).
[d] From Donnelly and Giliomee (1985a), total ant richness, A.J. Prins (pers. comm.).
[e] From Ofer et al. (1978).
[f] From Tohmé (1969).
[g,h] From Hunt (1977).

Greenslade (1985 and unpublished data) sampled ants in open forest in the Mt. Lofty Ranges near Adelaide, South Australia while Majer (1980a) assessed the ant fauna in 10 open forest plots near Wagerup, which is south of Perth, Western Australia. Majer et al. (1982) also investigated the ant fauna in heathland plots at Eneabba, to the north of Perth.

Table 15 shows the ant species richness recorded from each plot or overall study area. Where data are available, the estimated total species for each mediterranean region is also given.

There is considerable variation in ant species richness at the individual site level, even within a particular mediterranean region. However, if mean species per site are considered, the data indicate that Australian forests and heaths have the highest ant species richness, followed by the South African fynbos and then the Californian sites. The Israeli and Chilean sites have by far the fewest ant species per study site.

Comparison of the total ant species richness for each mediterranean region is complicated by the fact that they are of differing areas and not all have been completely surveyed for ants. With this in mind however, some tentative comments may be made (Table 15).

The high ant species richness of the Australian regions may be associated with the extreme antiquity of of the Australian Gondwanaland fauna allowing considerable time for speciation to take place. Coupled with this is the enriching effect of new faunal elements arriving from the north of Australia. This may explain the apparently higher species richness of ants in South Australia, where the arid and semi-arid fauna elements are increased by the presence of sub-humid elements associated with the nearby eastern seaboard, than in the south of Western Australia. This effect might not be present in the mediterranean part of Western Australia which has no adjacent sub-humid region.

In contrast to the relatively high site species richness in South Africa, the value for the total mediterranean region of the Cape is extremely low. This is at first unexpected in view of the fact that ancient Gondwanaland elements also exist here. It is probably explained by the small size and isolated nature of this mediterranean region of South Africa.

The relatively young age of the Mediterranean Basin, Californian, and Chilean ant faunas could in part explain their low richness when compared with Australia. However, the low richness of ants in Chile, when compared with California is surprising. Hunt (1973) has explained this phenomenon in terms of three components. These are the biogeographical setting and isolation of Chile and, perhaps more importantly, the low predictability and high amplitude of climatic phenomena in Chile imposing a higher rate of extinction on the ant fauna than that of California.

3.3.3 Ant generic composition

The distribution of genera collected in the bulked study sites described in Table 15 are shown in Table 16. This table also shows the total genera collected and the total genera within each subfamily. It should be stressed that these are mainly lists of the genera present in the study sites rather than complete lists for each mediterranean region. They are therefore a guide to the preponderance of genera in each region rather than a definitive list of the genera present.

The Australian sites contained the highest number of genera and the Chilean ones the least. The South African region had a relatively low number of genera while generic variety was intermediate in both the Californian and Mediterranean Basin areas.

The Myrmeciinae were confined to Australia while the Ponerinae were found mainly in Australia; one genus from this subfamily was also found in Lebanon and South Africa. Ecitoninae were only present in the Californian sites while the Dorylinae were confined to the South African and Lebanese mediterranean regions. Ants of the subfamily Pseudomyrmecinae were present in California and Chile but were not sampled elsewhere. The Myrmicinae, Dolichoderinae and Formicinae were present in all mediterranean regions although there were more species of Dolichoderinae in Australia than elsewhere. This is because of the high diversity of *Iridomyrmex* there.

The ground fauna was dominated by individuals from the genus *Iridomyrmex* in Australia, *Anoplo-*

Table 16. Number of ant species per subfamily and genus collected from study sites in mediterranean regions.

	Western Australian forest[a]	Western Australian heath[b]	South Australian forest[c]	South African fynbos[d]	Israeli vegetation transect[e]	Lebanon whole country[f]	Californian vegetation transect[g]	Chilean vegetation transect[h]
MYRMECIINAE	3	–	4	–	–	–	–	–
Myrmecia	3	–	4	–	–	–	–	–
PONERINAE	9	5	7	1	–	1	–	–
Amblyopone	1	–	1	–	–	–	–	–
Heteroponera	1	–	1	–	–	–	–	–
Rhytidoponera	2	2	1	–	–	–	–	–
Sphinctomyrmex	1	–	1	–	–	–	–	–
Cerapachys	–	1	+	1	–	–	–	–
Brachyponera	1	1	–	–	–	–	–	–
Trachymesopus	1	–	+	–	–	–	–	–
Ponera	–	–	–	–	–	1	–	–
Hypoponera	1	–	+	–	–	–	–	–
Leptogenys	1	–	–	–	–	–	–	–
Odontomachus	–	1	–	–	–	–	–	–
ECITONINAE	–	–	–	–	–	–	2	–
Neivamyrmex	–	–	–	–	–	–	2	–
DORYLINAE	–	–	–	1	–	1	–	–
Dorylus	–	–	–	1	–	1	–	–
PSEUDOMYRMECINAE	–	–	–	–	–	–	1	1
Pseudomyrmex	–	–	–	–	–	–	1	1
MYRMICINAE	24	22	15	27	31	58	21	8
Myrmica	–	–	–	–	–	–	1	–
Pogonomyrmex	–	–	–	–	–	–	2	3
Aphaenogaster	–	1	–	–	3	3	–	–
Messor	–	–	–	1	5	15	–	–
Veromessor	–	–	–	–	–	–	2	–
Oxyopomyrmex	–	–	–	–	1	1	–	–
Pheidole	3	–	2	4	2	3	4	–
Stenamma	–	–	–	–	–	–	2	–
Meranoplus	4	4	1	1	–	–	–	–
Podomyrma	1	–	+	–	–	–	–	–
Leptothorax	–	–	–	–	1	4	4	–
Tetramorium	4	5	–	11	8	13	–	–
Rhoptromyrmex	–	–	–	1	–	–	–	–
Monomorium	4	5	+	7	8	11	1	–
Chelaner	1	5	3	–	–	–	–	–
Nothidris	–	–	–	–	–	–	–	1
Solenopsis	–	–	+	–	–	1	2	4
Adlerzia	1	1	+	–	–	–	–	–
Anisopheidole	1	–	–	–	–	–	–	–
Ocymyrmex	–	–	–	1	–	–	–	–
Cardiocondyla	1	–	–	–	2	1	–	–
Crematogaster	2	1	2	–	1	6	3	–
Epopostruma	–	–	1	–	–	–	–	–
Mesostruma	–	–	+	–	–	–	–	–
Strumigenys	2	–	+	–	–	–	–	–
Smithistruma	–	–	–	1	–	–	–	–
DOLICHODERINAE	15	12	6	2	2	4	5	5
Dolichoderus	1	1	1	–	–	–	–	–
Liometopum	–	–	–	–	–	1	1	–
Iridomyrmex	13	10	5	1	–	–	1	–
Dorymyrmex	–	–	–	–	–	–	2	4
Bothriomyrmex	–	–	–	–	–	1	–	–
Tapinoma	1	1	–	–	2	2	1	1
Technomyrmex	–	–	–	1	–	–	–	–

Table 16. (Continued).

	Western Australian forest[a]	Western Australian heath[b]	South Australian forest[c]	South African fynbos[d]	Israeli vegetation transect[e]	Lebanon whole country[f]	Californian vegetation transect[g]	Chilean vegetation transect[h]
FORMICINAE	22	18	12	13	16	36	21	9
Melophorus	5	9	2	–	–	–	–	–
Notoncus	1	–	1	–	–	–	–	–
Prolasius	3	1	–	–	–	–	–	–
Lasiophanes	–	–	–	–	–	–	–	2
Brachymyrmex	–	–	–	–	–	–	–	2
Myrmelachista	–	–	–	–	–	–	–	2
Plagiolepis	–	–	+	5	2	1	–	–
Anoplolepis	–	–	–	1	2	–	–	–
Acantholepis	–	–	–	2	–	4	–	–
Stigmacros	4	2	+	–	–	–	–	–
Prenolepis	–	–	–	–	–	1	1	–
Paratrechina	–	–	+	–	1	2	–	–
Lasius	–	–	–	–	1	3	–	–
Acanthomyops	–	–	–	–	–	–	1	–
Myrmecocystus	–	–	–	–	–	–	3	–
Cataglyphis	–	–	–	–	4	8	–	–
Formica	–	–	–	–	–	1	2	–
Polyergus	–	–	–	–	–	–	1	–
Camponotus	8	5	5	5	5	16	13	3
Polyrhachis	1	1	+	–	1	–	–	–
Total species	73	57	>44	44	49	100	50	23
Total genera	29	19	28	16	17	23	21	10

[a] From Majer (1980a).
[b] From Majer *et al.* (1982).
[c] From P.J.M. Greenslade (1985 and unpublished data).
[d] From Donnelly and Giliomee (1985a).
[e] From Ofer *et al.* (1978).
[f] From Tohmé (1969).
[g,h] From Hunt (1977).

lepis and *Pheidole* in South Africa, *Camponotus*, *Tapinoma*, *Pheidole* and *Messor* in Israel and Lebanon, and *Liometopum* and *Formica* in California. Hunt (1977) states that no ground ants reached large colony sizes in the Chilean habitats that were studied.

3.3.4 Seasonality of the ant fauna
Seasonal variation in numbers of ant species and individuals for selected South and Western Australian and South African sites are shown in Fig. 6a–d.

The seasonal pattern is strikingly similar at all of the five sites with a pronounced winter trough and a spring-summer peak in both individuals and species of ants trapped per month. The amplitude of the seasonal variation in South Australia is greater than that exhibited in Western Australia, a phenomenon which may be associated with a greater seasonal fluctuation in climate in the former State.

The elevated position of the South African species richness curve is probably associated with the large area which was sampled. However, the amplitude of seasonal fluctuations in South Africa is similar to that at the South Australian sites.

3.3.5 Influence of fire on the ant fauna
Figures 7a and 7b show two months of pre-fire data and 9 months of post-fire data for a Western Australian and South Australian site. These are the same localities illustrated in Figures 6b and 6c so the differences between these graphs may largely be attributed to the effects of fire.

At both sites there was a temporary increase in the number of species and individuals of ants trapped following the fire although this stimulation in the catch declined after a period of six months. By nine months after the fire the Western Australian catch was lower than that of the unburnt plot

223

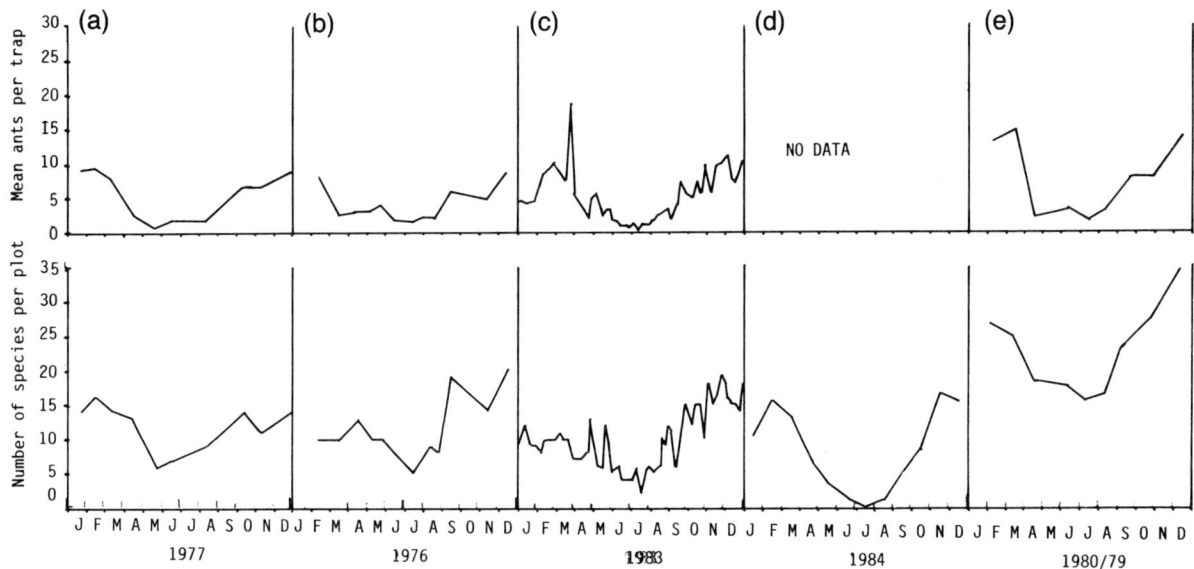

Fig. 6. Total ant species trapped per plot per month (or week) and mean number of ant individuals per pitfall trap per month (or week) in Australia and South Africa in: a) *Eucalyptus wandoo* woodland, Kojonup, Western Australia (33°50′ S, 116°52′ E) (from Majer 1980b); b) *Eucalyptus marginata* forest, Dwellingup, Western Australia (32°52′ S, 116°13′ E) (from Majer unpublished data); c) *Eucalyptus obliqua* forest, Belair, South Australia (35°00′ S, 138°40′ E) (from O'Dowd 1985a); d) *Eucalyptus obliqua/E. baxteri* forest, Bridgewater, South Australia (from P.J.M. Greenslade 1985); e) Fynbos, Jonkershoek Valley, South Africa (33°57′ S, 18°55′ E) (from Donnelly 1983).

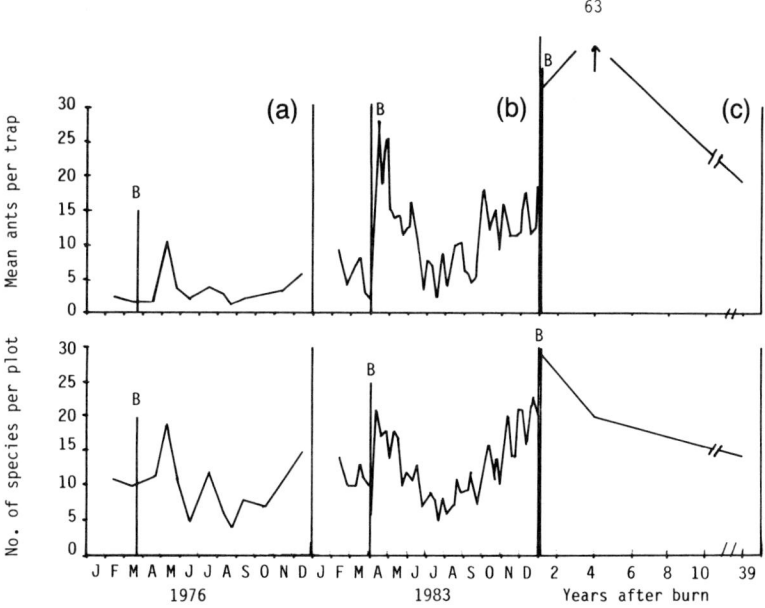

Fig. 7. Total ant species trapped per plot and mean number of ant individuals per pitfall trap obtained before, and at intervals of time after a fire in Australia and South Africa in: a) *Eucalyptus marginata* forest, Dwellingup, Western Australia (32°52′ , 116°13′ E) (from Majer unpublished data); b) *Eucalyptus obliqua/E. baxteri* forest, Belair, South Australia (35°00′ S, 138°40′ E) (from O'Dowd 1985b and unpublished data); c) Fynbos, Jonkershoek Valley, South Africa (33°57′ S, 18°55′ E) (from Donnelly and Giliomee 1985b). The vertical line (B) indicates the data of the fire.

while the South Australian catch continued to exceed that of the unburnt plot.

The South African data (Fig. 7c) were obtained by comparing a range of plots of differing age since the last burn. As with the Australian sites there was a stimulation of the variety and abundance of ants trapped following the fire. There was a gradual return to lower levels as the period since fire increased. The data from both continents suggest that the ant fauna of these mediterranean regions exhibits a high degree of resilience to fire.

4 Synthesis

This review has provided a certain amount of comparative information on the soil and litter invertebrates of mediterranean regions. However, for reasons of sampling design or season of sampling, the data were often not truly comparable.

Generally speaking, the fauna is dominated by mites, springtails and, to a lesser extent, ants. The data indicate that the fauna is strongly seasonal in its abundance or activity and it may be affected by altitude, degree of moisture, vegetation type and time since fire. Some of the taxa which were looked at in detail show a degree of resilience to fire. However, insufficient data are available to relate this resilience to that which is characteristic of other climatic regions.

We believe that it is premature to produce a detailed synthesis of soil and litter invertebrate data from mediterranean ecosystems. However, it is hoped that the data which we have provided will provide a baseline with which data from future studies may be compared.

5 References

Anon, 1963. Ecological study of the Mediterranean zone. Bioclimatic map of the Mediterranean zone. Recherches sur la zone Aride XXI, Paris.

Anon, 1978. Soil maps of the world. FAO-UNESCO, Paris.

Baguena, C.L., 1967. Scarabaeoidea de la fauna Ibero-Balear y Pirenaica. Consejo Superior de Investigaciones Cientifcas, Madrid.

Breytenbach, G.J., 1982. Small mammal responses to environmental gradients in the Groot Swartberg of the southern Cape. M.Sc. Thesis. University of Pretoria, Pretoria.

Briese, D.T. and Macauley, B.J., 1980. Temporal structure of an ant community in semi-arid Australia. Australian Journal of Ecology, 5, 121–134.

Cassagnau, P., 1986. Les écomorphoses des collemboles: 1. Deviation de la morphogenèse et perturbations histophysiologiques. Annales de la Société Entomologique de France, 22, 7–33.

Cody, M.L., Fuentes, E.R., Glanz, W., Hunt, J.H. and Moldenke, A.R., 1977. Convergent evolution in the consumer organisms of mediterranean Chile and California. In: Mooney, H.A. (ed.). Convergent evolution in Chile and California. I.B.P. Series, Dowden, Hutchinson and Ross, Inc., Pennsylvania.

Di Castri, F. and Vitaldi Di Castri, V., 1981. Soil fauna of Mediterranean-climate regions. In: Di Castri, F., Goodall, D.W. and Specht, R.L. (eds.). Mediterranean-type Shrubland. Elsevier Scientific Publishing Company, Amsterdam.

Donnelly, D., 1983. Comparison of community structure of the epigaeic fauna, with particular reference to ants and beetles, between fynbos sites under different management practices. M.Sc. Thesis, University of Stellenbosch.

Donnelly, D. and Giliomee, J.H. 1985a. Community structure of epigaeic ants (Hymenoptera: Formicidae) in fynbos vegetation in the Jonkershoek valley. Journal of the Entomological Society of Southern Africa, 48: 247–257.

Donnelly, D. and Giliomee, J.H., 1985b. Community structure of epigaeic ants in a pine plantation and in newly burnt fynbos. Journal of the Entomological Society of South Africa, 48, 259–265.

Du Merle, P., 1982. Ant visiting of shrubs and trees on a French mediterranean mountain. Insectes Sociaux, 29, 422–444.

Force, D.C., 1981. Post-fire insect succession in southern California chaparral. American Naturalist, 117, 575–582.

Greenslade, P., 1974. Ecological and zoogeographical notes on Collembola of Kangaroo Island, South Australia. Pedobiologia, 14, 256–265.

Greenslade, P., 1977, A re-examination of the genus *Corynephoria Absolon* (Collembola, Sminthuridae). Revue Ecologie et Biologie du Sol, 14, 241–256.

Greenslade, P., 1985. Phenology and diversity of epigaeic Collembola in the Mt. Lofty Ranges, South Australia. In: Soil and litter invertebrates of Australian mediterranean-type ecosystems. WAIT School of Biology Bulletin, No. 12, 60–62.

Greenslade, P., 1986. Small arthropods. In: Wallace, H.R. (ed.). The Ecology of the Forests and Woodlands of South Australia. Government Printer, Adelaide. pp. 144–153.

Greenslade, P. and Greenslade, P.J.M., 1971. The use of baits and preservatives in pitfall traps. Journal of the Australian Entomological Society, 10, 253–260.

Greenslade, P., and Majer, J.D., 1980. Collembola of rehabilitated mine sites in Western Australia. In: Soil biology as related to land use practices. Proceedings 7th International Colloquium of Soil Zoology, 397–408.

Greenslade, P. and Majer, J.D. (eds.), 1985. Soil and litter invertebrates of Australian mediterranean-type ecosystems. WAIT School of Biology Bulletin, No. 12, 93 pp.

Greenslade, P., and Rosser, G., 1984. Fire and soil-surface insects in the Mount Lofty Ranges, South Australia. In: Proceedings of the 4th International Conference on Mediterranean Ecosystems, 63–64.

Greenslade, P.J.M. 1985. Some effects of season and geographic aspect on ants (Hymenoptera: Formicidae) in the Mt. Lofty Ranges, South Australia. Transactions of the Royal Society of South Australia, 109, 17–23.

Greenslade, P.J.M., and Greenslade, P. 1973. Epigaeic Collembola and their activity in a semi-arid locality in southern Australia during summer. Pedobiologia, 13, 227–235.

Hunt, J.H., 1973. Comparative ecology of ant communities on mediterranean regions of California and Chile. Ph.D. Thesis, University of California, Berkely.

Hunt, J.H., 1977. Ants. In: Thrower, N.J.W. and Bradbury, D.E. (eds). Chile – California Mediterranean Scrub Atlas. I.B.P. Series, Dowden, Hutchinson and Ross, Inc., Pennsylvania.

Hutson, B.R. and Kirkby, C.A. 1985. Populations of Collembola and Acarina in litter and soil of South Australian forests and the effects of a wildfire. In: Soil and litter invertebrates of Australian mediterranean-type ecosystems. WAIT, School of Biology Bulletin, No. 12, 34–36.

Hutson, B.R. and Veitch, L.G., 1983. Mean annual population density of Collembola and Acari in the soil and litter of three indigenous South Australian forests. Australian Journal of Ecology, 8, 113–126.

Louw, J.F., 1968. Die ekologie van die malmier, *Anoplolepis custodiens* (Smith) (Hymenoptera: Formicidae), in die sentrale Oranje Vrystraat. I. Die bogrondse aktiviteitspeil van die werkers. Journal of the Entomological Society of Southern Africa, 31, 241–248.

Lumaret, J.P., 1977. Les scarabées coprophages de la Garrigue. Annales de la Société d'Horticulture et d'Histoire Naturelle de L'Herault, 117, 98–108.

Lumaret, J.P., 1980. Analyse des communautés de scarabeides coprophages dans le maquis Corse, et étude de leur rôle dans l'utilisation des excréments. Ecologia Mediterranea, 5, 51–58.

Majer, J.D., 1980a. A preliminary ecological survey of the Wagerup ant fauna. Alcoa Environmental Research Bulletin, No. 3, 19 pp.

Majer, J.D., 1980b. Report on a study of invertebrates in relation to the Kojonup fire management programme. W.A. Institute of Technology, School of Biology Bulletin, No. 2, 22 pp.

Majer, J.D., 1984. Short-term responses of soil and litter invertebrates to a cool autumn burn in Jarrah (*Eucalyptus marginata*) forest in Western Australia. Pedobiologia, 26, 229–247.

Majer, J.D., Sartori, M., Stone, R. and Perriman, W.S., 1982. Colonization by ants and other invertebrates in rehabilitated mineral sand mines near Eneabba, W.A. Reclamation and Revegetation Research, 1, 63–81.

Martin, P., 1982. Los Scarabaeinae (Coleoptera, Scarabaeoidea) de la Peninsula Iberica e Islas Baleares. Thesis Doctoral, Universidad Complutense, Madrid.

Matthews, E.G., 1972. A revision of the scarabaeine dung beetles of Australia. I. Tribe Onthophagini. Australian Journal of Zoology, Supplementary Series, 9, 1–330.

Matthews, E.G., 1974. A revision of the scarabaeine dung beetles of Australia. II. Tribe Scarabaeini. Australian Journal of Zoology, Supplementary Series, 24, 1–211.

Merritt, R.W. and Anderson, J.R., 1977. The effects of different pasture and rangeland ecosystems on the annual dynamics of insects in cattle droppings. Hilgardia, 45, 31–71.

Mostert, D.P., Siegfried, W.R. and Louw, G.N., 1980. Protea nectar and satellite fauna in relation to the food requirements and pollinating role of the Cape sugarbird. South African Journal of Science, 76, 409–412.

O'Dowd, D.J., 1985a. Seasonal patterns in the activity of ants in Belair Recreation Park, South Australia. In: Soil and Litter Invertebrates of Australian Mediterranean-type Ecosystems. WAIT School of Biology Bulletin, No. 12, 70–72.

O'Dowd, D.J., 1985b. Effects of a low-intensity fire on the activity of ants and other invertebrates at Belair Recreation Park, South Australia. In: Soil and Litter Invertebrates of Australian Mediterranean-type Ecosystems. WAIT, School of Biology Bulletin No. 12, 73–76.

Ofer, J., Shulov, A. and Noy-Meir, I., 1978. Associations of ant species in Israel: A multivariate analysis. Israel Journal of Zoology, 27: 199–208.

Ovalle, H.V. and Solervicens, J.A., 1980. Observaciones sobre la biologia de *Megathopa villosa* Eschscholtz, 1822. Boletin del Museo Nacional de Historia Natural, Chile, 37, 235–246.

Poinsot, N., 1968. Cas d'anhydrobiose chez le collembole *Subisotoma variabilis* Gisin. Revue Ecologie et Biologie du Sol, 4, 585–586.

Poinsot-Balaguer, N. and Tabone, E. 1985. Étude d'un écosysteme forestier méditerranéen. 1. Composition et structure des peuplements microarthropodiens du sol dans une forêt mixte de la region Provinçale. Bulletin Ecologie, 16, 149–160.

Poinsot-Balaguer, N. and Tabone, E., 1986. Étude d'un écosysteme forestier méditerranéen. 2. Les collemboles d'une forêt mixte de la region Provinçale. Bulletin Ecologie, 17, 87–95.

Postle, A.C., 1986. The effect of soil type on soil and litter invertebrates of Western Australian jarrah forest. Ph.D. Thesis. University of Western Australia.

Ridsdill-Smith, T.J., Weir, T.A. and Peck, S.B. 1983. Dung beetles (Scarabaeidae: Scarabaeinae and Aphodiinae) active in forest habitats in southwestern Australia during winter. Journal of the Australian Entomological Society, 22, 307–309.

Ridsdill-Smith, T.J. and Hall, G.P., 1984a. Dung beetles and mites attracted to fresh cattle dung in south-western Australian pastures. CSIRO Division of Entomology Report, No. 34, 1–29.

Ridsdill-Smith, T.J. and Hall, G.P. 1984b. Seasonal patterns of adult dung beetle activity in south-western Australia. Proceedings of the Fourth International Conference on Mediterranean Ecosystems. pp. 139–140.

Sáiz, F., 1971. Notas ecologicas sobre los estafilinidos (Coleoptera) del Parque Nacional 'Fray Jorge', Chile. Boletin del Museo Nacional de Historia Natural, Chile, 32, 67–97.

Sáiz, F., 1975a. Aspectos mesofaunisticos hipogeos en el Parque Nacional 'Fray Jorge'. I. Analisis comunitario. Anales Musea de Historia Natural de Valparaiso, Chile, 8, 29–50.

Sáiz, F., 1975b. Coleopteros epigeos del Parque Nacional 'Fray Jorge'. Aspectos ecologicos y biogeograficos. Boletin del Museo Nacional de Historia Natural, Chile, 34, 137–171.

Sáiz, F., 1977. Soil Beetles. In: Thrower, N.J.W. and Bradbury, D.E. (eds). Chile-California Mediterranean Scrub Atlas. I.B.P. Series, Dowden, Hutchinson and Ross Inc., Pennsylvania.

Sáiz, F. and Calderon, R., 1976. Investigaciones ecologicas sobre las aranas del Parque Nacional 'Fray Jorge', Chile. Anales del Museo de Historia Natural de Valparaiso, Chile, 9, 65–72.

Silva, F. and Sáiz, F., 1975. Investigaciones ecologicas de los diplopodos del Parque Nacional 'Fray Jorge'. Anales del Museo de Historia Natural de Valparaiso, Chile, 8, 17–28.

Skaife, S.H., 1961. The Study of Ants. Longmans, London.

Specht, M.M., 1985. Seasonality of aerial, litter and soil fauna in Dark Island heathland, South Australia. In: Soil and litter invertebrates of Australian mediterranean-type ecosystems. WAIT School of Biology Bulletin, No. 12, 39–42.

Springett, J.A., 1976. The effect of planting *Pinus pinaster* Ait. on populations of soil microarthropods and on litter decomposition at Gnangara, Western Australia. Australian Journal of Ecology, 1, 83–87.

Thiele, H.U., 1977. Carabid Beetles in their Environments. Springer Verlag, New York.

Tohmé, G., 1969. Repartition Géographique des Fourmis du Liban. Ph.D. Thesis. University of Toulouse.

Trabaud, L., 1984. Fire effects on soil of the Mediterranean Basin. In: 2nd International Rangeland Conference Working Papers, Adelaide.

Wallace, M.M.H., 1967. The ecology of *Sminthurus viridis* (L.) (Collembola). 1. Processes influencing numbers in pastures in Western Australia. Australian Journal of Zoology, 15, 1177–1206.

Wallace, M.M.H., 1968. The ecology of *Sminthurus viridis* (Collembola). II. Diapause in the aestivating egg. Australian Journal of Zoology, 16, 871–883.

Wallace, M.M.H. and Mahon, J.A. 1971. The ecology of *Sminthurus viridis* (Collembola). III. The influence of climate and land use on its distribution and that of an important predator, *Bdellodes lapidaria* (Acari: Bdellidae). Australian Journal of Zoology, 19, 177–188.

Whitford, W.G., Depree, D.J., Hamilton, P. and Ettershank, G., 1980. Foraging ecology of seed-harvesting ants, *Pheidole* spp. in a Chihuahuan desert ecosystem. The American Midland Naturalist, 105, 159–167.

Systematic index

Abies cephalonica 46
Abies cilicica 46
Abies maroccana 46
Abies nebrodensis 46
Abies pinsapo 46
Ablepharus kitaibelii 180, 194
Acacia sp. 96, 104, 126
Acacia acuaria 28
Acacia acuminata 15
Acacia aneura 94, 96
Acacia bivenosa 14
Acacia caven 45, 87, 113, 114, 179
Acacia dealbata 173
Acacia hemiteles 28
Acacia ixiophylla 28
Acacia longifolia 173
Acacia mearnsii 97
Acacia myrtifolia 16
Acacia papyrocarpa 96, 175
Acacia pycnantha 17
Acacia ramulosa 96
Acacia rupicola 20
Acacia spinescens 21
Acacia stricta 23
Acaena splendens 113
Acanthocyrtus sp. 215
Acantholepis sp. 222
Acantholepis capensis 206
Acanthomurus sp. 215
Acanthomyops sp. 222
Acaridida 206, 208
Acarina 199–205, 207, 208, 211–213, 219
Acer macrophyllum 177
Acer monspessulanum 77
Achillea sp. 126
Acinos alpinus 78
Acomys cahirinus 186
Acrotriche fasciculiflora 16
Acrotriche patula 20
Acrotriche serrulata 17
Actinostrobus acuminatus 31

Adenanthos cygnorum 31
Adenostoma sp. 86
Adenostoma fasciculatum 32, 35, 36, 38, 40, 72, 73, 177, 211
Adenostoma fasciculatum var. *obtusifolium* 73
Adenostoma sparsifolium 32, 36, 37
Adesmia arborea 44
Adesmia bedwelli 204
Adlerzia sp. 221
Aextoxicon punctatum 75, 204
Agama stellio 194
Agamidae 193
Agathosma sp. 91
Agonis sp. 104
Agonis hypericifolia 15
Agropyron scabrum 25
Alauda arvensis 192
Alaudidae 191
Alectoris chukar 191, 192
Alectryon oleifolius 175
Alkanna tinctoria 79
Alleculidae 210
Allocasuarina acutivalvis 15
Allocasuarina campestris 15
Allocasuarina humilis 71
Allocasuarina luehmannii 14, 97, 100
Allocasuarina muellerana 19
Allocasuarina pusilla 14, 18, 64
Allocasuarina verticillata 14
Alnus rhombifolia 177
Alticidae 208, 210
Amblyopone sp. 221
Amphipoda 200, 202, 213
Anisopheidole sp. 221
Anisotomidae 203
Annelida 212, 213
Anobiidae 208, 210
Anoplolepis sp. 220, 222
Anoplolepis custodiens 203, 206
Anthicidae 203, 210
Anthus pratensis 192
Anthyllis hermanniae 51, 78

Anugidae 9, 193
Aphaenogaster sp. 221
Aphodiinae 210, 216
Aphyllanthes monspeliensis 49
Apodemus mystacinus 180, 186
Apodemus sylvaticus 180, 186
Arachnida 199, 200, 202, 208, 210–213
Araneae 199–203, 205, 207, 208, 210–213
Arbutus unedo 47, 48, 52, 57, 75–78
Arbutus menziesii 32
Archaeognatha 202, 203, 213
Arctostaphylos sp. 32, 33
Arctostaphylos crustacea 73
Arctostaphylos glandulosa 35, 40, 72
Arctostaphylos glauca 72
Arctostaphylos hookeri 72
Arctostaphylos montereyensis 72
Arctostaphylos pringlei 35, 40
Arctostaphylos pumila 72
Arctostaphylos pungens 73
Arctostaphylos tomentosa 72
Arctostaphylos viscida 73
Arctotheca calendula 25
Ardeidae 192
Ardeola ibis 192
Argania sp. 46
Arisarum vulgare 79
Aridius subfasciatus 208
Aristida sp. 32, 91, 126
Aristotelia chilensis 75
Aristotelia macqui 75
Artemisia sp. 33, 126
Artemisia californica 34, 178
Arthrocnemum sp. 126
Arthropleona 202, 203, 205
Aspalathus sp. 91
Asparagus acutifolius 49
Asphodelus aestivus 79
Asphodelus sp. 126
Aster haplopappus 42
Atheta obscuripennis 208
Astroloma conostephioides 17
Atriplex sp. 33, 67, 84, 94
Atriplex atacamensis 114
Atriplex rhagodioides 96
Atriplex stipitata 67
Atriplex suberecta 22
Atriplex vesicaria 82, 96, 175
Aulax cneorifolia 80
Avena fatua 178
Azara dentata 75
Azara lanceolata 75

Baccharis concava 42, 204
Baeckea crassifolia 21, 69
Baeckea maidenii 15

Baeckea priessiana 71
Bahia ambrosioides 42, 74
Ballota acetabulosa 51, 78
Balsamocarpon brevifolium 114
Banksia sp. 84, 101, 104, 173, 199
Banksia aemula 65
Banksia attenuata 15, 29, 31, 71
Banksia hookeriana 31
Banksia littoralis 15
Banksia marginata 14, 16, 18, 63, 64
Banksia menziesii 15, 29, 31
Banksia micrantha 30
Banksia oblongifolia 65
Banksia ornata 18, 64
Banksia robur 65
Banksia sphaerocarpa 71
Banksia tricuspis 30
Beaufortia elegans 29, 31
Berberis chilensis 44
Berzelia lanuginosa 80
Beyeria lechenaultii 20
Blattodea 199–203
Boidae 9, 193
Bolitobius seriaticollis 208
Borya constricta 28
Bothriomyrmex sp. 221
Brabejum stellatifolium 80
Brachylaena sp. 91
Brachyloma ericoides 21, 69
Brachymyrmex sp. 222
Brachypodium phoenicoides 89
Brachypodium ramosum 89
Brachypodium retusum 50
Brachyponera sp. 221
Brassica campestris 178
Briza maxima 79
Briza minor 45
Bromus sp. 178
Bupleurum fruticosum 75
Buxus sempervirens 50
Byrrhidae 203, 208, 210

Cactaceae 32
Calandrinia calyptrata 26
Calicotome spinosa 77
Calicotome villosa 46, 52–54, 56, 78
Callitris sp. 82
Callitris preissii 14, 15
Calluna vulgaris 60, 77, 131
Calothamnus quadrifidus 15
Calothamnus sanguineus 30
Calothamnus torulosus 30
Calytrix alpestris 18, 64
Calytrix flavescens 71
Calytrix tetragona 17, 21, 69
Camponotus sp. 222

Camponotus maculatus 204, 206
Canis familiaris dingo 173
Canis latrans 178
Cantharidae 203, 210
Canthium sp. 91
Canthon simplex 218
Carabidae 202, 203, 210, 211
Cardiocondyla sp. 221
Carduelis carduelis 191, 192
Carduelis chloris 191, 192
Carduelis spinus 192
Carduus pycnocephalus 79
Carex hallerana 50
Cassia sp. 96
Cassia brogniartii 114
Cassine sp. 91
Cassinia laevis 66
Cassinia longifolia 173
Castanopsis chrysophylla 72
Casuarina sp. 82, 104
Casuarina cristata 94, 175
Cataglyphis sp. 222
Catopidae 210
Ceanothus sp. 32, 86
Ceanothus crassifolius 73, 177, 211
Ceanothus dentatus 72
Ceanothus greggii 33, 35–37, 40, 72
Ceanothus spinosus 73
Cedrus atlantica 46
Cedrus brevifolia 46
Cedrus libani 46
Centaurea spinosa 46
Cerambycidae 203, 208
Cerapachys sp. 221
Ceratonia sp. 126
Ceratonia siliqua 46, 54, 78
Cercidium sp. 33
Cercis siliquastrum 78
Cercocarpus betuloides 33, 36, 73, 177
Cercotrichas galactotes 191
Chalcides guentheri 194
Chalcides ocellatus 194
Chamaeleo chamaeleon 194
Chelaner sp. 221
Chenopodium nitrariaceum 82
Chiliadenus iphionoides 56
Chilopoda 199–202, 208, 210, 212, 213
Chilorus ovallenus 207
Chrysomelidae 202, 203, 210
Chuquiragua oppositifolia 44, 113, 114
Cichorium spinosum 78
Cicindelinae 203
Cistus sp. 46
Cistus albidus 75, 77
Cistus creticus 53, 55
Cistus crispus 77

Cistus incanus (syn. *C. villosus*) 47, 51, 78
Cistus ladaniferus 57, 77
Cistus libanotis 59
Cistus monspeliensis 47–49, 57, 76–78
Cistus parviflorus 78
Cistus salviifolius 47, 53–55, 76–78
Clarkia tenella 45
Clematis microphylla 19
Cleridae 210
Cliffortia sp. 91
Clutia sp. 91
Coccinellidae 203
Coleoptera 199–203, 205, 207, 208, 210, 212, 213
Collembola 198–202, 204, 207, 208, 210–215, 219
Colliguaya integerrima 74
Colliguaya odorifera 43, 74
Colliguaya salicifolia 44, 74
Coluber jugularis 194
Coluber najadum 194
Coluber ravergieri 180, 194
Colubridae 9, 193
Columbidae 191, 192
Colydiidae 210
Conospermum stoechadis 71
Convolvulus althaeoides 79
Coprini 216–218
Corvidae 191, 192
Corynephoria sp. 215
Coturnix coturnix 192
Crassula sieberana 26
Crataegus aronia 53
Crataegus laevigata 78
Crematogaster sp. 221
Cricetulus migratorius 186
Crinodendron patagua 75
Crioceridae 210
Crustacea 199, 202, 208, 210, 212, 213
Cryptocarya alba 43, 74, 113, 114
Cryptocephalidae 208, 210
Cryptophagidae 210
Cryptopygus sp. 215
Cucujidae 203
Cupressus forbsii 32
Cupressus sempervirens 46, 78
Curculionidae 202, 203, 207, 208, 210, 211
Cussonia sp. 91
Cyrtodactylus kotschyi 194

Danthonia sp. 65
Danthonia auriculata 27
Danthonia pallida 23
Danthonia racemosa 25
Daphne gnidium 57
Daviesia incrassata 71
Daviesia striata 30
Daviesia virgata 173

Dendrocopos syriacus 191, 192
Dermaptera 199–202, 205, 211–213
Dermestidae 208
Deuterosminthurus sp. 215, 216
Dianella tasmanica 23
Dictyoptera 205
Diplopoda 199–202, 207, 208, 210–213
Diplura 199, 208, 212
Diptera 199–202, 207, 208, 210–214
Discolomidae 210
Dodonaea humilis 20
Dolichoderinae 220, 221
Dolichoderus sp. 221
Dorycnium pentaphyllum 49
Dorylinae 220, 221
Dorylus sp. 221
Dorymyrmex sp. 221
Drepanura sp. 215
Drimys winteri 75, 204
Dryandra cynaroides 71
Dryandra nivea 15, 30
Dryandra sessilis 30, 71

Ecballium elaterium 79
Ecdeiocolea monostachya 29
Echium vulgare 78
Ecitoninae 220, 221
Eirenis lineomaculata 194
Eirenis rothi 194
Elapidae 9, 193
Elateridae 203, 210
Elegia parviflora 80
Elytropappus sp. 91
Emberiza caesia 192
Emberiza calandra 191
Emberizidae 191, 192
Embioptera 199, 211, 212
Empodisma minus 65
Entomobrya sp. 215, 216
Entomobrya unostrigata 215
Entomobryidae 215
Entomobryomorpha 207, 208
Epacris impressa 14
Ephemeroptera 201
Epopostruma sp. 221
Eragrostis sp. 126
Eremaea beaufortioides 31
Eremaea pauciflora 29, 71
Eremophila sp. 96
Erica sp. 91
Erica arborea 46–48, 52, 76, 77
Erica ciliaris 60
Erica cinerea 77
Erica corifolia 80
Erica pulchella 80
Erica scoparia 60, 77

Erica verticillata 52
Eriogonum fasciculatum 34, 178
Eriogonum wrightii 39
Eriophyllum confertiflorum 40
Erithacus rubecula 192
Erodium sp. 178
Erodium botrys 65
Erotylidae 210
Eryngium rostratum 27
Escallonia illinita 44
Eucalyptus sp. 82, 101, 187, 199, 212, 214
Eucalyptus baxteri 14, 16, 63, 97, 100, 101, 223
Eucalyptus behriana 14, 26, 66, 94, 100
Eucalyptus calophylla 15
Eucalyptus camaldulensis 14, 94, 100
Eucalyptus cneorifolia 97
Eucalyptus cosmophylla 14, 199
Eucalyptus cypellocarpa 14
Eucalyptus diversifolia 14, 20, 68, 96, 97, 100
Eucalyptus dives 14, 94, 100
Eucalyptus drummondii 71
Eucalyptus dumosa 14, 94
Eucalyptus fasciculosa 14, 17, 97
Eucalyptus flocktoniae 104
Eucalyptus foecunda 14, 15, 21, 69
Eucalyptus globulus 178
Eucalyptus gomphocephala 15
Eucalyptus goniocalyx 17, 97
Eucalyptus gracilis 14, 15, 22, 96, 97, 100
Eucalyptus incrassata 14, 21, 69–71, 96, 100, 173
Eucalyptus largiflorens 14
Eucalyptus leptopoda 15
Eucalyptus leucoxylon 14, 19, 65, 97
Eucalyptus loxophleba 15
Eucalyptus macrorhyncha 14, 23, 24, 94, 100, 101
Eucalyptus marginata 15, 198, 223
Eucalyptus microcarpa 14, 25, 65, 94, 97, 100, 101
Eucalyptus obliqua 14, 16, 24, 63, 97, 100, 101, 173, 199, 223
Eucalyptus obtusiflora 15
Eucalyptus occidentalis 84
Eucalyptus odorata 14
Eucalyptus oleosa 15, 20, 22, 104
Eucalyptus ovata 97
Eucalyptus polyanthemos 14, 100, 101
Eucalyptus porosa 14, 97
Eucalyptus radiata 14, 23, 94, 100
Eucalyptus regnans 14
Eucalyptus rossii 94
Eucalyptus rudis 15
Eucalyptus salmonophloia 15
Eucalyptus sieberi 173
Eucalyptus socialis 14, 67, 71, 94, 96, 97, 100
Eucalyptus todtiana 29, 31
Eucalyptus viminalis 14, 65, 101
Eucalyptus viridis 14, 100
Eucalyptus wandoo 15, 28, 223

Euclea sp. 91
Eudera sculptilis 208
Eumeces schneideri 194
Euphorbia sp. 91
Euphorbia acanthothamnos 51, 78
Euphorbia characias 78
Euphorbia helioscopia 79
Eustachys sp. 91
Exocarpos cupressiformis 199

Felis catus 175
Felis concolor 179
Festuca pallescens 113
Ficedula hypoleuca 192
Ficus carica 78
Flotovia diacanthoides 75
Flourensia thurifera 42, 74, 114
Folsomides sp. 215
Formica sp. 222
Formicidae 199–203, 205, 208, 210–213, 219
Formicinae 220, 222
Fringilla coelebs 192
Fringillidae 191, 192
Fuchsia lycioides 42, 204

Gahnia deusta 68
Galerida cristata 191, 192
Galium aparine 79
Gamasina 206, 207, 208
Garrulus glandarius 191, 192
Garrya flavescens 37
Gaultheria shallon 72
Geckonidae 193
Genista acanthoclada 46, 51, 78
Genista scorpius 49
Geranium core-core 45
Gerbilus dasyurus 186
Gnaphalium californicum 36
Gochnatia fascicularis 74
Gonioma sp. 91

Hakea sp. 199
Hakea brachyptera 31
Hakea falcata 15
Hakea gibbosa 65
Hakea megalosperma 30
Hakea muellerana 21, 69
Hakea obliqua 31
Hakea ruscifolia 71
Halimium commutatum 59
Halimium halimifolium 59
Halticinae 202, 203
Haplopappus sp. 204
Haplopappus foliosus 42, 74
Haplopappus glutinosus 44, 74
Haplopappus squarrosa 34

Helenium aromaticum 45
Helianthemum nummularium 51, 78
Helichrysum picardii 59
Helichrysum stoechas 78
Heliotropium stenophyllum 114
Hemidactylus turcicus 194
Hemiptera 200, 201, 202, 205, 211–213
Heteromeles arbutifolia 178
Heteroponera sp. 221
Heteroptera 199, 208, 210
Hibbertia sp. 199
Hibbertia acerosa 30
Hibbertia riparia 18, 64
Hibbertia sericea 16–18, 64
Hibbertia stricta 24
Hippolais sp. 191
Homoptera 199, 208, 210
Hymenoptera 199–203, 205, 208, 210–213, 219
Hyparrhenia sp. 91
Hypericum empetrifolium 51
Hyphaene sp. 126
Hypocalymma xanthopetalum 30
Hypodiscus cristatus 80
Hypogastruridae 215
Hypolaena fastigiata 18, 64
Hypoponera sp. 221

Iguanidae 193
Insecta 199, 200, 202, 210–213
Iridomyrmex sp. 220, 221
Iridomyrmex humilis 204, 206
Isopoda 199–202, 207, 208, 210, 212, 213
Isopogon ceratophyllus 24
Isoptera 199, 201, 202, 213
Isotomidae 215

Jepsonia parryi 38
Juglans californica 178
Juniperus sp. 33, 46
Juniperus oxycedrus 50, 77, 78
Juniperus phoenicea 57, 75, 78

Kageneckia angustifolia 44, 74
Kageneckia oblonga 43, 74, 204
Keckiella antirrhinoides 34, 40
Kingia australis 30

Lacerta laevis 180, 194
Lacerta trilineata 180, 194
Lagriidae 203
Lagurus ovatus 79
Lambertia multiflora 30
Laniidae 191, 192
Lanius excubitor 191, 192
Lanius nubicus 191
Lanius senator 191

Larrea divaricata 33
Lasiopetalum behrii 68
Lasiophanes sp. 222
Lasius sp. 222
Lathridiidae 207, 208, 210
Laurelia sempervirens 113
Laurus nobilis 78
Lavandula sp. 46
Lavandula stoechas 59, 76–78
Lavatera cretica 79
Lepidobrya sp. 215
Lepidocyrtoides sp. 215
Lepidocyrtus sp. 215, 216
Lepidoptera 199, 201, 202, 208, 210–213
Lepidosperma sp. 199
Lepidosperma carphoides 18, 64
Lepidosperma laterale 18, 21, 64, 69
Lepidosperma semiteres 16, 17, 24, 63
Leptocarpus hyalinus 80
Leptogenys sp. 221
Leptospermum sp. 173, 199
Leptospermum coriaceum 173
Leptospermum erubescens 15, 71
Leptospermum juniperinum 187
Leptospermum myrsinoides 14, 16, 18, 63, 64, 173, 187
Leptospermum phylicoides 173
Leptothorax sp. 221
Leucadendron laureolum 80
Leucadendron salignum 80
Leucopogon conostephioides 29, 71
Leucospermum sp. 91
Leucospermum truncatulum 80
Leysera sp. 91
Liometopum sp. 221, 222
Lithobiomorpha 202
Lithraea caustica 43, 74, 113, 114
Lobelia tupa 42
Lolium sp. 126
Lolium rigidum 25
Lomandra sp. 96
Lomandra dura 14
Lomandra effusa 14, 28
Lomandra filiformis 23
Lomatia ilicifolia 23
Loncovilius discoideus 208
Lonicera implexa 50, 75
Lonicera subspicata 37, 39
Lotus scoparius 38
Loxocarya fasciculata 31
Luscinia svecica 192
Lynx rufus 179

Mabuya vittata 194
Macropidia fuliginosa 30
Maireana sp. 67
Maireana aphylla 82

Maireana pentatropis 22
Maireana pyramidata 82, 175
Maireana sedifolia 82, 96, 175
Majorana syriaca 53, 54, 56
Malachiidae 203
Malpolon monspessulanus 194
Malva silvestris 79
Mantodea 201, 213
Maytenus sp. 46, 91
Maytenus boaria 113
Mecoptera 213
Medicago arborea 78
Medicago orbicularis 79
Megathopa villosa 218
Melaleuca sp. 84, 173
Melaleuca acuminata 22
Melaleuca cordata 15
Melaleuca lanceolata 14, 20, 66, 67, 97
Melaleuca nodosa 65
Melaleuca rhaphiophylla 15
Melaleuca sieberi 65
Melaleuca squarrosa 187
Melaleuca trichophylla 71
Melaleuca uncinata 14, 15, 21, 69, 70, 96, 97
Melandryidae 203, 210
Melanophthalma australis 208
Meloidae 203
Melophorus sp. 222
Mentophilus hollandiae 217
Meranoplus sp. 221
Meriones tristrami 180, 186
Mesomelena stygia 31
Mesomelaena tetragona 30
Mesostruma sp. 221
Messor sp. 221, 222
Metalasia sp. 91
Microcoryphia sp. 205
Micromeria nervosa 78
Microtus guentheri 180, 186
Mimulus longiflorus 34
Monomorium sp. 221
Monomorium delagoensis 206
Monomorium schultzei 204, 206
Monticola solitarius 191, 192
Mordellidae 203
Motacilla alba 192
Motacillidae 192
Muehlenbeckia hastulata 43
Mulinum spinosum 44, 113
Murmidiidae 210
Mus musculus 186
Muscari commutatum 79
Muscari comosum 79
Muscicapa striata 191
Muscicapidae 191, 192
Myoporum platycarpum 175

Myrceugenia correaefolia 204
Myrica faya 129
Myrmecia sp. 221
Myrmeciinae 221
Myrmecocystus sp. 222
Myrmelachista sp. 222
Myrmica sp. 221
Myrmicinae 220, 221
Myrtus communis 48, 57, 78

Nectarinia osea 191, 192
Nectariniidae 191, 192
Neivamyrmex sp. 221
Nematoda 199, 213
Nerium oleander 78
Neuroptera 202, 213
Nilionidae 210
Nitidulidae 203
Nothidris sp. 221
Nothofagus sp. 179
Nothofagus glauca 113
Nothofagus obliqua 113
Nothofagus pumilio 113
Notoncus sp. 222

Ocotea sp. 91
Ocymyrmex sp. 221
Odonata 213
Odontomachus sp. 221
Olea sp. 61
Olea europaea 46, 78, 131
Olea europaea var. *oleaster* 52
Olea europaea var. *sylvestris* 57
Olearia muelleri 28
Olearia revoluta 28
Oligodesmus michelbacheri 207
Oniticellini 216–218
Onitini 216–218
Onobrychis ebenoides 79
Onosma frutescens 78
Onthophagini 216, 218
Onthophagus lemur 217
Onthophagus minutus 217
Onthophagus rupicapra 217
Onychophora 213
Ophiomorus latastii 194
Ophisaurus apodus 180, 194
Ophisops elegans 194
Opiliones 199
Opilionida 202, 210–213
Orchesella sp. 215, 216
Oribatida 204, 206–208
Orthoptera 199–203, 205, 210, 212, 213
Oryctolagus cuniculus 175
Ostomidae 210
Oxalis parvifolia 44

Oxylobium parviflorum 28
Oxyopomyrmex sp. 221

Paratrechina sp. 222
Paridae 191, 192
Parus major 191, 192
Passer domesticus 191, 192
Passer hispaniolensis 192
Passerina sp. 91
Pauropoda 199, 208
Paussinae 203
Pedilinae 208
Penstemon centranethifolius 37
Peromyscus maniculatus 178
Persea lingue 75, 113
Petrophile drummondii 31
Petrophile media 71
Petrophile striata 30
Peumus boldus 42, 75, 113, 114
Phagnalon graecum 51, 78
Phagnalon rupestre 56
Phalacridae 203
Phasianidae 191, 192
Phasmatodea 202, 213
Pheidole sp. 221, 222
Pheidole capensis 206
Phengodidae 210
Phillyrea sp. 46
Phillyrea angustifolia 48, 49, 57, 75–77
Phillyrea latifolia 47, 50, 76, 78
Phillyrea media 52, 77
Phlomis fruticosa 51, 78
Phlomis viscosa 56
Phoenicurus ochruros 192
Phoenicurus phoenicurus 192
Phoridae 201, 202, 203, 205
Phylloscopus bonelli 192
Phylloscopus collybita 192
Phyllota remota 18, 64
Phymatocarpus porphyrocephalus 31
Picidae 191, 192
Pinus brutia 46
Pinus calabra 46
Pinus clusiana 46
Pinus coulteri 177
Pinus dalmatica 46
Pinus edulis 33
Pinus halepensis 46, 55, 58, 78, 88, 89, 131
Pinus italica 46
Pinus jeffreyi 32
Pinus laricio 46
Pinus mauretanica 46
Pinus mesogeensis 46
Pinus muricata 32
Pinus nigra 46
Pinus pallasiana 46

Pinus pinaster 88, 129
Pinus pinea 46
Pinus ponderosa 32, 33
Pinus radiata 32
Pinus sabiniana 32
Pinus salzmanni 46
Pinus torreyana 32
Pistacia sp. 126
Pistacia lentiscus 46, 47, 54, 55, 57, 58, 75, 76, 78
Pistacia paleastina 53, 55, 56
Pistacia terebinthus 50
Plagiolepis sp. 222
Plantago varia 27
Platanus orientalis 78
Platanus racemosa 177, 178
Platylobium obtusangulum 16
Ploceidae 191, 192
Poa sp. 32
Podanthus mitiqui 42
Podolepis capillaris 28
Podolepis lessonii 28
Podomyrma sp. 221
Poduromorpha 207, 208
Pogonomyrmex sp. 221
Polydesmidae 210
Polyergus sp. 222
Polyrhachis sp. 222
Polyxenidae 210
Pomaderris obcordata 20
Ponera sp. 221
Ponerinae 221
Populus sp. 126
Porlieria chilensis 113, 204
Prasium majus 78
Prenolepis sp. 222
Prinia gracilis 191, 192
Prolasius sp. 222
Prorastriopes sp. 215, 216
Prosopis sp. 33, 126
Prosopis chilensis 113
Prostigmata 204, 206, 208
Protea sp. 91, 200
Protea neriifolia 80
Protea nitida 80
Protea repens 204
Protura 199, 207, 208, 212
Proustia cuneifolia 42, 179
Proustia pungens 204
Pselaphidae 203, 207, 208, 210
Pseudomyrmecinae 220, 221
Pseudomyrmex sp. 221
Pseudoscorpionida 199, 202, 210, 212, 213
Pseudotsuga macrocarpa 177
Pseudotsuga menziesii 33
Psocoptera 199, 200, 202, 207, 208, 210–213
Psoralea bituminosa 78

Pteridium esculentum 14
Pterocelastrus sp. 91
Ptiliidae 207, 208, 210
Ptinidae 207, 210
Ptyodactylus hasselquistii 194
Pultenaea daphnoides 16
Pultenaea myrtoides 65
Pultenaea tenuifolia 21, 69
Punica granatum 78
Puranius sp. 208
Putterlickia sp. 91
Puya berteroniana 43
Puya chilensis 42
Pycnonotidae 191, 192
Pycnonotus barbatus 191, 192
Pygopogidae 9, 193
Pyrus amygdaliformis 78
Pythidae 203

Quercus sp. 32, 126
Quercus aegilops 46
Quercus agrifolia 32, 38, 177, 178
Quercus arizonica 33
Quercus boissieri 46, 53
Quercus calliprinos 46, 53–55
Quercus canariensis 129
Quercus cerris 46
Quercus chrysolepis 211
Quercus coccifera 46, 49, 52, 57, 58, 75, 77, 78, 88, 89
Quercus douglasii 32
Quercus dumosa 32, 35–38, 40, 73, 177
Quercus emoryi 33
Quercus faginea 46, 129
Quercus ilex 46, 47, 50, 52, 76, 77, 88, 89, 211
Quercus infectoria 46
Quercus ithaburensis 46, 56
Quercus lobata 32
Quercus pubescens 46
Quercus pyrenaica 129
Quercus robur 129
Quercus rotundifolia 129
Quercus suber 46, 77, 88, 129, 211
Quercus turbinella 33
Quercus wislizenii 32, 37
Quillaja saponaria 43, 74, 113, 114

Rapanea sp. 91
Rattus norvegicus 186
Rattus rattus 186
Relhamia sp. 91
Restio sp. 28, 91
Restio egregius 80
Retama sp. 126
Rhagodia parabolica 26
Rhamnus alaternus 47, 75, 77
Rhamnus palaestina 54, 55

Rhamnus punctata 53
Rhaphithamnus spinosus 75
Rhododendron ponticum 129
Rhoptromyrmex sp. 221
Rhus sp. 91
Rhus integrifolia 38
Rhus laurina 34
Rhus ovata 40, 72
Rhynchocalamus melanocephalus 194
Rhytidoponera sp. 221
Rosa canina 78
Rosmarinus officinalis 46, 48, 57, 59, 75, 79, 89
Rubia peregrina 49
Rubus canescens 78
Rubus ulmifolius 49
Ruscus aculeatus 50

Salix sp. 177
Salpingidae 210
Salvia sp. 32
Salvia apiana 34
Salvia mellifera 34, 73
Salvia triloba 54, 55
Sarcopoterium spinosum 46, 51, 53–56, 79, 88
Satureja gilliesii 43, 74
Satureja thymbra 55
Saxicola torquata 192
Scaphidiidae 207, 208, 210
Scarabaeidae 202, 203, 216
Scarabaeinde 216, 218
Scarabaeini 216–218
Scarabaeus laticollis 217
Scarabaeus rugosus 217
Schinus montanus 44
Schoenus sp. 65
Schoenus breviculmis 27
Schottia sp. 91
Scincidae 193
Scolopacidae 192
Scolopax rusticola 192
Scolopendromorpha 202
Scolytinae 203, 210
Scorpionida 210, 213
Scutigeromorpha 202
Scydmaenidae 203, 208, 210
Senecio vulgaris 79
Sequoia sempervirens 32
Sequoiadendron giganteum 32
Serinus serinus 192
Sideroxylon sp. 61, 91
Siphonaptera 201, 202, 210, 213
Siphonotidae 210
Sisyrinchium junceum 45
Skytanthus acutus 114
Smilax aspera 50, 78
Sminthuridae 215

Sminthurinus sp. 215
Sminthurus viridis 215
Smithistruma sp. 221
Solenomelus pedunculatus 45
Solenopsis sp. 221
Solidago californica 39
Solpugida 202, 203, 205, 210, 213
Sphaeridia sp. 215
Sphaerotrichopidae 207, 210
Sphinctomyrmex sp. 221
Spyridium spathulatum 17
Spyridium subochreatum 64
Staberoha cernua 80
Staphylinidae 202, 203, 207, 208, 210, 211
Stauracanthus genistoides 59
Stellaria cuspidata 45
Stenamma sp. 221
Stenotaphrum sp. 91
Stigmacros sp. 222
Stipa sp. 32, 84, 100
Stipa eremophila 19
Stipa nitida 22
Stipa pulchra 40
Stirlingia latifolia 29, 71
Strepsiptera 202, 213
Streptopelia decaocto 191, 192
Streptopelia senegalensis 191
Streptopelia turtur 191
Strumigenys sp. 221
Styrax sp. 126
Styrax officinalis 54, 56
Suaeda sp. 126
Sylvia atricapilla 192
Sylvia communis 191
Sylvia curruca 192
Sylvia hortensii 191
Sylvia melanocephala 191, 192
Sylvia rupelli 192
Sylviidae 191, 192
Symphyla 199, 202, 208, 213
Symphypleona 202, 203, 205, 208
Sympieza articulata 80

Talguenea quinquenervia 43
Tamarix sp. 126
Tapinoma sp. 221, 222
Tarsonemini 204, 206, 208
Tauschia parishii 36
Technomyrmex sp. 221
Teiidae 8, 193
Templetonia retusa 20
Tenebrionidae 203, 207, 210, 211
Testudo graeca 194
Tetraclinis articulata 46, 131
Tetramorium sp. 221
Tetraria sp. 91

Tettigonioidea 199
Teucrium chamaedrys 49, 50
Teucrium divaricatum 51
Teucrium polium 51, 79
Thamnochortus sp. 91
Thamnochortus dichotomus 80
Thapsia garganica 79
Themeda sp. 91
Themeda triandra 14, 23, 27, 100
Thymelaea hirsuta 79
Thymelaea tartonraira 79
Thymus capitatus 46, 51, 79
Thysanoptera 199–202, 208, 210, 212, 213
Thysanura 200, 202, 208, 210, 213
Tomocerus sp. 215, 216
Tordylium apulum 79
Trachymesopus sp. 221
Trachypogon sp. 91
Tragopogon sp. 79
Trevoa trinervis 43, 74
Trichocereus chilensis 43
Trichoptera 201
Trifolium stellatum 79
Triodia sp. 82
Triodia irritans 22, 71
Tristachya sp. 91
Trogidae 203, 210
Turdidae 191, 192
Turdus merula 191, 192
Turdus philomelos 192
Typhlopidae 9, 193
Typhlops simoni 194

Ulex parviflorus 77
Umbellularia californica 177
Umbilicus rupestris 79
Upupa epops 191, 192
Upupidae 191, 192
Urginea maritima 79
Uropodina 204, 206–208

Vaccinium obovalis 73
Vaccinium ovatum 72
Valenzuelia trinervis 44, 74
Varanidae 9, 193
Verbascum undulatum 79
Veromessor sp. 221
Viburnum tinus 77
Vicia villosa 79
Vulpes vulpes 175

Wahlenbergia stricta 27
Willowsia sp. 215

Xanthorrhoea sp. 199
Xanthorrhoea australis 14, 18, 24, 64
Xanthorrhoea drummondii 15
Xanthorrhoea resinosa 65
Xanthorrhoea semiplana 16, 17
Xanthorrhoea tateana 14
Xenylla sp. 215
Xylomelum angustifolium 31

Ziziphus sp. 46, 126
Zygophyllum sp. 126
Zygophyllum ovatum 22

General index

abundance, ants 219
Adelaide, S. Aust. 97–99
Adelong, N.S.W. 94, 95
Adrall, Spain 131–133
Adrar, Algeria 120, 124
Afghanistan 144, 145
Aflou, Algeria 119, 123
Africa 141, 145, 216, 218
Afromontane forest 91
Agrinion, Greece 125, 126
Ahmadi, Kuwait 120, 124
Ahwaz, Iran 120, 124
Ain El Ksar, Algeria 119, 123
Akko, Israel 126–128
Aknoul, Morocco 118, 122
Al Hufuf, Saudi Arabia 120, 124
Albany, W. Aust. 104–106
Albertacce, Corsica 118, 122
Albury, N.S.W. 94–96
Alcobaca, Port. 129, 130
Aleppo, Syria 119, 123
Alexandroupolis, Greece 125, 126
Alicante, Spain 131–133
Allone Abba, Israel 88
Allonim, Israel 88
Almeria, Spain 118, 122
Alvalade, Port. 129, 130
Amazon Basin, Brazil 159
Amman, Jordan 119, 123
amphibians 8
amphiphyte 5, 16–60
Anamur, Turkey 119, 123
Andes Mtn, Chile 44, 159
Angol (El Verg), Chile 113, 115, 116
Angorichina, S. Aust. 96, 98, 99
Aniane, France 89
Anoghia, Crete 125, 126
Antioch, Cal. 107–109
Antofagasta, Chile 114–116
ants 204, 214, 218–223
Aoulef El Ared, Algeria 120, 124

aquatic mammals 8, 171–194
Arak, Iran 120, 124
Ararat, Vic. 100, 102, 103
arboreal mammals 8, 171–194
Argostolion, Greece 125, 126
arid climate 4, 140
arid fynbos 61
Aridity Index (Budyko) 142
Arizona, U.S.A. 33, 110–112
Aschad, U.S.S.R. 120, 124
Ashdot Yaaqov, Israel 126–128
Asia 141, 145
Astakos, Greece 78
Athens, Greece 51, 125, 126
Atlantic Ocean 143, 146
Attica, Greece 78
Auburn Dam, Cal. 106, 108, 109
Augusta District, W. Aust. 85
Australia – mediterranean 149–155
Australia – subtropical 152, 153
Australia – tropical 152–154
autotrophe 6, 16–60
Avedat, Israel 126–128
Avenal, Cal. 107–109
Averroes, Morocco 118, 122
Avoca, Vic. 100, 102, 103
Awbari, Libya 120, 124
axyle 7, 16–60

Bab-Bou-Idir, Morocco 118, 122
Backus Ranch, Cal. 107, 108, 110
Badajoz, Spain 118, 122
Badgingarra, W. Aust. 85, 104, 105
Badjaling N.R., W. Aust. 183, 189
Baget, Spain 131–133
Baghdad, Iraq 120, 124
Bahrain 119, 123
Baja California, Mex. 86
Balkans 145
Balladonia, W. Aust. 104–106
Ballan, Vic. 100, 102, 103

Ballarat, Vic. 100, 102, 103
Balmaceda, Chile 113, 115, 116
Balranald, N.S.W. 67, 82, 94, 95
Baños de Jahuel, Chile 113, 115
Banyuls, France 211
Barcelona, Spain 131–133
bark consistency 7, 16–60
bark shedding 7, 16–60
bark shedding – seasonality 7, 16–60
bark thickness 7, 16–60
Bartlett Dam, Ariz. 110–112
Bas Languedoc, France 89
Baskale, Turkey 119, 123
Basra, Iraq 120, 124
batha 88, 126
bats 8, 171–194
Baviaanskloof Mts, S. Afr. 90
Bayramic, Turkey 119, 123
beaded short roots 7, 16–60
Beaufort West, S. Afr. 134–136
Beaumont Pump Plant, Cal. 107–109
Beechworth, Vic. 100, 102, 103
Beer Sheva, Israel 118, 122, 126–128
Beerwah, Qld 65
beetles, dung 216–218
Beja, Port. 129, 130
Belair, S. Aust. 97–99, 214
Belair R.P., S. Aust. 183, 188
Bell Canyon Saddle, Cal. 73
Benalla, Vic. 100, 102
Bencubbin, W. Aust. 104, 105
Bendigo, Vic. 100, 102, 103
Beniatjer, Spain 131–133
Berkane, Morocco 118, 122
Berri, S. Aust. 97–99
Berryessa Lake, Cal. 106, 108, 109
Bet Dagan, Israel 126–128
Bet Netofa, Israel 126–128
Bet Oren, Israel 184
Bet Qama, Israel 126–128
Bet Sehan, Israel 119, 123
Beulah, Vic. 100, 102
Billiatt C.P., S. Aust. 175, 183, 188
Billyacatting Hill N.R., W. Aust. 183, 189
Bingol, Turkey 120, 124
Bird Island, S. Afr. 134–136
birds 8, 171–194
Birecik, Turkey 120, 124
Black River Pumps, Ariz. 110–112
Blanchetown, S. Aust. 67
Boca, Cal. 107–109
Boghari, Algeria 119, 123
Boica, Port. 129, 130
Bontevok N.P., S. Afr. 91
Boort, Vic. 100, 102, 103
Booylgoo, W. Aust. 104–106

Bordertown, S. Aust. 68, 97, 98
Braganca, Portugal 118, 122
Brannan Island, Cal. 106, 108, 109
Brazil 159–165
Briançon, France 117, 121
Bridgetown, W. Aust. 104–106
Brisbane Ranges, Vic. 24, 83
Broken Hill, N.S.W. 94–96
Brookfield C.P., S. Aust. 22, 174, 183, 185, 188, 193
Browns Valley, Cal. 218
Bucak, Turkey 119, 123
bulb 7, 16–60
Bunbury, W. Aust. 104–106
Bundaleer F.R., S. Aust. 97, 98
Bungalbin, W. Aust. 183, 185, 189, 193
Burlingame, Cal. 107–109
Burtkraal, S. Afr. 91
Bushire, Iran 119, 123
bushveld 134
Butler Valley Ranch, Cal. 106–108

Cabo Carvoeiro, Port. 129, 130
Cachagua, Chile 42
Cachuma Lake, Cal. 107–109
Cairo, Egypt 119, 123
Caldera, Chile 114–116
California, U.S.A. 32, 34–40, 72, 73, 86, 106–110, 159–165,
    176, 183, 185, 190, 193, 198, 202, 211, 212, 214, 216–218
Caltrans Vernal Pool, Cal. 38
Camp Pardee, Cal. 107–109
Camp Pendleton, Cal. 107–109
Camperdown, Vic. 100, 102, 103
campina 160
Canberra, A.C.T. 94–96
Capdella, Spain 131–133
Cape Agulhas, S. Afr. 134–136
Cape Borda, S. Aust. 97–99
Cape Columbine, S. Afr. 134–136
Cape de Couedic, S. Aust. 97–99
Cape Naturaliste, W. Aust. 104–106
Cape Nelson, Vic. 100, 102, 103
Cape Northumberland, S. Aust. 97–99
Cape Otway, Vic. 100, 102, 103
Cape Province, S. Afr. 61, 80, 91, 134–136, 160, 180, 184, 216
Cape Riche, W. Aust. 85
Cape Schanck, Vic. 101–103
Cape St Blaize, S. Afr. 134–136
Cape St Francis, S. Afr. 134–136
Cape Town, S. Afr. 91, 134–136, 159
Caramulo, Port. 129, 130
Carmel, Cal. 86
Carnamah, W. Aust. 104–106
carnivore 8, 171–194
carnivorous plant 6, 16–60
carrascal 57
Castellfort, Spain 131–133

Castellón, Spain 131–133
Castelo Branco, Port. 129, 130
Castlemaine, Vic. 101–103
cat, feral 175
Catalonia, Spain 58
caudex 7, 16–60
Cauquenes, Chile 113, 115, 116
Cauro, Corsica 47, 76, 88
Cazevieille, France 217
Ceduna, S. Aust. 96, 98
cellulose 63–80, 160–165
Central Valley, Chile 45
Cerro La Campana, Chile 160
Cerro Moreno, Chile 114, 115
chamaephyte 5, 16–60
Chañaral, Chile 114, 115
chaparral 32, 35–38, 40, 72, 73, 86, 158–165, 202, 219
Charlton, Vic. 100, 102, 103
Chelva, Spain 131–133
chenopod shrubland 82
Cheyne Beach, W. Aust. 85
Chico Univ. Farm, Cal. 106, 108, 109
Chile 41–45, 74, 75, 87, 113–116, 159–165, 179, 183, 185, 193, 198, 204, 208–210, 214, 216, 217
Chile Chico, Chile 113, 115
Chillán, Chile 113, 115, 116
Chula Vista, Cal. 107–109
Cihanbeyli, Turkey 119, 123
Citrusdal, S. Afr. 134–136
cladode 6, 16–60
Clare, S. Aust. 97–99
climagram – Emberger 4, 140
climatic type – Emberger 3, 4, 94–136, 140–146
climatic type – Köppen 3, 4, 94–136
closed shrubland 88
cloud forest 204, 208
Coast Range, Chile 159
coastal fynbos 61
coastal sage scrub 32, 34, 86
coastal succulent scrub 32, 86
Cofrentes, Spain 131–133
Coimbra, Portugal 118, 122
Colac, Vic. 100, 102, 103
Colina, Chile 45, 113, 115
Collie, W. Aust. 104–106
Concepción, Chile 113, 115, 116
Concon, Chile 217, 218
Condobolin, N.S.W. 94, 95
Constitución, Chile 113, 115, 116
Contulmo, Chile 113, 116
contractile roots 7, 16–60
Cook, S. Aust. 96, 98, 99
Cootamundra, N.S.W. 94–96
Copiapo, Chile 114, 115
coppice forest 88, 89
Corackerup N.R., W. Aust. 85

Cordillera de la Costa, Chile 75, 160
Cordillera de los Andes, Chile 74, 160
Corfu, Greece 125, 126
corm 7, 16–60
Corsica, France 47, 48, 76, 88, 216, 217
Coyhaique, Chile 113, 115, 116
Cranbourne, Vic. 173, 183, 187
Creswick, Vic. 101, 102
Crete 126
crown diameter 6, 16–60
cryptocotyly 8, 16–60
cryptophyte 5, 16–60
Cullinco, Chile 113, 115, 116
Curicó, Chile 113, 115, 116
Curtin University (South Bentley), W. Aust. 104, 105
Cuyamaca Lake, Cal. 107–109
cyperoid rootlets 7, 16–60

Danger Point, S. Afr. 134–136
Dark Island Soak, S. Aust. 18, 21, 64, 69, 83
Dasseneiland, S. Afr. 134–136
dauciform rootlets 7, 16–60
Davis Dam, Ariz. 110–112
Davis Expt. Farm, Cal. 106, 108, 109
Death Valley, Cal. 107, 108, 110
deciduous leaves 157–159
Deepwalls, S. Afr. 134–136
Dehibat, Tunisia 119, 123
Del Mar, Cal. 38
Denia, Spain 131–133
Deniliquin, N.S.W. 82, 94–96
Denmark Res. Stn, W. Aust. 104, 105
Descanso, Cal. 160
Désert des Agriates, Corsica 48, 88
Desulo, Sardinia 119, 123
dietary type 8, 171–194
dimorphic leaves 78
dingo 173
Diyarbakir, Turkey 120, 124
Doha, Qatar 120, 124
Donald, Vic. 100, 102
Doñana N.P., Spain 59, 60
Dongolocking N.R., W. Aust. 183, 185, 189, 193
Donnybrook, W. Aust. 104–106
Dookie, Vic. 100, 102, 103
Douglas, Ariz. 110–112
Dubbo, N.S.W. 94–96
dune fynbos 61, 91
dung beetles 216–218
Durdiwarrah, Vic. 100, 102, 103
Durokoppin N.R., W. Aust. 28, 183, 193
Duttons Landing, Cal. 106, 108, 109
Dwellingup, W. Aust. 104–106, 198, 199

East London, S. Afr. 134–136
East Yuna N.R., W. Aust. 85, 183, 193

Echo Valley, Cal. 72, 107–109
Echuca, Vic. 100, 102, 103
Eclipse Island, W. Aust. 104–106
ecomorphological characters 5–8, 16–60
ecomorphosis 215
Egypt 145
El Adem, Egypt 118, 122
El Belloto, Chile 113, 115
El Bosque, Chile 114, 115
El Kom, Syria 120, 124
El Meghaim, Algeria 120, 124
El Roble, Chile 179, 183, 185, 193
El Rosario, Mex. 86
El Sharqiyah, Libya 119, 123
El Teniente, Chile 113, 115, 116
El Verg (Angol), Chile 115, 116
Elandsberg Mts, S. Afr. 91
Elat, Israel 126–128
Elazig, Turkey 120, 124
Elgin, S. Afr. 134–136
Emberger climatic type 4, 94–136, 140
Emberger-Giacobbe Index 140, 142
Emberger Pluviothermic Quotient 3, 4, 94–136, 139, 142
En Nebek, Syria 118, 122
En Shemer, Israel 126–128
Eneabba, W. Aust. 31, 85, 104, 105, 220
Enguera, Spain 131–133
Ensenada, Mex. 86
epicormic buds 7, 16–60
epigaeic invertebrates 9, 197–226
epigeal 8, 16–60
Eregli, Turkey 118, 122
Erez, Israel 126–128
ericoid heathland 157–167
Erivan, U.S.S.R. 120, 124
Erzigan, Turkey 120, 124
Eslida, Spain 131–133
Esperance, W. Aust. 104–106
espinal 45, 87
Estany Gento, Spain 131–133
ether extractives 63–80, 160–165
Eucla, W. Aust. 104–106
Eudunda, S. Aust. 97, 98
Europe 142, 216, 218
Euston, N.S.W. 67, 94, 95
evaporative coefficient (k) 4, 94–136, 149–154
Even Sapir, Israel 126–128
evergreen leaves 157–167

Faizabad, Afghanistan 119, 123
Fall River Mills, Cal. 107–109
Farina, S. Aust. 96, 98, 99
Faro, Port. 129, 130
Ferndale, Cal. 106, 108, 109
Ferries-McDonald C.P., S. Aust. 70
Fethiye, Turkey 119, 123

Fez, Morocco 119, 123
Figueira de Castelo Rodrigo, Port. 129, 130
Figueres, Spain 131–133
fire 69, 77, 173, 222
Fleming, Cal. 107–109
Flicker Ridge, near Berkeley, Cal. 73
Flix, Spain 131–133
Florida, U.S.A. 159–165
flowering – pyrogenic 8, 16–60
flowering – seasonality 7, 16–60
foliage projective cover 5, 150–154
foliar nutrients – calcium 8, 63–80, 157–167
foliar nutrients – magnesium 8, 63–80, 157–167
foliar nutrients – nitrogen 8, 63–80, 157–167
foliar nutrients – phosphorus 8, 63–80, 157–167
foliar nutrients – potassium 8, 63–80, 157–167
Folsom Dam, Cal. 107–109
foraging zone 8, 171–194
Forbes, N.S.W. 94–96
forest 32, 33, 39, 46, 50, 75, 77, 82, 84, 88, 91, 117, 207, 211, 213, 219
forest – cloud 204, 208
forest – deciduous 41, 46, 117, 211
forest – open 14–16, 23, 24, 63, 82, 83, 88, 89, 117
forest – sclerophyll 14–16, 24, 41, 63, 83, 84
forest – tall open 14, 15
Forrest, W. Aust. 104, 105
Fort Valley, Ariz. 110–112
Forty Oaks, Israel 88
fossorial mammals 8, 171–194
Fowlers Bay, S. Aust. 96, 98, 99
fox 175
France 49, 50, 75–77, 88, 89, 142, 143, 159–165, 198, 211–213, 216, 217
Frankston, Vic. 101–103
Fray Jorge N.P., Chile 208–210
Fremantle, W. Aust. 104–106
Friant Govt. Camp, Cal. 107–109
fruit dehiscence 8, 16–60
Fundo Santa Laura, Chile 43, 74, 87, 160
funnel extraction 9, 212, 214
fynbos 90, 91, 134, 159–165, 181, 198, 200–203, 213, 219, 220
fynbos – arid 61, 90
fynbos – coastal 61, 134
fynbos – dune 91
fynbos – grassy 91
fynbos – mountain 61, 80, 91, 134, 160

Gabes, Tunisia 119, 123
Gafsa, Tunisia 119, 123
Galil/Galilee, Israel 53–56, 179, 180, 184, 186, 190, 191, 193, 194
Gamtoos, S. Afr. 91
Garraf, Spain 58
garrigue 46, 49, 58, 88, 89, 126, 217, 218
Gat, Libya 118, 122

Gaza, Israel 118, 122
Geelbeck, S. Afr. 217
Geelong, Vic. 100, 102, 103
Gellibrand Hill, Vic. 25, 83
geophyte 5, 16–60
George, S. Afr. 90, 134–136
Georgetown, S. Aust. 97, 98
Geraldton, W. Aust. 85, 104–106
Gerar, Israel 126–128
Gerona, Spain 131–133
Gilat, Israel 126–128
Gilboa Mts, Israel 179, 184, 186, 190, 191, 194
Gilet, Spain 131–133
Gjirokaster, Albania 119, 123
Gonabad, Iran 119, 123
Goolgowi, N.S.W. 70
Graaf Reinet, S. Afr. 134–136
Grahamstown, S. Afr. 91, 134–136
Grand Canyon, Ariz. 110–112
grassland 14, 89, 117, 131
grassland – tussock 27
Grasspatch, W. Aust. 84
Great Fish Point, S. Afr. 134–136
Greece 51, 52, 78, 79, 88, 125, 126, 142
Green Hill, S. Aust. 65
Grenfell, N.S.W. 94, 95
Griffith, N.S.W. 94–96
Grizzly Island, Cal. 107–109
Groot Drakenstein, S. Afr. 134–136
Groot Swartberg, S. Afr. 200
ground mammals 8, 171–194
Growth Indices 4, 5, 94–136
Guardamar, Spain 131–133
Guercif, Morocco 119, 123

H.4, Jordan 119, 123
Hadim, Turkey 119, 123
Hai, Iraq 120, 124
Haleakala N.P., Maui, Hawaii, 160
Halkidiki, Greece 52
Hamilton, Vic. 100, 102, 103
Hawaiian Islands 159, 160
Hawker, S.A. 96, 98
Hawley Lake, Ariz. 107–109
Hay, N.S.W. 94–96
Hay Plain, N.S.W. 82
heathland 60, 61, 84, 85, 90, 157–167, 219
heathland – arid 61
heathland – dry 14, 18, 64, 83
heathland – open 18, 64, 90
heathland – shrub 15
heathland – wet 65, 83
Hefzi Bah Gilib, Israel 119, 123
height 6, 16–60
Hellinikon (Athens), Greece 51, 126
hemicryptophyte 5, 16–60

Herault, France 77
herbaceous leaves 5, 16–60
herbivore 8, 171–194
Hermitage, S. Afr. 134–136
Hetch Hetchy, Cal. 107–109
Hinis, Turkey 120, 124
holoxyle 7, 16–60
Horsham, Vic. 100, 102, 103
Horta, Azores 117, 121
Hula Farm, Israel 126–128
Hulata, Israel 126–128
Humansdorp, S. Afr. 91
Hume Reservoir, N.S.W. 94–96
humid climate 140–146
hygrophilous forest 75, 159, 160
Hymettus, Greece 51
hypogeal 8, 16–60

Ibiza, Spain 131–133
Ierapetra, Crete 125, 126
Ifrane, Morocco 118, 122
Igualada, Spain 131–133
Ile D'Yeu, France 117, 121
India 142
Indices, Growth 4, 5, 94–136
Indio Date Garden, Cal. 107, 108, 110
Inherm, Morocco 119, 123
Innes N.P., S. Aust. 172, 174, 183, 185, 189, 193
insectivore 8, 171–194
invertebrates 9, 197–226
invertebrates, epigaeic 9, 197–226
invertebrates, litter 9, 197–226
invertebrates, soil 9, 197–226
Iran 145
Iraq 145
Isernia, Italy 118
Isfahan, Iran 120, 124
Israel 53–56, 88, 126–128, 179, 184, 186, 190, 191, 193, 194, 219
Italy 142
Ivanhoe, N.S.W. 94–96
Izmir, Turkey 119, 123

Jafr, Jordan 119, 123
Jansenville, S. Afr. 134–136
jarrah forest 84, 198, 199
Jarrahdale, W. Aust. 84
Jask, Iran 118, 122
Jaspe, Portugal 57
Jeddah, Saudi Arabia 118, 122
Jerilderie, N.S.W. 82
Jijona, Spain 131–133
Jonkershoek, S. Afr. 80, 134–136, 159, 160, 201, 203, 205–207, 219
Jordan Valley 219
Juncal, Chile 44, 114, 115

Jurien Bay, W. Aust. 29, 104, 105

Kadina, S. Aust. 97, 98
Kaffrarian thicket 91
Kaffrarian succulent thicket 91
Kalamata, Greece 125, 126
Kalamunda, W. Aust. 104–106
Kalbarri N.P., W. Aust. 85
Kalgoorlie, W. Aust. 104–106, 176
Kandhila, Greece 78
Kapunda, S. Aust. 97–99
Karatas, Turkey 119, 123
Karistos, Greece 78
karoid shrubland 90, 134
Karragullen, W. Aust. 199
karri tall open-forest 183, 185, 189, 193
Karridale, W. Aust. 104–106, 183, 185, 189, 193
Kasbah Tadla, Morocco 119, 123
Katanning, W. Aust. 104–106
Kavalla, Greece 125, 126
Kazalinsk, U.S.S.R. 120, 124
Kefallinia, Greece 126
Kefar Yehezqel, Israel 126–128
Keith, S. Aust. 18, 19, 21, 64, 68, 69, 83, 97–99
Keles, Turkey 118, 122
Kellerberrin, W. Aust. 104–106
Kerang, Vic. 100, 102, 103
Kettleman City, Cal. 107–109
Kew, Vic. 101–103
Khanaqin, Iraq 120, 124
Ki Ki, S. Aust. 68
Kimba, S. Aust. 96, 98
Kimi, Greece 125, 126
Kinchega N.P., N.S.W. 82
King William's Town, S. Afr. 134–136
Kingscote, S. Aust. 97–99
Kishon Reservoir, Israel 126–128
Knights Ferry, Cal. 107–109
Knysna, S. Afr. 90, 134
Konitsa, Greece 125, 126
Köppen climatic type 4, 94–136
Kodj Kodjin N.R., W. Aust. 28
Koeberg, S. Afr. 91
Kogelberg, S. Afr. 80, 159, 160
Konya, Turkey 119, 123
Koolau Mtns, Oahu, Hawaii 160
Kozani, Greece 125, 126
Kuitpo, S. Aust. 199, 200, 214
Kunduz, Afghanistan 120, 124
Kurkes, Albania 119, 123
Kwam Umbu, Egypt 118, 122
kwongan 30, 31, 71, 84
Kyancutta, S. Aust. 96, 98, 99
Kybybolite, S. Aust. 97–99
Kyeema, S. Aust. 199, 200, 214

La Coruna, Spain 117, 121
La Serena, Chile 114–116
Laghouat, Algeria 119, 123
Laguna Beach, Cal. 73
Lahav, Israel 126–128
Lake Grace, W. Aust. 104, 105
Lake Solano, Cal. 107–109
Lake Spaulding, Cal. 107–109
Lake Victoria, N.S.W. 94–96
Lakeshore, Cal. 107–109
Lameroo, S. Aust. 97, 98
Langgewens, S. Afr. 134–136
Larache, Morocco 118, 122
Laverton, Vic. 100, 102, 103
Laverton, W. Aust. 104–106
leaf angle 6, 16–60
leaf area 6, 16–60, 150, 151
leaf area index 150–152
leaf area: assim. stem area 6, 16–60
leaf colour 6, 16–60
leaf consistency 6, 16–60
leaf duration 6, 16–60, 158
leaf glands 6, 16–60
leaf length 6, 16–60
leaf margin 6, 16–60
leaf – number per 10 cm stem 8, 16–60, 152
leaf nutrients 63–80, 157–167
leaf seasonality 6, 16–60
leaf size 6, 16–60
leaf specific area 68–80
leaf stomata 6, 16–60
leaf structure 8, 157–167
leaf surface resins 6, 16–60
leaf thickness 63–80
leaf tomentosity 6, 16–60
leaf width 6, 16–60
Lebanon 219
leptophyll 6, 16–60
Lérida, Spain 131–133
Lesvos, Greece 125, 126
life duration 6, 16–60
Light Index 151
lignin 63–80, 160–165
lignotuber 7, 16–60
Lime Lake, W. Aust. 85
Linares, Chile 113, 115, 116
Lisbon, Port. 129, 130
litter 9
Little Panoche Dam, Cal. 107–109
Little Desert N.P., Vic. 70
littoral mediterranean climate 140–146
Llay-Llay, Chile 114, 115
Lluch, Spain 131–133
Lod, Israel 126–128
Lodi, Cal. 107–109
Lonquimay, Chile 113, 115, 116

Los Andes, Chile 114–116
Los Angeles, Cal. 73, 86, 177
Los Angeles, Chile 113, 115, 116
Los Banos, Cal. 107–109
Los Cerrillos, Chile 114, 115
Los Dominicos, Chile 179, 184, 186, 190, 193
Lovedale, S. Afr. 134–136
Lower Munroe Canyon, Cal. 73
Lucindale, S. Aust. 97, 98

Macã, Port. 129, 130
macchia 131, 134
macrophyll 6, 16–60
macrotherm 152
Mafraq, Jordan 118, 122
Mahón, Spain 131–133
Maian, Jordan 119, 123
Maitland, S. Aust. 97, 98
malacophyll 6, 16–60
Malakiyen, Syria 120, 124
Malazgirt, Turkey 120, 124
Malibu Canyon, Cal. 73
mallee 14, 20–22, 26, 66–71, 82, 84
mallee – broombush 21, 69, 70, 83
Mallee Cliff N.P., N.S.W. 82
mammals 8
Manaus, Brazil 159, 160
Mandeville Island, Cal. 107–109
Mandura Motel, W. Aust. 104, 105
Manjimup, W. Aust. 104–106
Manteca, Cal. 107–109
Many Farms, Ariz. 110–112
maquis 46–48, 52, 75, 78, 88, 126, 159, 160, 211, 213, 216, 217
maquis arboré 76, 77, 88
Markley Cove, Cal. 107–109
Marree, S. Aust. 96, 98, 99
Maryborough, Vic. 101–103
Marseille, France 75, 159, 160
Mas d'Abeilles, France 211, 212, 216
Mascara, Algeria 118, 122
Mashad, Iran 119, 123
Mashed Soleyman, Iran 120, 124
Masirah Island, Oman 118, 122
Matakana, N.S.W. 82
Matmata, Tunisia 119, 123
matorral 41, 43, 46, 74, 157–167, 204, 208, 211, 219
matorral – coastal 42, 74, 87, 159, 160, 165
matorral – montane 44, 74, 87, 159, 160, 165
matorral – sub-Andean 159
matorral – typical 87, 160, 165
Matroosberg, S. Afr. 134–136
Mazari Sharif, Afghanistan 120, 124
McDermid Rock, W. Aust. 183, 185, 189, 193
McNary, Ariz. 110–112
Medina, Saudi Arabia 119, 123
Mediterranean Basin 46, 88, 89, 117–124

mediterranean climate 140–146
mediterraneaneity 139
Medrignac, France 117, 121
megaphanerophyte 5, 16–60
megaphyll 6, 16–60
megatherm 152
Melbourne, Vic. 101–103
Melipilla, Chile 184, 190
Melton, Vic. 26, 66
Menia, Egypt 119, 123
Meningie, S. Aust. 97, 98
Menzies, W. Aust. 104–106
Merbein, Vic. 100, 102, 103
Merredin, W. Aust. 104–106
Mesa Expt. Stn, Ariz. 110–112
mesophanerophyte 5, 16–60
mesophyll 6, 16–60
mesotherm 5, 152
Mesudiye, Turkey 118, 122
Methoni, Greece 125, 126
Mevo Betar, Israel 126–128
Mexico 86
microphanerophyte 5, 16–60
microphyll 6, 16–60
microtherm 5, 152
Middleback Station, S. Aust. 175, 183, 188
Mildura, Vic. 67, 100, 102, 103
Millau, France 117, 121
mineral nutrients 63–80
Miniyah, Egypt 119, 123
Mishmar Ayyalon, Israel 126–128
Mishmar Ha Negev, Israel 126–128
Mizpe Ramon, Israel 126–128
Moisture Index (M.I.) 4, 5, 94–136, 149–154
Mojave, Cal. 107, 108, 110
Molina, Chile 113, 115, 116
Monchique, Port. 129, 130
Monemvasia, Greece 78
Montagu, S. Afr. 134–136
Monte Real, Port. 129, 130
Montemór-o-Velho, Port. 129, 130
Monterey Co. Cal. 72, 159
Monterey Pen., Cal. 160
Monticello Dam, Cal. 107–109
Montpellier, France 88, 89, 142
Montseny, Spain 131–133
Montserrat, Spain 131–133
Moorlands, S. Aust. 67
Moraga Ridge, Cal. 73
Mornington, Vic. 101–103
Mostaganem, Algeria 118, 122
mountain grassland 131
mountain fynbos 80
Mt Auriol (Beziers), France 77
Mt Barker, S. Aust. 97–99
Mt Barker, W. Aust. 104–106

Mt Burr Forest, S. Aust. 97–99
Mt Carmel, Israel 88, 179, 180, 184, 186, 190, 193, 194
Mt Crawford Forest, S. Aust. 97, 98
Mt Dandenong, Vic. 23, 83
Mt de la Gardiole, France 88
Mt Diablo, Cal. 86
Mt Eliza, Vic. 101–103
Mt Gambier, S. Aust. 97–99
Mt Gilboa, Israel 88, 180
Mt Hope, N.S.W. 95, 96
Mt Laguna, Cal. 107–109
Mt Lesueur, W. Aust. 30, 85
Mt Lofty Ranges, S. Aust. 65, 174, 199, 220
Mt Lofty Summit, S. Aust. 16, 63, 83
Mt Maures, France 88
Mt Meron, Israel 88, 184
Mt Ney, W. Aust. 84
Mt Rescue C.P., S. Aust. 175, 183, 188
Muhraqa, Israel 88, 184
Mungo N.P., N.S.W. 82
Munroe Canyon, Cal. 73
Murcia, Spain 118, 122
Muresk, W. Aust. 104–106
Murray Bridge, S. Aust. 97, 98
Murzuq, Libya 120, 124
Mus, Turkey 120, 124
Myall Lakes N.P., N.S.W. 172, 174, 183, 185, 193
Mycorrhizal rootlets 7, 16–60
Myponga, S. Aust. 97, 98
Myrtleford, Vic. 100, 102, 103

N-fixation 6, 16–60
Nacimiento Dam, Cal. 107–109
Nadgee N.R., N.S.W. 172, 183, 185, 187, 193
Nahal Shiqma, Israel 126–128
Nahariyya, Israel 126–128
Najaf, Iraq 120, 124
nanophanerophyte 5, 16–60
nanophyll 6, 16–60
nanotherm 5, 152
Nanya, N.S.W. 82
Naracoorte, S. Aust. 97, 98
Narcut (Comet Bore) C.P., S. Aust. 183, 188
Narrandera, N.S.W. 94, 95
Narrogin, W. Aust. 104, 105
Narweena, N.S.W. 82
Negba, Israel 126–128
Net Photosynthetic Index 151
Neve Ya'ar, Israel 88
New Melones Dam, Cal. 107–109
New South Wales 66, 67, 70, 82, 83, 94–96
Newark, Cal. 107–109
Nhill, Vic. 100, 102, 103
Nicosia, Cyprus 119, 123
Nir Yizhaq, Israel 126–128
No Mans Land, N.S.W. 82

nodules, root 7, 16–60
Nogales, Ariz. 110–112
Nord-Salang, Afghanistan 118, 122
Nornalup N.P., W. Aust. 85
North Tarin Rock Res., W. Aust. 183, 185
Northam, W. Aust. 104, 105
Nowa Nowa S.F., N.S.W. 172, 173, 183, 185, 187, 193
Nugadong N.R., W. Aust. 183, 189
Numurkah, Vic. 100, 102, 103
Nuriootpa, S. Aust. 97, 98
nutrient content (leaf) 63–80, 157–167

oak forest 46, 77, 107, 125, 131
oak woodland 32, 33, 86, 88
Oakdale Woodward Dam, Cal. 107–109
oakland 32, 129
oakland – deciduous 129
Oleo – Ceratonion 129
Oliana, Spain 131–133
Olot, Spain 131–133
omnivore 8, 171–194
Oodnadatta, S. Aust. 96, 98
open scrub 14, 15, 20–22, 26, 59, 61, 66–71, 90
open shrubland 88
Orange Co., Cal. 86
Orange Free State, S. Afr. 204
organs shed – seasonality 6, 16–60
organs periodically shed 6, 16–60
Orlando, Florida 160
Orotava, Canaries 118, 122
Osorno, Chile 113, 115, 116
Oudtshoorn, S. Afr. 134–136, 200
Outeniqua Mts, S. Afr. 90
Ouyen, Vic. 100, 102
Ovalle, Chile 114, 115
overstorey – Foliage Projective Cover 151
overstorey – species richness 153

Page, Ariz. 110–112
Pakistan 142, 145
Palaearctic Region 139–148
Palermo, Sicily 118, 122
Palma de Mallorca, Spain 131–133
Palomar Mountain, Cal. 40
Pan Ban Lake, N.S.W. 82
Panimávida, Chile 113, 115, 116
Papudo, Chile 42, 87
Para Wirra R.P., S. Aust. 17, 83, 183, 187
parasitic plant 6, 16–60
Parkes, N.S.W. 94, 95
Paros, Greece 78
Paso Marchant, Chile 44, 87
Pec, 121
Peloponnisos, Greece 126
Pemberton Forest, W. Aust. 104, 105
Peñablanca, Chile 114, 115

Penhas douradas, Port. 129, 130
Penmarch, France 117, 121
per-humid climate 4, 140
Perth, W. Aust. 104–106
Perup, W. Aust. 183, 185, 189, 193
phanerocotyly 8, 16–60
phanerophyte 5, 16–60
photosynthesis 159
photosynthetic organs 6, 16–60
phrygana 46, 51, 78, 79, 88
phyllode 6, 16–60
Pichidangui, Chile 74, 160, 184, 190
pine forest 32, 33, 46, 125, 131
pinewood 129
Pingelly, W. Aust. 104, 105
Pisa, Italy 118, 122
pitfall traps 9, 201, 211, 214
Placerville IFG, Cal. 107–109
plumule position 8, 16–60
Pluviothermic Quotient (Emberger) 3, 4, 94–136, 139
Poitiers, France 117, 121
Pollensa, Spain 131–133
Polpaico, Chile 45, 87
Ponta Delgada, Azores 118, 122
Port Augusta, S. Aust. 96, 98, 99
Port Elizabeth, S. Afr. 134–136
Port Lincoln, S. Aust. 96, 98, 99
Port Pirie, S. Aust. 97–99
Portland, Vic. 100, 102, 103
Porto/Serra P., Port. 129, 130
Portugal 57, 129, 130
post-fire succession 69, 77, 173
Potrerillos, Chile 114–116
protein-crude 63–80
proteoid rootlets 7, 16–60
proteoid shrubland 90, 157–167
Province, France 77
Puchuncavi, Chile 184, 190
Puech-du-Mas-du-Juge, France 49
Puechabon, France 50, 88, 89
Puigcerda, Spain 131–133
Pulletop N.R., N.S.W. 82
Pulumur, Turkey 119, 123
Punta Angeles, Chile 113, 115, 116
Punta Carranza, Chile 113, 115, 116
Punta Lavapié, Chile 113, 115, 116
Punta Tortúga, Chile 114–116
Punta Tumbes, Chile 113, 115, 116

Qadmous, Syria 118, 122
Quebrada de la Plata, Chile 179
Queenscliffe, Vic. 100, 102, 103
Queensland, Aust. 65
Quercetum cocciferae 77
Quercetum ilicis 77
Quillota, Chile 114–116

Quintero, Chile 114, 115
Qumren, Jordan 118, 122

rainforest 207, 208
Ramat David, Israel 119, 123
Ramat Yishay, Israel 126–128
Rancagua, Chile 113, 115, 116
Rancho Bernardo, Cal. 34
Rankins Springs, N.S.W. 83
Rapha, Saudi Arabia 120, 124
Ras Tammurah, Saudi Arabia 120, 124
Ravensthorpe, W. Aust. 84
Rawlinna, W. Aust. 104–106
Refresco, Chile 114–116
Reggane, Algeria 120, 124
renewal buds 5, 16–60
Rengo, Chile 113, 115
Renmark, S. Aust. 97, 98
renosterveld 61, 90, 91, 134
reptiles 8, 171–194
Requena, Spain 131–133
resilience 215
restioid heath 90, 157–167
restioid rootlets 7, 16–60
rhizode 7, 16–60
rhizome 7, 16–60
Rhodes, Greece 125, 126
Rinconada, Chile 184, 186, 193
Rio Bueno, Chile 113, 115, 116
Rio Cisnes, Chile 113, 115, 116
Riversdale, S. Afr. 134–136
Riverside Citrus Expt. Station, Cal. 107–109
Robe, S. Aust. 97–99
Rodhos, Greece 125, 126
Rome, Italy 118, 122
Roosevelt, Ariz. 110–112
root depth 7, 16–60
root modification 7, 16–60
root morphology 7, 16–60
root spread 7, 16–60
rootlet – cyperoid 7, 16–60
rootlet – dauciform 7, 16–60
rootlet – mycorrhizal 7, 16–60
rootlet – proteoid 7, 16–60
rootlet – restioid 7, 16–60
rootlet modification 7, 16–60
Roseworthy, S. Aust. 97–99
Rottnest Island, W. Aust. 104–106
Rutherglen, Vic. 100, 102, 103

S. Julián de Vilat, Spain 131–133
Saad, Israel 126–128
Sabugal, Port. 129, 130
Sabzevar, Iran 120, 124
sage, coastal 32
Sahara 145

Saint Antoine, Egypt 119, 123
Saint Pons, France 117, 121
Sakakah, Saudi Arabia 119, 123
Sallum, Egypt 118, 122
Salmon Gums, W. Aust. 104, 105
salt complex 84
Salt Springs, Cal. 107–109
Salter Springs, S. Aust. 83
saltmarsh 84
San Antonio, Chile 113, 115
San Bernardino, Cal. 86
San Carlos Res., Ariz. 110–112
San Diego, Cal. 86
San Diego Co., Cal. 34, 38, 40, 159, 160
San Dimas Expt. For., Cal. 73, 176, 183, 185, 190, 193, 211
San Dimas (Tanbark Flat), Cal. 107–109, 211
San Fernando, Chile 113, 115, 116
San Gabriel Mts, Cal. 176
San Jacinto Mts, Cal. 86
San José de Maipo, Chile 113, 115
San Jose Hills, Cal. 177
San Luis Dam, Cal. 107–109
San Mateo, Spain 131–133
San Quintin, Mex. 86
Sandy Creek C.P., S. Aust. 183, 188
Santa Barbara, Cal. 86
Santa Laura, Chile 179, 184, 185, 193
Santa Rosa Mts, Cal. 86
Santa Ynez Mts, Cal. 86
Santiago, Chile 114–116
saprophyte 6, 16–60
Sartene, Corsica 118, 122
Sasa, Israel 126–128
savanna woodland 14, 19, 25, 28, 32, 65, 83
Scadden, W. Aust 84
sclerophyll 6, 14, 16–60, 157–165
sclerophyll index 8, 63–80, 157–165
sclerophylly 157–165
scrub 160
scrub – coastal 32, 42
scrub – coastal sage 32, 34
scrub – open 14, 15, 20–22, 26, 59, 61, 66, 67–71, 90
scrub – sclerophyll 43
scrub – thorny 204, 207, 208
seasonality – ants 214, 222
Sede Moshe, Israel 126–128
Sednaya, Syria 119, 123
Sedom Pans, Israel 126–128
seed dissemination 8, 16–60
seedling – germination type 8, 16–60
Sefid Dast, Iran 119, 123
semi-arid climate 140–146
semi-continental climate 140–146
semixyle 7, 16–60
Semman, Iran 120, 124
Serra da Arrábida, Portugal 57

Serviceton, Vic. 100, 102, 103
Sesimbra, Port. 129, 130
Sewell, Chile 113, 115, 116
Seymour, Vic. 101–103
Shalahaddin, Iraq 120, 124
Shasta Dam, Cal. 107–109
Sheluhot, Israel 126–128
Shepparton, Vic. 100, 102, 103
Shiraz, Iran 119, 123
shoot growth 151–155
shoot growth – seasonality 7, 16–60
shoot length 150–153
shrub communities 32, 33
shrub grassland 88
shrubland 15, 73, 77, 82, 84, 88, 90
shrubland – chenopod 82
Sidi Barani, Egypt 118, 122
Sidi-Bel-Abbes, Algeria 118, 122
Sierra Ancha, Ariz. 110–112
Signal Hill, S. Afr. 91
Sinjar, Iraq 120, 124
Sintra/Pena, Port. 129, 130
Siros, Greece 125, 126
Siverek, Turkey 120, 124
Skiathos Island, Greece 88
Sky Oaks, Cal. 35–39
Slenfer, Syria 118, 122
snakes 8, 171–194
Snowflake, Ariz. 110–112
Snowtown, S. Aust. 97–99
sobol 7, 16–60
soil cores 9, 198
Soil Fertility Index 151
Solsona, Spain 131–133
Somerset East, S. Afr. 134–136
South Africa 61, 80, 90, 91, 134–136, 159–165, 180, 184, 186, 198, 200, 212, 214
South Australia 14, 16–22, 63–65, 67–71, 83, 96–99, 174, 183, 185, 188, 189, 193, 198, 212–214, 219
Southern Cross, W. Aust. 104–106
Soviet Union 145
Spain 58–60, 131–133, 142, 216, 217
species richness 5
species abundance – ants 219
species richness – ants 219
species richness – overstorey 153–155
species richness – plants 5, 81–91, 153–155
species richness – small mammals 8, 153–155, 171–194
species richness – understorey 153–155
specific leaf weight 63–80, 160
Spes Bona, S. Afr. 134–136
spinescence 7, 16–60
St Albans, Vic. 27
St Arnaud, Vic. 100, 102, 103
St Clément, France 88
St Chinian, France 77

St Gély-du-Fesc, France 49, 77, 89
Sta Cruz de Palma, Canaries 118, 122
Sta Cruz de Tenerife, Canaries 118, 122
Star Swamp, W. Aust. 84
Stavros, Greece 52
Stawell, Vic. 100, 102, 103
Stellenbosch, S. Afr. 159
stem – diameter 7, 16–60
stem – height 7, 16–60
stem lignification 7, 16–60
stem – number from base 7, 16–60
steppe 87, 117
Stewart Mtn, Ariz. 110–112
stilt root 7, 16–60
Stirling Range N.P., W. Aust. 85
Stirling West, S. Aust. 97–99
Stockton Mowry Bridge, Cal. 107–109
stomata 6, 16–60
strandveld 61, 91, 134
Strathalbyn, S. Aust. 97–99
stratum – ground 6, 16–60
stratum – mid 6, 16–60
stratum – upper 6, 16–60
Streaky Bay, S. Aust. 96, 98, 99
sub-humid climate 140–146
sub-mediterranean climate 140–146
succession – post-fire 69, 77, 173
succulent 6, 16–60
succulent karoo 90, 134
summer drought 139–148
Sutherland, S. Afr. 134–136
SW Cape, S. Afr. 61, 80, 91, 134–136, 160
swamp 84
Swan Coastal Plain, W. Aust. 199
Swan Hill, Vic. 100, 102, 103
Swartberg, S. Afr. 90, 180, 184, 186
Swartberg Mts, S. Afr. 90, 180
Swartboschkloof, S. Afr. 204
Swellendam, S. Afr. 91
Syria 145

T.3, Syria 120, 124
Table Mountain, S. Afr. 134–136
Tagounite, Morocco 120, 124
Tahoe City, Cal. 107–109
Tal Shahar, Israel 126–128
Talca, Chile 113, 115, 116
Talme Yafe, Israel 126–128
Taltal, Chile 114, 115
Tarcoola, S. Aust. 96, 98, 99
Tareena, N.S.W. 82
Tarragona, Spain 131–133
Tavira, Port. 129, 130
Tazerbo, Libya 119, 123
Tefia, Canaries 118, 122
Temora, N.S.W. 94–96

Tempe Citrus Expt. Stn, Ariz. 110–112
Temuco, Chile 113, 115, 116
thermal continentality 140
Thermal Index (T.I.) 4, 5, 94–136, 151
Thermi, Greece 52
therophyte 5, 16–60
Thessaloniki, Greece 125, 126
thicket 84, 91
Til Til, Chile 43
Tirat Zevi, Israel 126–128
Titograd, Yugoslavia 119, 123
Tizi-Ouzou, Algeria 118, 122
Tlemcem, Algeria 141, 142
Tocumwal, N.S.W. 94, 95
Tomarsa, Turkey 119, 123
Torndirrup N.P., W. Aust. 85
tortoise 8
Trabzon, Turkey 118, 122
Tracey Pumping Plant, Cal. 107–109
Traiguén, Chile 113, 115, 116
tree mallee 84
Trinity Dam Hatchery, Cal. 106, 108, 109
Truro, S. Aust. 67
Tsitsikama Mts, S. Afr. 90
tuart woodland 84
tuber 7, 16–60
Tucson, Ariz. 110–112
Tulelake, Cal. 106, 108, 109
Tullgren-type funnel 9, 212, 214
Turkestan 145
Turkey 88, 145
Turntable Creek, Cal. 107–109
Tutanning, W. Aust. 71, 84, 85
Twitchell Dam, Cal. 107–109
Two People Bay N.R., W. Aust. 85, 183, 189
Tygerberg, S. Afr. 91

underground stems 6, 16–60
understorey – species richness 155
Urfa, Turkey 120, 124

Valencia, Spain 131–133
Valladolid, Spain 118, 122
Vallenar, Chile 114, 115
valley bushveld 90
Valmy, France 211
Valparaiso, Chile 113, 115, 116
Van, Turkey 119, 123
Vassilika, Greece 52
vegetative multiplication 7, 16–60
vegetative regeneration after fire 7, 16–60
vertebrates 8, 171–194
Viana do Castelo, Port. 129, 130
Victoria, Aust. 23–27, 66, 67, 70, 83, 100–103, 173, 174, 183, 185, 187
Victoria, Chile 113, 115, 116

Vicuña, Chile 114, 115
Vidauban, France 77
Viella, Spain 131–133
Vila Real, Port. 129, 130
Vila Real de St Antònio, Port. 129, 130
Vlore, Albania 118, 122
Volos, Greece 78, 125, 126
Voorhis Ecological Reserve, Cal. 176, 177, 183, 185, 190, 193

waboomveld 90
Wagerup, W. Aust. 220
Wagga, N.S.W. 94–96
Wahweap, Ariz. 110–112
Waikeri, S. Aust. 71
W.A.I.T., now Curtin Univ., W. Aust. 104, 105
Waite Institute, S. Aust. 97–99
Walnut Grove, Cal. 107–109
Walpeup, Vic. 100, 102, 103
Wandering, W. Aust. 104, 105
Wangaratta, Vic. 100, 102, 103
Warm Springs Dam, Cal. 106, 108, 109
Warner Springs, Cal. 107, 108
Warooka, S. Aust. 97, 98
Warracknabeal, Vic. 100, 102
Warrenben C.P., S. Aust. 18, 83
Warrnambool, Vic. 100, 102, 103
Watermans, W. Aust. 84
Watheroo, W. Aust. 217
Wellington, N.S.W. 94–96
Wentworth, N.S.W. 82, 94–96
Werribee, Vic. 100, 102, 103
West Bendering N.R., W. Aust. 183, 185
West Wyalong, N.S.W. 66
Western Australia 15, 28–31, 71, 84, 85, 104–106, 175, 183, 185, 189, 193, 198, 212–214, 217, 219
Whiskey Town, Cal. 107–109
Whiteriver, Ariz. 110–112
Whyalla, S. Aust. 96, 98
William Creek, S. Aust. 96, 98, 99
Willow Creek, Cal. 106, 108, 109
Willows, Cal. 107–109
Wilroy N.R., W. Aust. 183, 185

Wilson's Prom., Vic. 83
Wingfield, S. Afr. 134–136
Winkelman, Ariz. 110–112
Wongan Hills, W. Aust. 104, 105
woodland 33, 82, 84, 86, 88, 89
woodland – evergreen 43
woodland – low 29, 84
woodland – savanna 14, 15, 19, 25, 28, 65, 83
woodland – sclerophyll 14, 15, 17, 29
Woodline Hills, W. Aust. 183, 185, 189, 193
Wyalong, N.S.W. 94, 95
Wycheproof, Vic. 100, 102, 103
Wyperfeld N.P., Vic. 172–174, 183, 185, 187, 193

xeromorphy 157
xerophytic 207, 208

Yarrawonga, Vic. 100, 102, 103
Yathong N.R., N.S.W. 82
Yavne, Israel 126–128
Yenda, N.S.W. 94–96
Yongala, S. Aust. 96, 98, 99
York, W. Aust. 104–106
Yorke Pen., S. Aust. 20
Yorkrakine Rock N.R., W. Aust 183, 189
Yornaning N.R., W. Aust. 183, 189
Yosemite N.P., Cal. 73
Young, N.S.W. 94, 95
Yudnapinna, S. Aust. 96, 98, 99
Yugoslavia 144
Yuma Citrus Stn, Ariz. 110–112
Yunak, Turkey 119, 123
Yunta, S. Aust. 96, 98

Zabol, Iran 120, 124
Zanjan, Iran 119, 123
Zapallar, Chile 42, 113, 115, 116, 179, 184, 186, 193
Zara, N.S.W. 82
Zeelim, Israel 126–128
Zora, Israel 126–128
Zucaina, Spain 131–133

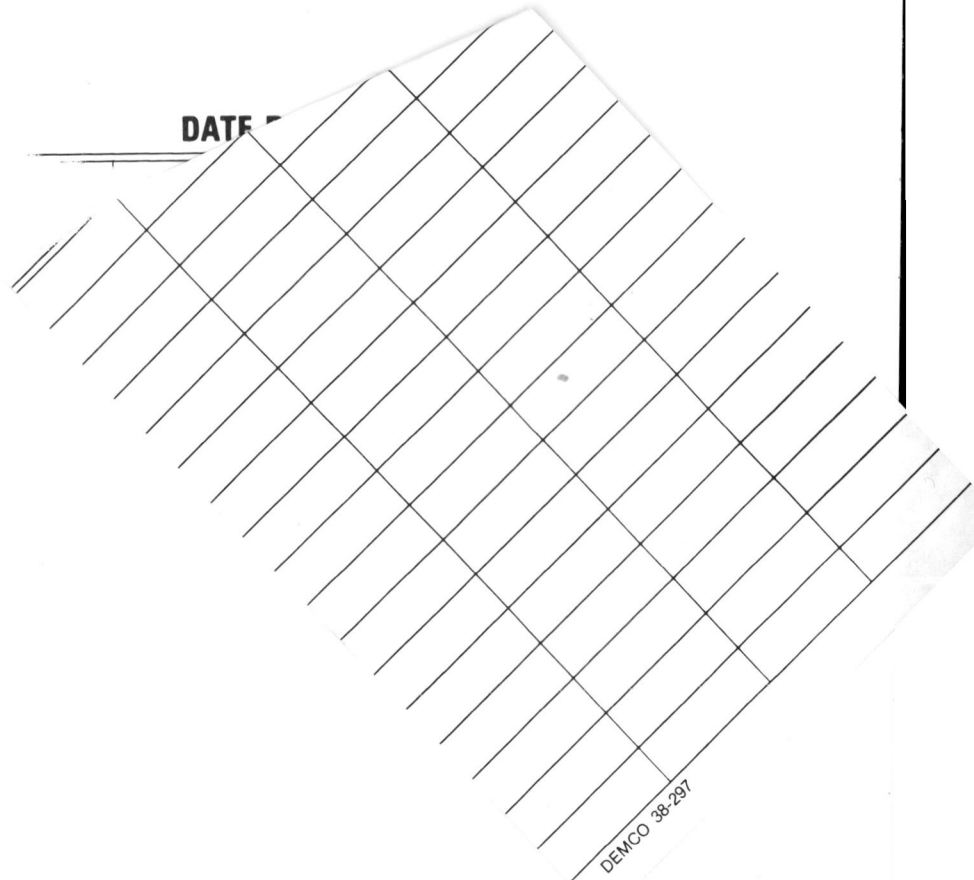